AF698551

Pain and Headache

Vol. 9

Series Editor
Philip L. Gildenberg, Houston, Tex.

KARGER

Basel · München · Paris · London · New York · New Delhi · Singapore · Tokyo · Sydney

Neurotransmitters and Pain Control

Neurotransmitters and Pain Control

Volume Editors
H. Akil, Ann Arbor, Mich.
J. W. Lewis, Ann Arbor, Mich.

37 figures and 17 tables, 1987

KARGER

Basel · München · Paris · London · New York · New Delhi · Singapore · Tokyo · Sydney

Pain and Headache

Drug Dosage

The authors and the publisher have exerted every effort to ensure that drug selection and dosage set forth in this text are in accord with current recommendations and practice at the time of publication. However, in view of ongoing research, changes in government regulations, and the constant flow of information relating to drug therapy and drug reactions, the reader is urged to check the package insert for each drug for any change in indications and dosage and for added warnings and precautions. This is particularly important when the recommended agent is a new and/or infrequently employed drug.

All rights reserved.

No part of this publication may be translated into other languages, reproduced or utilized in any form or by any means, electronic or mechanical, including photocopying, recording, microcopying, or by any information storage and retrieval system, without permission in writing from the publisher.

© Copyright 1987 by S. Karger AG, P.O. Box, CH-4009 Basel (Switzerland)
Printed in Switzerland by Druckerei 'Der Bund', Bern
ISBN 3-8055-4579-7

Contents

Classical Neurotransmitters Role in Pain Control

Opioid Systems and Analgesia

Non-Opioid Peptides and Analgesia: Focus on Cholecystokinin

Activation of Endogenous Mechanisms of Analgesia

Acknowledgments

We would like to thank Mrs. Adele Henry for her invaluable assistance in the editing of this book. We also thank Dr. Alfred Mansour for providing the receptor autoradiogram on the cover and Dr. S.J. Watson for his input throughout the preparation and planning of this volume. Most importantly, we are indebted to the authors who have given of their valuable time to contribute the chapters which constitute this work. Obviously, without them, it would not be. This work was supported, in part, by NIDA Grant No. RO1 DA02265 and NIMH Grant No. PO1 MH422251.

Huda Akil
James W. Lewis

Preface

The identification of the neurotransmitters of pain is perhaps one of the most significant obstacles impeding the development of effective analgesia-inducing drugs or therapeutic modalities. Although many advances in our understanding of this topic have been made in recent years, we still are in search of a magic bullet against pain. Our purpose in this volume has been to bring together the knowledge of various experts in the field of neurochemistry of pain and analgesia. Using biochemical, anatomical, and pharmacological tools, these researchers have attempted to understand the response to noxious stimuli (what we term pain) or the inhibition of the pain response (what we term analgesia). As the reader might expect, no one transmitter or modulator is uniquely responsible for either pain perception or pain inhibition. Indeed, almost every known brain transmitter when studied in relation to pain has been found to have some role. In some cases, such as opioid peptides, the case for an important role in pain modulation is reasonably strong and has face validity. For other substances, such as some of the newly discovered peptides, the case is less convincing, mostly because the investigation is in its early stages.

It is of course impossible to cover in one volume all the accumulated knowledge on the neurochemistry of pain modulation. Rather, we have attempted to give the reader an overview of the potential modulatory role of three main classes of neurochemical substances: (1) Classical neurotransmitters, such as catecholamines, serotonin and acetylcholine. (2) The opioid peptide family with the multiple products deriving from the opioid genes. (3) The nonopioid peptides, with a focus on one peptide in particular, cholecystokinin. In each of these three areas, we have included at least one chapter with a general overview of the field, as well as some chapters which detail a more discrete body of work focused on a particular subset of current questions. In this fashion we hope to provide the reader with two levels of information, a general appreciation of the current knowledge about each of these three classes of substances, and more specific examples of the type of research currently being performed in the area. Some of these examples reach paradoxical conclusions, may even be controversial, but they do challenge any illusion of a neat and tidy understanding of the biochemistry of pain control.

The last section of the book, rather than using specific neurochemicals as its starting point, focuses on known analgetic phenomena: acupuncture analgesia, stress-induced analgesia and stimulation-produced analgesia. The authors of these chapters, by summarizing work on the biology of analgetic phenomena, clearly point to the complexity of pain modulation, involving multiple anatomical sites of regulation, as well as multiple neurochemical systems. In so doing, they show us, from a different angle, both the complexity and dynamism involved in the mammalian response to painful stimuli.

There is a great deal this book does not deal with. Most notable is the question of the anatomical substrates integrating pain modulation, a topic ably covered in this series by the monograph of Dr. Willis. In addition, implications of the current biochemical information for human work and for the development of novel pharmaceutical agents is also left for future volumes in this series.

We are most grateful to the authors who have contributed to this collection. It is our hope that by presenting as one unit this overview of the biochemistry of pain modulation, we can allow a sharper perception of both the limitations and the power of this level of discourse. This, in turn, may ultimately cumulate in a better integration of all basic research on pain, and result in novel ways of controlling it.

H. Akil
J.W. Lewis

Classical Neurotransmitters Role in Pain Control

Pain Headache, vol. 9, pp. 1–63 (Karger, Basel 1987)

Cholinergic and Catecholaminergic Modulation of Nociceptive Reactions

Interactions with Opiates

Agu Pert

Biological Psychiatry Branch, National Institute of Mental Health, Bethesda, Md., USA

Cholinergic Modulation

Over the past 50 years, considerable evidence has accumulated in support of an important role for acetylcholine in nociceptive processes. Three basic suppositions appear to underlie most of the studies that have attempted to evaluate the antinociceptive properties of cholinergic compounds and their interactive effects with opiate analgesics: (1) that cholinergic transmission is either involved in the transmission or inhibition of nociceptive information, and, therefore, the use of appropriate cholinergic drugs should reveal the operation of such cholinergic mechanisms; (2) that opiate analgesics somehow mediate at least part of their antinociceptive effects through cholinergic mechanisms; thus, cholinergic drugs should alter opiate analgesia; and (3) that cholinergic drugs may exert their antinociceptive effects through opiate activation.

Davis et al. [1932] were probably the first to discover that cholinergic compounds might have antinociceptive properties. These investigators found that 0.1 mg/kg of nicotine was able to abolish nociceptive reactions to the distention of the gall bladder in cats. This effect appeared to be specific to visceral pain, since reactions to limb pinches were not altered. One year later, Pellandra [1933] reported that the anticholinesterase, physostigmine, increased the threshold for pain elicited by radiant heat applied to the forehead of human subjects.

Anticholinesterases

As noted above, one of the first cholinergic compounds to be assessed for antinociceptive activity was physostigmine, an anticholinesterase which potentiates the actions of endogenous acetylcholine by preventing its enzymatic degradation by acetylcholinesterase. Following the demonstration by Pellandra [1933] that physostigmine may have analgesic properties, Slaughter and Gross [1933] reported that some of the pharmacologic effects of opiates could be potentiated by anticholinesterases. This prompted Slaughter and Munsell [1940] to consider whether anticholinesterases might also potentiate the antinociceptive properties of opiates. In a study employing tail pressure to cats as a nociceptive stimulus, they found that doses of neostigmine, which had no apparent effects by themselves, did potentiate the effects of morphine on the reactions to tail pinch. The concurrent administration of atropine (a muscarinic blocking agent) appeared to reverse the synergistic action.

Several other studies in man have presented evidence to suggest that anticholinesterases may have analgesic properties. Floodmark and Wrammer [1945] reported that neostigmine, as well as physostigmine, increased pain thresholds in man as defined by the application of radiant heat to the forehead. Christensen and Gross [1948], using similar procedures, also found that neostigmine had antinociceptive properties in man. More recently, Sitaram et al. [1977] have demonstrated antinociceptive properties for physostigmine in normal volunteers, employing a four-point scale to judge the aversive qualities of 31 intensities of electric shock. The same investigators also found that physostigmine decreased the P_{100} component of shock-elicited averaged evoked cortical potentials in man. They postulated that acetylcholine-induced reticular formation arousal may somehow serve to dampen the emotional component of pain. The finding that anticholinesterases have antinociceptive properties is not universal, however. Andrews [1942], using a design similar to that employed by Floodmark and Wrammer [1945] and Christensen and Gross [1948], were not able to demonstrate antinociceptive properties for even larger doses of neostigmine. At any rate, the reported antinociceptive effects of neostigmine are somewhat puzzling. They are most likely due to peripheral actions, since this compound does not penetrate the brain to any appreciable degree. In fact, Floodmark and Wrammer [1945] have suggested that neostigmine may have produced apparent analgesic effects to radiant heat in their study by increasing the blood flow through the skin. A similar mechanism could also account for the antinociceptive effects of physostigmine in other studies that have utilized radiant heat as a nociceptive stimulus. The effects reported by Sitaram, on the other hand, might be centrally mediated, since the subjects were pretreated with methyl atropine, which protected the peripheral muscarinic cholinergic sites. This still does not rule out the possible participation of peripheral nicotinic cholinergic receptors.

Antinociceptive properties have been demonstrated also for anticholinesterases in a number of animal species and tasks. Harris et al. [1969], Howes et al. [1969] and Lipman and Spencer [1980] reported that systemic administration of physostigmine increased response latencies in the tail-flick or tail-withdrawal tasks in mice. These effects were probably mediated centrally, since neostigmine (which does not penetrate the blood-brain barrier readily) administered intraventricularly (i.c.v.) has also been shown to be effective in inhibiting this reaction [Pedigo et al., 1975]. Surprisingly, however, intraventricular physostigmine has not been found to be effective [Lipman and Spencer, 1980], possibly due to the relatively low dose (1/100 of the active peripheral dose) employed. Mudgill et al. [1974] also failed to elicit antinociceptive effects in the tail-flick test following i.c.v. injections of 3 nmol of physostigmine. The ability of centrally administered neostigmine to increase tail-flick reaction latencies [Pedigo et al., 1975] also suggests that anticholinesterases may exert their effects, at least partially, by activating descending inhibitory mechanisms to the spinal cord. Physostigmine has also been found to increase reaction times of mice on the hot plate at similar doses to those used above [Cox and Tha, 1972; Pleuvry and Tobias, 1971], although Chen [1958] failed to demonstrate antinociceptive effects following 0.5 mg/kg in the Haffner tail-clip test with mice. Cox and Tha [1972] also examined the antinociceptive properties of dyflas (an anticholinesterase) and found it to be ineffective in altering hot plate reactions, even though it also decreased brain cholinesterase levels by the same degree as physostigmine. This could imply that cholinesterase inhibition is not critical for the effects of physostigmine on the hot plate or, alternatively, that the predominant effect of dyflas is on cholinesterase rather than acetylcholinesterase.

Antinociceptive effects have also been reported for other anticholinesterases. Koehn and Karczmar [1978] and Koehn et al. [1980) reported that DFP produced marked antinociception in rats and mice measured by the hot-plate or tail-flick procedures. Interestingly, Koehn and Karczmar [1978] also found that the effects of DFP were blocked by atropine sulfate (a muscarinic blocker that acts both centrally as well as peripherally) and not by the peripherally acting atropine methyl nitrate. Harris et al. [1969] reported similar interactive effects following administration of physostigmine. These findings suggest that the antinociceptive actions of physostigmine are mediated centrally through the muscarinic cholinergic receptor. A central site of action for the antinociceptive effects of anticholinesterase compounds is further supported by findings that neostigmine, administered peripherally in moderate doses, does not generally alter nociceptive reactions in cats [Slaughter and Munsell, 1940], rats [Saxena, 1958], mice [Pleuvry and Tobias, 1971] or primates [Pert, 1975]. At high doses (0.165 mg/kg s.c.), neostigmine has been reported to alter the reactions of cats to tail pressure [Slaughter and Munsell, 1940]. However, assessment of the pain reaction was complicated at this dosage level by violent side reactions including profuse salivation, muscular twitching and muscular weakness. The authors concluded that the animals were 'so sick that they did not care about the pain'. de Jongh [1954] also observed alterations in reactions of guinea pigs to radiant heat by neostigmine. The relationship of this effect to cholinesterase inhibition is questionable, however, since low doses of atropine also produced similar actions.

Herz and Yacoub [1964], using shock-elicited vocalization in the rat to assess nociception, also reported antinociceptive properties for mintocal, a cholinesterase inhibitor, but noted that the effective dose range was very near the toxic level for this compound. Ireson [1970] found that physostigmine (0.3 mg/kg) was inactive as an antinociceptive agent in tail-shock induced vocalization, but that it did inhibit writhing reactions induced by s.c. injections of phenylbenzoquinone.

Catecholaminergic and Serotonergic Involvement in Anticholinesterase Antinociception

One apparent unstated assumption behind studies that have evaluated the antinociceptive properties of anticholinesterases is that increasing the availability of endogenous acetylcholine at the cholinergic synapse should be sufficient to induce an analgesic state. This suggests that pathways carrying nociceptive information may be under a tonic inhibitory influence of acetylcholine, which is further enhanced by anticholinesterases. While such an influence may be direct, there is also evidence to suggest the participation of other neurohumoral systems in the antinociceptive effects of anticholinesterases following such cholinergic perturbations. Pleuvry and Tobias [1971], for example, evaluated and compared the antinociceptive activity of both physostigmine and morphine following depletion of brain amines and indoleamines with various drugs. They found that pretreatment with reserpine, which depletes catecholamines and serotonin, significantly reduced the antinociceptive activity of both physostigmine and morphine on hot plate reactions in mice. Pretreatment with diethyldithiocarbamate, a dopamine beta-hydroxylase inhibitor that reduces the synthesis of norepinephrine, as well as *p*-chlorophenylalanine (PCPA), a tyrosine hydroxylase inhibitor that reduces the synthesis of serotonin, also decreased the antinociceptive activity of both compounds. On the other hand, pretreatment with 5-hydroxytryptophan increased the antinociceptive effects of both morphine and physostigmine. These findings suggest that catecholaminergic and serotonergic systems are critical for the expression of physostigmine-induced analgesia. Activation of descending serotonergic and noradrenergic pathways, of course, has been shown to be critical for the expression of opiate analgesia [Yaksh, 1979]. Further support for the involvement of serotonin in anticholinesterase-induced analgesia comes from the observation that DFP increases serotonin levels in the medulla, mesencephalon, and diencephalon [Koehn and Karczmar, 1978]. The same authors, however, were not able to antagonize DFP-induced analgesia with PCPA and concluded that the antinociceptive effects of DFP may not depend on serotonin.

Opiate Involvement in Anticholinesterase Antinociception

While the role of serotonin and catecholamines in anticholinesterase antinociception is somewhat equivocal, it is also possible that the activation of endogenous opiate systems may somehow underlie the actions of these compounds. Harris et al. [1969] were the first to systematically study the interactive effects between anticholinesterases and opiate antagonists. They found that at all doses employed (0.5 to 20 mg/kg), naloxone effectively antagonized the ability of physostigmine to raise tail-flick latencies in mice. Koehn and Karczmar [1978] reported that 5 mg/kg of naloxone was as effective in antagonizing DFP-induced antinociception in the hot-plate test as atropine sulfate. Koehn et al. [1980] extended these findings by demonstrating that (−)isomers of 5,9 alpha-diethyl-2(3-furylmethyl)-2-hydroxy-6,7-benzomorphan and

2-allyl-2′-hydroxy-9 B-methyl-5-phenyl-6,7-benzomorphan HCl were effective in antagonizing both morphine and DFP-induced antinociception, while the corresponding (+)isomers were ineffective against either compound. Naloxone was also effective at 5.0 and 0.5 mg/kg, but not at 0.05 mg/kg. Ramabadran and Jacob [1978] also were able to antagonize the antinociceptive effects of physostigmine with naloxone and the (−)isomer MR 2266, an active antagonist. The (+)isomer MR 2267 was inactive. While these findings seem to implicate an opiate mediation of DFP and physostigmine antinociception, it is not clear whether the opiate antagonists were exerting their effects through pharmacological or physiological antagonistic actions. No attempt was made to fully evaluate the actions of these compounds by themselves on aversive thresholds. It has been shown by others that opiate antagonists do possess hyperalgesic properties [Jacob et al., 1974; Frederickson, 1977; Grevert and Goldstein, 1977; Jacob and Ramabadran, 1978; Amir and Amit, 1978; Ramabadran and Jacob, 1979]. Koehn et al. [1980] also found that animals made tolerant to morphine did not exhibit cross-tolerance to DFP. This finding seems to suggest that perhaps the antinociceptive actions of anticholinesterases may not be related to endogenous opiate activation. In addition, not all studies have reported that opiate antagonists are effective antagonists of anticholinesterases. Lipman and Spencer [1980], for example, did not observe antagonism of physostigmine analgesia in the tail-immersion test following 1 mg/kg of naloxone.

Confounding Side Effects of Anticholinesterases. While the findings above appear to suggest that anticholinesterases enjoy antinociceptive properties, it is still unclear whether these effects are related to an attenuation of pain or to other factors. Anticholinesterases produce severe side effects in animals and man. The muscarinic effects include extreme salivation, involuntary defecation and urination, sweating, bradycardia, and hypotension. Nicotinic actions, mediated predominantly through actions at the neuromuscular junction, include fatigability and generalized weakness, involuntary twitchings, and eventual paralysis and respiratory depression.

Considering that all of the animal nociceptive tests cited above utilize motor reactions as an endpoint, it seems likely that at least some of the antinociceptive effects of anticholinesterases may have been due to the motor debilitating actions of these compounds. It is surprising that this factor has been addressed in only a few reports as a possible confounding variable. Koehn and Karzmar [1978] measured locomotor activity of rats concurrently with antinociception following the administration of DFP. They found that DFP, at doses that increased reaction latencies in the hot-plate or tail-flick tests, also decreased locomotor activity by more than 50%. However, they argued that the antinociceptive actions of DFP could not be due to its depression of locomotor activity for several reasons. First, they noted that there was no relationship between the antinociceptive activity of other compounds (e.g. morphine) and their effects on locomotor behavior. 'DFP and morphine produced similar analgetic effects yet differed diametrically with respect to motor activity.' Second, atropine antagonized the antinociceptive actions of DFP but not its depressant actions, demonstrating a dissociation between the two processes. Third, PCPA, like DFP, produced locomotor depression in the rat yet decreased the reaction latencies in the hot-plate and tail-flick tests. They concluded that locomotor depression is not a sufficient determinant of antinociception in laboratory animals. In a subsequent study from the same laboratory, Koehn et al. [1980] reported that lower doses of DFP (0.1 mg/kg) produced antinociception that was not accompanied by locomotor effects. In addition, higher doses of DFP (1.5 mg/kg) produced antinociceptive effects that outlasted its effects on locomotor activity by at least 4 h. Interestingly, naloxone also completely antagonized DFP-induced antinociception while it was totally ineffective in altering DFP-induced attenuation of locomotor output. Other investigators, however, have not been able to determine definitively whether the side effects of anticholinesterases contribute to their apparent antinociceptive actions. As noted before, Herz and Yacoub [1964] found that mintocal inhibited both tail-shock-induced vocalization and conditioned avoidance responses in the rat only at doses that were very near the toxic level. Slaughter and Munsell [1940] were only able to induce an antinociceptive effect with neostigmine at a dose that also produced violent side reactions including muscular weakness, twitching, and salivation. In this case, the effects (including antinociception) had to be peripheral, since this compound does not penetrate the brain readily. Pert [1975] also reported that high doses of physostigmine increased the thresholds of rhesus monkeys to electrical shocks in a shock-titration paradigm. However, the author noted that the highest doses seemed to depress all behavior and, therefore, it was difficult to separate true antinociceptive effects of this compound from its general depressant properties. In this study, the animals became unresponsive to shock immediately following i.v. injections. This effect generally lasted only 15 minutes, at which point there was an immediate return to baseline levels of responding. At the highest dose used, it was not possible to initiate responding even when the current was suddenly increased to the maximal level at which undrugged animals normally exhibited profound reactions.

Interactions of Anticholinesterase with Opiate Analgesics. One belief that was partially responsible for initiating and maintaining an interest in the potential for anticholinesterases to induce analgesia was that some of the effects of opiates may be related to their inhibition of cholinesterases [Bernheim and Bernheim, 1936]. This effect of morphine was assumed by a number of early investigators to underlie at least some of its antinociceptive effects. If opiates were, in fact, producing antinociception by inhibiting cholinesterases, then it would be likely that the addition of anticholinesterases to opiates would potentiate their actions. There are numerous studies that have examined this interaction. The earliest studies were conducted in man with neostigmine. Andrews [1942] was the first to demonstrate that neostigmine increased both the intensity and duration of morphine, meperidine, and methadone-induced analgesia in man as assessed by radiant heat. Abaza and Gregoire [1952], Christensen and Gross [1948] and Floodmark and Wrammer [1945] reported similar findings for neostigmine. The potentiation of opiate analgesia in these studies presumably was the result of peripheral cholinesterase inhibition by neostigmine and did not represent a direct central interaction. It is possible, for example, that neostigmine exerts its interactive effects with opiates by altering the disposition or metabolism of these compounds [Komlos and Komlos-Szarz, 1954].

The interactive effects between opiates and anticholinesterases has been studied extensively in laboratory animals. Physostigmine has been almost universally found to enhance the antinociceptive activity of opiate analgesics. Bhargawa and Way [1972] found that both DFP and physostigmine potentiated morphine-induced antinociception as measured in the tail-flick test. Howes et al. [1969], Harris et al. [1969], as well as Takemori [1975] reported similar findings for physostigmine in mice using a similar task. Harris et al. [1969] also found that mixed agonist-antagonist opiates, which had little or no activity by themselves, increased reaction times significantly in the presence of nearly inactive doses of physostigmine. They postulated that the narcotic-antagonist analgesics are acting on supraspinal cholinergic mechanisms inhibitory to the tail-flick reflex. Physostigmine has also been shown to potentiate morphine as well as nalorphine in the hot-plate test [Pleuvry and Tobias, 1971], while Ireson [1970) has found potentiation of morphine in both the tail-shock-induced vocalization and phenylbenzoquinone writhing tests.

The potentiating effects of anticholinesterases are probably centrally mediated in these animal studies, since potentiation of opiate analgesia has also been reported following intraventricular injections of even small quantities of physostigmine. Further support for this assumption comes from the observation that neostigmine has generally been found to be ineffective in potentiating opiates in studies with laboratory animals [Saxena, 1958; de Jongh, 1954; Pleuvry and Tobias, 1971]. Pert [1975] also did not observe any interactive effects between morphine and neostigmine in the rhesus monkey. In contradiction to the findings above, Slaughter and Munsell [1940] reported that neostigmine potentiated morphine-induced inhibition of tail-pinch reactions in cats.

Interaction of Anticholinesterases with Other Antinociceptive Agents. While the majority of studies have examined the antinociceptive effects of anticholinesterases with opiate analgesics, there are several studies that also have evaluated interactive effects between anticholinesterases and other antinociceptive compounds. Lipman and Spencer [1980], for example, reported that physostigmine potentiated not only morphine-induced analgesia, but also the antinociceptive effects of clonidine in the tail immersion test. Kamat et al. [1972] found that physostigmine decreased the ED_{50} of dipyrone, a non-opiate analgesic, on the hot-plate test. The authors proposed that cholinesterase-induced potentiation of opiate analgesia might reflect cholinergic involvement in central pain mechanisms per se rather than a specific involvement in narcotic analgesia.

While many of the early studies that examined the interactive effects between opiates and anticholinesterases were based on the premise that opiate analgesics may induce some of their analgesic effects through cholinesterase inhibition, it seems that this position is no longer tenable. Komlos et al. [1950], for example, reported that one of the most potent anticholinesterases, tetraethylpyrophosphate, was incapable of potentiating the effects of any analgesic substances. A similar finding was reported by Porszasz et al. [1981] while Saxena [1958] was able to potentiate morphine-induced antinociception with neostigmine but not with DFP, which inhibits brain as well as peripheral cholinesterases and is more potent than the former. In the same study, atropine failed to antagonize the potentiating effects of neostigmine, which suggests that, at least for this anticholinesterase inhibitor, muscarinic cholinergic activation does not appear to be critical for its ability to potentiate morphine. Further evidence against a role for the mediation of opiate-induced analgesia through cholinesterase inhibition comes from studies that have examined the relationship between the analgesic potency of opiates and their ability to inhibit acetylcholinesterase. Knoll and Komlos [1951] not only failed to find a relationship between the pain relieving effects and degree of cholinesterase inhibition produced in the cerebral cortex by several analgesics, but also observed that the simultaneous injection of morphine and an anticholinesterase did not increase the inhibitory activity of the former in the brain. Young et al. [1956] were also unable to demonstrate a relationship between the analgesic actions of a group of opiates and their anticholinesterase activity. Furthermore, levalorphan, which has little analgesic activity if any, had a greater anticholinesterase activity than any of the active opiates. It seems clear that cholinesterase inhibition plays a relatively minor role (if any) in the antinociceptive effects of opiate analgesics.

Cholinomimetic Compounds

The antinociceptive effects of cholinomimetic compounds have not been evaluated in man, undoubtedly because of their severe side effects. Several studies in primates, however, have failed to find that cholinomimetic drugs elevate shock thresholds. Houser and Houser [1973] did not find alterations in shock thresholds in the squirrel monkey following injections of pilocarpine, a cholinomimetic with predominantly muscarinic cholinergic agonist actions. Using a similar paradigm, Pert [1975] also was unable to demonstrate antinociceptive properties in the rhesus monkey for pilocarpine or arecoline, a cholinomimetic that also has nicotinic actions in addition to prominent muscarinic effects. In contrast to these negative findings in primates, there are numerous studies that have reported antinociceptive effects in the rodent following administrations of cholinomimetics. Chen [1958] was the first to demonstrate antinociceptive actions for tremorine in mice using the Haffner tail-clip test. Lenke [1958] confirmed these findings. Keranen et al. [1961] were also able to inhibit the reactions of mice to the application of a tail-clip with tremorine. Furthermore, Herz and Yacoub [1964] reported that tremorine decreased vocalizations induced by electrical stimulation of the tail root in rats, while van Eick and Bock [1971] found that tremorine decreased the reactions of mice on a hot plate. The antinociceptive actions of oxotremorine, another very potent cholinomimetic with predominantly muscarinic actions, has been more extensively investigated. Harris et al. [1969] reported that oxotremorine was 250 times more potent than morphine in blocking the tail-flick reflex in mice. In a subsequent study, the same laboratory [Howes et al. 1969] found a positive relationship between the brain levels of acetylcholine and the degree of inhibition of the tail-flick reaction for various doses of oxotremorine. It is not clear, however, if the effects of oxotremorine on brain acetylcholine have any relevance for its analgesic actions, since, in the same study, this relationship did not hold up for other compounds and drug combinations. Tripathi et al. [1980] have also found that oxotremorine produced dose-related increases in acetylcholine levels which reached statistical significance at the ED_{84} antinociceptive dose. The increased steady-state level of acetylcholine was temporarily correlated with the peak antinociceptive activity of oxotremorine. Acetylcholine turnover also decreased with increasing doses of oxotremorine. In the same study, mice injected with oxotremorine twice a day for 4 days developed tolerance to its actions in the tail-flick test. Metys et al. [1969] and Lipman and Spencer [1980] also found that oxotremorine increased tail withdrawal reactions in mice. Metys et al. [1969] also reported that oxotremorine was effective in attenuating the hot-plate, tail-clip, and writhing reactions in mice, tail-shock-induced vocalizations in rats, and tooth pulp-elicited nociceptive reactions in rabbits. Inhibition of tail-shock-induced vocalizations in rodents by oxotremorine has also been reported by Leslie [1969] and Ircson [1970]. Oxotremorine also inhibited phenylbenzoquinone-induced writhing reactions in the latter study. Not surprisingly, considering its effects in other nociceptive tests, cholinergic activation by oxotremorine also has been found to increase reaction times in the hot-plate test [Bite and Gyorgy, 1974; Cox and Tha, 1972; Pleuvry and Tobias, 1971].

Pilocarpine, a cholinomimetic that has a dominant muscarinic action but also causes anomalous cardiovascular responses, has not been found to produce antinociceptive effects as conclusively as oxotremorine. Pert [1975] failed to observe elevations in shock-titration thresholds in the primate following i.v. administration of pilocarpine. Saxena [1958] also failed to find antinociceptive effects for pilocarpine (up to 0.2 mg/kg) in the Haffner tail-clip test. However, Houser and van Hart [1973], using a spatial preference test, found that the aversive threshold was raised significantly following administrations of pilocarpine (doses of 2.5 mg/kg or above were found to be effective). In a subsequent study, Houser [1976] also found an elevation in aversive thresholds in the same test following injections of 10 and 20 mg/kg of pilocarpine.

While the antinociceptive effects of pilocarpine are somewhat equivocal, the antinociceptive effects of arecoline, a cholinomimetic with mixed nicotinic and muscarinic activity, are somewhat more consistent in the rodent. Herz [1962] found that arecoline inhibited reactions of mice to tail pressure in doses that did not cause tremor. In a subsequent study, Herz and Yacoub [1964] reported that arecoline inhibited the reactions elicited by electrical stimulation of the tail root in rats. The conditioned avoidance response was also inhibited at somewhat higher doses in the same study. Metys et al. [1969] also reported that arecoline inhibited a number of different nociceptive reactions in mice, rats and rabbits.

The central actions of cholinomimetic-induced analgesia are supported by studies that have characterized the antinociceptive effects of acetylcholine following intraventricular injections. Pedigo et al. [1975] found that acetylcholine injected i.c.v. inhibited the tail-flick reactions as well as phenylquinone writhing in mice. I.c.v. neostigmine also elevated tail-flick latencies and, at lower doses, reduced the ED_{50} of acetylcholine.

Systemically administered atropine, but not methyl atropine, blocked the antinociceptive effects of i.c.v. acetylcholine. Pedigo and Dewey [1981a, b] confirmed these findings for mice in the tail-flick test. Dewey et al. [1975] also studied the effects of i.c.v. acetylcholine and acetylcholine analogues on writhing reactions induced by intraperitoneally administered acetylcholine and found that 8 μg was the ED_{50} for i.c.v. acetylcholine. It was also reported that i.c.v. injections of *alpha* substituted acetylcholine were inactive in inhibiting writhing responses, while both the racemate and the S(+)isomers of the *beta* methyl substituted acetylcholine were active.

Interestingly, it is the S(+) conformation of the *beta* methyl substituted acetylcholines which resemble the preferred conformation of acetylcholine [Canepa et al., 1966]. Furthermore, the S(+)isomer of *beta* substituted acetylcholine is more potent than the R(−)isomer at peripheral muscarinic sites [Cocolas et al., 1971]. The S(−)isomer of three analogs of carbamyl beta-methylcholine also had a higher potency in the writhing test following i.c.v. injections than the R(−)isomers. The antinociceptive effects of these isomers also corresponded with their efficacy in the guinea pig ileum preparation. Metys et al. [1969] also reported the inhibition of nociceptive reactions in rodents following i.c.v. injections of carbachol and oxotremorine.

Further support for the possibility that cholinomimetic activation produces antinociceptive effects through a central mechanism comes from studies that have compared the abilities of peripherally and centrally acting anticholinergic compounds to reverse cholinomimetic-induced analgesia and from studies that have evaluated the antinociceptive actions of peripherally active agonists. Leslie [1969] found that oxotremorine-induced antinociception in the tail-flick test was antagonized by a number of centrally acting anticholinergic drugs (e.g. atropine, benactyzine, benztropine, scopolamine, promethazine). Quaternary atropine-like drugs that do not penetrate the blood-brain barrier to any significant degree proved to be ineffective antagonists. Oxotremorine-induced inhibition of tail-shock-induced vocalization and phenylbenzoquinone writhing has also been antagonized by atropine [Ireson, 1970]. Atropine and other tertiary atropine-like drugs, but not methyl atropine, eliminated the attenuation of nociceptive reactions by arecoline in the rat [Herz, 1962]. Neither Houser [1976] nor Metys et al. [1969], using a variety of nociceptive tests, was able to demonstrate antinociceptive effects for carbachol, a peripherally acting cholinomimetic. These studies seem to suggest that cholinomimetic compounds exert their antinociceptive effects through a central mechanism.

Possible Anatomical Foci that May Mediate the Antinociceptive Actions of Cholinomimetics. Metys et al. [1969] reported that intraventricular injections of carbachol, as well as oxotremorine, produced an increase in the threshold for eliciting licking reactions in the rabbit by tooth-pulp stimulation. The authors concluded that cholinomimetics must induce their antinociceptive effects by acting on structures adjacent to the ventricles. In order to define further the anatomical foci that mediated these effects, they injected small quantities of carbachol (0.3 μg) into various regions of the rat brain and measured the changes in nociceptive reactions following such injections. They found that injections of this cholinomimetic into the septal area, as well as the midbrain reticular formation and central gray matter, produced dramatic increases in the stimulation thresholds needed to evoke vocalization afterdischarges in rats. Injections into the caudate nucleus or dorsal hippocampus also appeared to elevate thresholds, but to a much lesser degree. Injections into the medial thalamic nuclei were even less effective. Similar effects were found following injections of arecoline and oxotremorine. The efficacy of oxotremorine, which has no nicotinic properties, as well as the ability of atropine to reverse the antinociceptive effects of these cholinomimetics, suggest a rather specific muscarinic mechanism underlying these central actions.

Katayama et al. [1984a, b] have very elegantly characterized sites in the cat pons which appear to mediate antinociceptive as well as motoric effects of cholinomimetics. A variety of studies have reported that injections of cholinomimetics into the rostral pontine tegmentum produce postural atonia [Mitler and Dement, 1974; Von Dongen et al., 1978; Velasco et al., 1981]. Other reports have noted that animals are also unresponsive to external stimuli, even those with noxious characteristics following pontine injections [Baxter, 1969; George et al., 1964; Hernandez-Peon et al., 1963]. More recently, Katayama et al. [1984a] have carefully mapped the rostral pontine tegmentum for somatomotor, visceromotor as well as nociceptive somatosensory effects of carbachol (0.4 μg in 0.2 μl of saline). Injections of carbachol into an area caudal to the ventral tegmental nucleus produced suppression of both postural somatomotor and various visceromotor functions in cats. The reduction of sympathetic tone following intracerebral carbachol may have been responsible for producing a number of visceromotor effects, including relaxation of nictating membranes, miosis, reduction of resting blood pressure and heart rate as well as the failure of external stimuli to produce changes in respiration and heart rate. Injections of carbachol into this region also produced an abolition of responsiveness to various external stimuli and a profound reduction of the resting tone of postural muscles. Injections of carbachol into an area ventral to the locus coeruleus produced predominantly a suppression of postural somatomotor functions. Injections of carbachol into an area surrounding the lateral half of the brachium conjunctivum, on the other hand, appeared to suppress nociceptive somatosensory functions selectively. Injections of carbachol into this area suppressed the tail-flick reaction as well as defensive/affective reactions in response to noxious stimuli. In a subsequent study [Katayama et al., 1984b], the same authors characterized more fully the antinociceptive effects of carbachol injections into the pons. They reported that injections into the dorsal part of the parabrachial region of the cat pons produced a profound suppression of nociceptive responses which were not accompanied by generalized sensory, emotional, or motor deficits, indicating a specific effect on nociceptive input. Injections of carbachol into the ventral part of the parabrachial region produced a slight suppression of motor responses in addition to antinociception. The authors proposed that the antinociceptive effects of carbachol in this region of the cat brain may involve descending inhibitory influences on the sensory input to the spinal cord. Since atropine sulfate but not mecamylamine, a nicotinic antagonist, was able to reverse carbachol-induced antinociception, it appears that muscarinic but not nicotinic receptors are involved in mediating these cholinomimetic effects. Opiate mechanisms also are not apparently involved, since naloxone failed to antagonize the antinociceptive effects of carbachol, and injections of morphine into carbachol-sensitive sites failed to produce analgesia.

Endogenous acetylcholine in the pons may play a role in modulating nociception. Katayama et al. [1984c] have found that injections of atropine into the parabrachial region attenuate antinociceptive effects elicited by the exposure of cats to novel environmental changes without affecting their general responsiveness to environmental stimulation. This suggests that certain external stimuli are capable of activating this cholinergic mechanism to produce an attenuation of nociceptive reactions.

Since the effects of cholinomimetics in the parabrachial region appear to be mediated through the muscarinic cholinergic receptor, it is interesting to note that the density of muscarinic receptors is quite high in the dorsal parabrachial nucleus [Wamsley et al., 1981]. The ventral parabrachial nucleus also appeared to have a rather high concentration of autoradiographic grains in the pontine levels which diminished somewhat in the rostral extent. Since [^{3}H]QNB binding has also been found to be high in the nucleus tractus spinal nervi trigemini and the substantia gelatinosa trigemini, muscarinic cholinomimetics may be expected to play an important role in modulating trigeminal nociceptive transmission as well. The periaqueductal gray matter has been reported to have intermediate to moderate concentrations of QNB binding [Wamsley et al., 1982; Rotter et al., 1979; Nanaka and Moraji, 1984]. It is conceivable that muscarinic receptors in this area represent an additional target for cholinomimetics in the induction of analgesia [Metys et al., 1969].

Role of Other Neurohumors in Cholinomimetic-Induced Antinociception. While the antinociceptive effects of cholinomimetic compounds have generally been thought to be mediated directly through the cholinergic receptor interactions, there is some evidence to implicate the participation of other neurohumoral systems as well. Bite and Gyorgy [1971] demonstrated that the antinociceptive effects of oxotremorine could be potentiated by tetrabenazine, a catecholamine depleter. This effect of tetrabenazine could be antagonized by a concomitant administration of desipramine, amphetamine, and enhanced by 5-hydroxytryptophan, a serotonin precursor. Phenoxybenzamine, a potent alpha-adrenergic blocker, also potentiated the antinociceptive effects of oxotremorine in the hot-plate test. The authors concluded that the potentiation of oxotremorine by tetrabenazine is related to decreased catecholaminergic activity produced by the compound. Pleuvry and Tobias [1971] also reported similar findings. In this study, pretreatment with alpha-methyl-*p*-tyrosine (AMPT), a tyrosine hydroxylase inhibitor, and diethyldithiocarbamate potentiated the antinociceptive effects of oxotremorine. Conversely, pretreatment with *L*-DOPA decreased the antinociceptive efficacy of oxotremorine. Oxotremorine antinociception was unaffected by pretreatment with *p*-chloro-phenylalanine or 5-hydroxytryptophan. Sethy et al. [1971], on the other hand, reported changes in tremorine-induced antinociception by diethyldithiocarbamate, reserpine, and *p*-chloro-phenylalanine that are at variance with the above results. To add further confusion to this issue, Pedigo and Dewey [1981a, b] reported that both reserpine and tetrabenazine antagonized the antinociceptive effects of i.c.v. acetylcholine. More selective depletors like AMPT and diethyldithiocarbamate were ineffective while *p*-chloro-phenylalanine methyl ester increased the ED_{50} of oxotremorine. The authors concluded that the effects of tetrabenazine and reserpine could be due to their multiple effects on central biogenic amines or to changes in other brain systems such as calcium or various intracellular mediators. One possible variable that may account for the disparate findings with reserpine is the amount of time allowed to elapse following treatment. Howes et al. [1969] reported that reserpine given simultaneously with oxotremorine potentiated its antinociceptive effects, while administration 16 h before antagonized its effects.

Role of Endogenous Opiate Mechanisms in the Antinociceptive Effects of Cholinomimetics. While some studies have tried to relate the actions of cholinomimetics to brain amines, other studies have postulated an involvement of opiate mechanisms. Howes et al. [1969] and Harris et al. [1969] reported that naloxone and nalorphine were effective antagonists of oxotremorine-induced antinociception. Howes et al. [1969], however, failed to find cross-tolerance between oxotremorine and morphine. This finding appears to be inconsistent with the observation by Pedigo and Dewey [1981] that mice made tolerant to morphine exhibited significant cross-tolerance to oxotremorin. Pedigo et al. [1975] also found that the antinociceptive effects of acetylcholine were blocked by five narcotic antagonists in the same rank order potency in which they antagonized morphine. Interestingly, the stereospecificity of pentazocine and cyclazocine was reserved in antagonizing acetylcholine with respect to their antagonistic actions against morphine. Dewey et al. [1976] also found similar effects with benzomorphan antagonists (i.e. the *d*-isomers were more potent in reversing the effects of acetylcholine, while the *l*-isomers were more potent in reversing the effects of morphine). These data differ from the results of Ramabadran and Jacob [1978] who showed a similar stereospecificity for opiate antagonists in reversing antinociceptive effects of morphine or systematically administered arecoline. Pedigo and Dewey [1981] proposed that this discrepancy may be due to the different routes of administration used for the cholinergic agonists. Finally, it must be noted that Pleuvry and Tobias [1971] and Lipman and Spender [1980] failed to find that naloxone or nalorphine had any ability to reverse analgesia induced by oxotremorine.

One additional observation makes it unlikely that i.c.v. acetylcholine-induced antinociception is mediated through a specific opiate mechanism. Pedigo and Dewey [1981] constructed modified Schild plots for naloxone inhibition of acetylcholine and morphine-induced antinociception. The apparent pA2 value for morphine was quite similar to previously reported values [Tulunay and Takemori, 1971] and the slope was close to unity. In contrast, the pA2 for acetylcholine was much lower and the slope was significantly greater than one. These data suggest that naloxone is acting in a noncompetitive manner against i.c.v. acetylcholine and probably at a different site from that at which it inhibits morphine antinociception. Thus, it appears that acetylcholine and morphine produce their antinociceptive effects through different if not entirely independent mechanisms.

Cholinomimetics and Opiate Interactions. Considering the ability of cholinomimetics compounds to alter nociceptive thresholds, it is not surprising that a number of them have been found to potentiate morphine antinociception. Saxena [1958] found that doses of pilocarpine, which had no antinociceptive activity by themselves, did potentiate morphine-induced antinociceptive effects in the Haffner tail-clip test. Lipman and Spencer [1980], using the tail-withdrawal test, found that oxotremorine produced antinociceptive effects that were additive with morphine as well as clonidine, while Pleuvry and Tobias [1971] reported potentiation of the antinociceptive effects of morphine with oxotremorine in the hot-plate test. Howes et al. [1969] found potentiation of morphine antinociceptive effects in the tail-flick test with oxotremorine. A central site of action for such interactive effects is likely, since i.c.v. acetylcholine also has been found to potentiate morphine antinociception in the mouse [Pedigo et al., 1975]. Mudgill et al. [1974], however, have found that 15 nmol of acetylcholine produced a significant antagonism of morphine analgesia when tested 30 min after injection. No significant decrease was observed at 15 min after treatment nor following injections of 30 nmol of acetylcholine. Surprisingly, in the same study, i.c.v. physostigmine also decreased morphine analgesia, an observation that is in opposition to numerous other reports reviewed above which demonstrated potentiation of morphine by systemically administered physostigmine.

Confounding Effects of Cholinomimetics. One of the most noticeable behavioral effects of cholinomimetics like tremorine or oxotremorine is the production of muscular tremors. In fact, this characteristic of cholinomimetics was originally used to evaluate the potential efficacy of anti-parkinsonian drugs [Blockus and Everett, 1957]. At high doses, cholinomimetics like arecoline, pilocarpine, and tremorine, produce catalepsy in mice [Zetler, 1968]. The cataleptic effects are apparently mediated through muscarinic receptors, since atropine and scopolamine were effective antagonists of this action. Tang and Yim [1965] studied the effects of arecoline on monosynaptic reflexes from the lumbar spinal cord in spinal cats. Following intra-arterial injections into the spinal cord blood supply, arecoline and oxotremorine were found to depress both monosynaptic and polysynaptic reflex discharges recorded from the ventral root following dorsal root stimulation. Atropine was considerably more potent than methylatropine nitrate in blocking the effects of arecoline. The authors concluded that arecoline-induced depression of the monosynaptic discharge was likely the result of an action of arecoline on Renshaw cells through muscarinic rather than the classical nicotinic receptors. It is possible, however, that the effects of arecoline may have been mediated through muscarinic receptors located on other interneurons in the spinal cord.

While cholinomimetics exert dramatic effects on somatomotor as well as spinal reflex activity, it is again surprising that few studies have considered these effects when evaluating the antinociceptive actions of such cholinergic compounds. Katayama et al. [1984a, b] have demonstrated the type of analysis that can and should be performed on somatomotor and somatosensory functions when evaluating the antinociceptive actions of drugs, especially cholinomimetics.

Cholinolytics

Since cholinomimetics and anticholinesterases generally appear to attenuate nociceptive reactions, cholinolytics should exert the opposite effect. Again, the assumption underlying this postulation is that cholinergic mechanisms are tonically active in suppressing nociceptive transmission or integration. Unfortunately, the findings from studies that have evaluated the actions of anticholinergics on nociceptive behaviors are somewhat equivocal. Christensen and Gross [1948] failed to find that atropine and scopolamine altered the aversive nature of radiant heat applied to the forehead in human subjects. Dundee et al. [1961], however, reported that i.v. injections of atropine and scopolamine increased the sensitivity of human volunteers to somatic pain induced by tibial pressure. Migdal and Frumin [1963], in a preliminary report, found that when a combination of atropine and benzotropine were administered prior to the induction of anesthesia in patients undergoing elective surgery, 4 out of 14 subjects required no analgesics postoperatively while 10 needed only an initial dose of 50 mg of meperidine on the average of 5 h postoperatively. This finding appears to suggest that anticholinergics may have analgesic effects in man. A number of studies in laboratory primates appear to support this possibility. Houser and Houser [1973] found that scopolamine raised aversive thresholds in squirrel monkeys in the shock-titration paradigm. Pert [1975] and Pert and Maxey [1975] confirmed these findings using a similar task. As little as 0.05 mg/kg of scopolamine was sufficient to raise shock thresholds in rhesus monkeys, while even 0.20 mg/kg of methscopolamine was ineffective. While the effects of anticholinergic compounds on shock thresholds are relatively consistent in the laboratory primate, the findings concerning the effects of these compounds on nociceptive reactions in the rodent are less than equivocal. Initially, Houser and van Hart [1973, 1974] did not observe any effects of scopolamine on shock threshold as measured by the spatial preference test. However, when scopolamine and atropine were given to rats during their initial exposure to the spatial preference procedure (when thresholds were high), both decreased aversive thresholds [Houser, 1976]. Since methyl atropine and methyl scopolamine were about as effective as the tertiary analogues, it appears that these effects are not predominantly central in nature. Two other studies in rodents have reported that anticholinergics increase nociceptive thresholds. de Jong [1954] reported that atropine raised reaction thresholds to thermal radiation in guinea pigs. van Eick and Bock [1971] also studied the actions of scopolamine on the hot-plate reactions of mice. While scopolamine was found to delay the hot-plate licking reactions of the animals, careful observation brought to light that under the influence of anticholinergics, another response pattern appeared at the moment when the licking reaction could be expected to occur (short rapid trembling of one of the hind feet). Morphine was observed to reduce the frequency of both the paw-lick and the trembling reactions, while scopolamine reduced the frequency of the paw-lick reactions but increased the frequency of the trembling reactions. It appears that scopolamine had simply prevented the appearance of one type of reaction, primarily by interfering with its motor expression.

Cholinolytic and Opiate Interactions. Slaughter and Munsell [1940] were probably the first to notice that anticholinergic drugs may antagonize opiate analgesia. Smith et al. [1944] also noted a tendency for scopolamine to abolish the analgesic effects of opiates. Quinine, however, appeared to have the same effect. Two studies in man also have reported a tendency for anticholinergics to decrease opiate analgesia. Christensen and Gross [1948] found that 0.3 mg/kg atropine injected i.v. 40 min after methadone decreased the intensity and duration of analgesia. Dundee et al. [1961] found that scopolamine decreased the analgesic index of pethidine. The findings from more recent studies, however, are not as consistent. While Takemori et al. [1975] found that atropine raised the ED_{50} of morphine-induced antinociception in the tail-flick test, de Jongh [1954] and Denisenko [1965] reported that anticholinergics potentiated the antinociceptive effects of morphine in the rodent. Pert [1975] also found a potentiation of morphine-induced increases in the shock thresholds of rhesus monkeys by scopolamine. In this study, scopolamine appeared to combine in a multiplicative fashion with morphine. This interaction was taken to suggest that the two compounds are exerting their antinociceptive actions through different systems or mechanisms. This possibility received further support in a subsequent study from the same laboratory. Pert and Maxey [1975] reported that tolerance to morphine attenuated the response to scopolamine, but tolerance to scopolamine had no effect on morphine-induced antinociception. In addition, the shock threshold was not modified by injections of scopolamine or arecoline into brain sites that had been found previously to be responsive to morphine injections. These findings again were interpreted to imply that morphine and scopolamine do not exert their antinociceptive effects through identical neural substrates. On the other hand, the fact that tolerance to morphine attenuated the effect of a second administration of scopolamine does indicate that there may be some degree of overlap between the substrates affected by either compound, whether at a biochemical or neurophysiological level.

Nicotinic Activation and Antinociception. Although nicotine was one of the first cholinergic compounds for which antinociceptive actions were described [Davies et al., 1932], considerably less research has been directed at characterizing the effects of this class of compounds on nociceptive reactions than muscarinic agonists. Mattila et al. [1965] reported that nicotine as well as ethylnornicotine increased the reaction latencies of mice on the hot-plate. The same doses that increased hot-plate latencies, however, also decreased significantly the motor activity of the animals and caused them to fall down from a rotating rod.

Mecamylamine, a tertiary nicotinic blocker which penetrates the blood-brain barrier, effectively antagonized the antinociceptive as well as locomotor depressant actions of nicotine. Several years later, the same laboratory described a more comprehensive study that carefully analyzed the antinociceptive and sedative actions of nicotine [Mattila et al., 1968]. Following an injection of 2.5 mg/kg of nicotine s.c., the antinociceptive as well as locomotor depressant actions reached their peak value in 10 min and persisted for as long as 40 min. In general, there seemed to be a positive correlation between the antinociceptive and locomotor depressant actions. In rabbits, i.v. injections of 50–200 μg/kg produced restlessness, intense panting, and a decrease in reactions to tooth pulp stimulation. The antinociceptive effects appeared within two minutes and lasted only 5–10 min following injection. Longer lasting actions were observed following s.c. injections of 2.5 μg/kg of nicotine. The tertiary blocking agent, mecamylamine, did not alter the nociceptive threshold in rodents by itself, but did effectively antagonize the antinociceptive effects of nicotine without blocking nicotine-induced respiratory stimulation. Scopolamine, the muscarinic blocker, and hexamethonium, a quaternary nicotinic blocker with only peripheral actions, were ineffective in antagonizing the effects of nicotine. These findings suggest that nicotine produces its antinociceptive actions through nicotinic activation at a central site. When nicotine was administered chronically to mice, tolerance developed to its antinociceptive effects. In addition, some cross-tolerance was found between morphine and nicotine. Mansner [1972] has also found tolerance to the antinociceptive effects of nicotine on the hot plate in mice. Phan et al. [1973] also described the antinociceptive effects of nicotine in mice and rats. In mice, an injection of 7.5–10 μg/kg produced a rapid rise in hot-plate latencies, an effect that also subsided rapidly. Ten to 15 min after an i.p. injection, a marked effect was found on the hot-plate reaction which dissipated rapidly over time, being no longer observable 30 min later. No tachyphylaxis was seen following repeated nicotine injections at intervals of 60 min. Tachyphylaxis did, however, develop to nicotine-induced motor impairment when measured with the roto rod test. On the other hand, daily treatment of mice with nicotine for 4 days did reduce its antinociceptive actions. In this study, mecamylamine had no effect itself but did antagonize the antinociceptive effects of nicotine. Atropine was entirely ineffective as an antagonist. High doses of nicotine were also found to antagonize acetic acid-induced writhing in mice. A central site of action for the antinociceptive effects of nicotine was suggested by the finding that intraventricular injections of nicotine (at somewhat excessive doses, i.e. 125 μg) in rats also elevated the latencies for hot-plate reactions. Again, mecamylamine antagonized this effect of nicotine. Sahley and Bernston [1979] also evaluated the antinociceptive effects of nicotine in mice and rats. In these studies, injections of 0.5 μg/kg of nicotine produced a rapid rise in tail-flick latencies in rats, an effect which lasted for about 40 min. While nicotine increased tail-flick latencies to radiant heat, surprisingly it also appeared to render the animals hyper-responsive to touch on pinches of the tail. It is not entirely clear why the actions of nicotine would be specific to radiant heat stimulus. It is interesting to note that nicotine has not been found to produce antinociceptive effects in other studies that have utilized pressure or electrical shock as nociceptive stimuli [Kamat et al., 1972; Pert, 1975; Rogers, 1979]. These findings are somewhat puzzling but do seem to suggest that motoric deficits may not be sufficient to account for the antinociceptive effects of nicotine in the tail-flick task. Intraventricular injections of nicotine were also found to elevate tail-flick latencies in the study by Sahley and Bernston [1979]. Nicotine-induced antinociception was antagonized by the centrally active nicotinic blocker mecamylamine but not by the quaternary blocking agent, hexamethonium. Surprisingly in this study, in contrast to previous studies, the muscarinic blocker, scopolamine, was also effective in antagonizing the effects of nicotine. Hexamethonium and scopolamine methylnitrate, which do not pass into the brain, had no effect on nicotine-induced antinociception. The authors concluded that nicotine may exert its antinociceptive actions through the muscarinic cholinergic system. Since nicotine is known to be a potent releaser of acetylcholine [Armitage and Hall, 1967; Armitage et al., 1966; Essman, 1973], it is possible that the antinociceptive effects of nicotine are mediated in part by the release of acetylcholine which may produce antinociception through muscarinic cholinergic receptors.

More recently, Tripathi et al. [1982] have also found antinociceptive effects for nicotine. As described by previous reports, 1 μg/kg s.c. nicotine had a rapid onset of action in the tail-flick test. The effect also dissipated rapidly, being barely detectable 10 min later. A good correspondence was observed between brain and plasma levels of nicotine and its antinociceptive actions in the mouse but not the rat. Although mecamylamine was inactive by itself on the tail-flick response, it did antagonize the effects of nicotine. Large doses of hexamethonium were also partially effective. Neither atropine nor scopolamine was able to block the effects of nicotine. Nicotine has also been shown to decrease reactions to radiant heat in the dog. Kamerling et al. [1982] found that nicotine administered intravenously increased the skin twitch latency in dogs to radiant heat. This effect of nicotine was also accompanied by increased cardiac and respiratory rates, hypothermia, and behavioral restlessness.

As noted above, the antinociceptive effects of nicotine appear to be mediated through a central site of action. Generally speaking, mecamylamine has invariably been found to antagonize the antinociceptive effects of nicotine, but hexamethonium, a quaternary blocker that acts peripherally, has either been ineffective or less effective than mecamylamine. Furthermore, direct injections of nicotine into the ventricles has also been shown to produce potent antinociceptive actions. Aceto et al. [1983] have evaluated the relative contributions of peripheral and central sites of action of nicotine by comparing the antinociceptive effects of a series of quaternary methiodides of nicotine with nicotine. They reported that following peripheral administration, all the methiodides were inactive in the tail-flick procedure, whereas nicotine was active in both tests. Intraventricular injections of the methiodides, on the other hand, were effective in inhibiting the tail-flick reflex. Kamerling et al. [1982] were also able to demonstrate an inhibition of skin twitch reactions in dogs following intraventricular injections of nicotine. The effects, however, were not as consistent as following systemic administration. While it is clear that the predominant effects of nicotine on nociceptive reactions are through central mechanisms, there is evidence that in some tasks peripheral actions may also play a role. Aceto et al. [1983] found that systemically administered nicotine pyrralidine methiodide was about one-third as active as nicotine in inhibiting writhing reactions induced by s.c. injections of phenylquinone. Interestingly, hexamethonium partially blocked nicotine and the methiodide derivative, whereas mecamylamine blocked nicotine completely. The authors concluded that nicotine may have both central and peripheral actions in the writhing test.

The central effects of nicotine are undoubtedly determined by the interactions of this compound with specific nicotinic cholinergic receptors in brain. Recently, Clarke et al. [1984, 1985] have described the autoradiographic distribution of high affinity binding sites for ^{3}H-nicotine in the rat brain. Especially high densities of binding were found in the thalamus (especially the so-called 'relay nuclei') and layers 3 and 4 of the cerebral cortex. These two regions may represent important foci for the antinociceptive effects of nicotine, although it is not clear how an initial effect at either focus may be translated into an inhibition of spinal nociceptive reflex activity. One major focus for the inhibitory effects of nicotine on the tail-flick reflex does, in fact, appear to be at the level of the spinal cord. Aceto et al. [1986] have reported that injections of nicotine into the subarachnoid space of mice produced a potent inhibition of the tail-flick. In fact, they found that it required 20–30 times less nicotine to inhibit the reflex following subarachnoid injections than following intraventricular injections. Furthermore, at the time of antinociception, almost all of the ^{3}H-nicotine was localized at the site of injection with very little recovered from supraspinal structures. It might be assumed that a nicotinic activation of Renshaw cells in the cord underlies the inhibition of the tail-flick reflex. It is unclear, however, whether this mechanism can also account for the inhibitory effects of nicotine on other nociceptive responses.

Although ^{3}H-nicotine binding has also been found to be quite high in the interpeduncular nucleus, superior colliculus, medial habenula, substantia nigra and the molecular layer of the dentate gyrus [Clarke et al., 1984, 1985], it is unlikely that the antinociceptive effects of nicotine are mediated through any of these structures, since they have not been implicated in nociception. It should be noted that portions of the periaqueductal gray matter also appeared to have binding that was clearly above background. However, it is also unlikely that this is an important site of action for nicotine, since Kamerling et al. [1982] failed to elicit antinociceptive effects in the dog following direct injections of nicotine into this structure.

As reviewed above, cholinomimetics have been reported by many to increase nociceptive thresholds. There is little evidence, however, that any of the antinociceptive effects of these compounds are mediated through nicotinic activation. Herz [1962] reported that mecamylamine was ineffective in antagonizing the antinociceptive effects of arecoline, a cholinomimetic with both nicotinic and muscarinic properties. Pedigo et al. [1975] failed to block the antinociceptive effects of intraventricular acetylcholine with mecamylamine. Dewey et al. [1975] have also presented evidence to suggest that the actions of i.c.v. acetylcholine are mediated through muscarinic and not nicotinic activation. They found that the alpha substituted acetylcholines were inactive in inhibiting the writhing reactions in mice, while the beta substituted acetylcholines were effective. The alpha substituted cholines have been shown to have a specific activity for nicotinic receptors while the beta substituted acetylcholines are potent at peripheral muscarinic sites.

While nicotine appears to possess potent antinociceptive properties, it also produces a number of untoward side effects that may alter nociceptive reactions without modifying pain perception per se. For example, in the peripheral neurons system, the primary action of nicotine is a transient stimulation followed by a more persistent depression of all automatic ganglia. At the neuromuscular junction the initial stimulant action is rapidly replaced by paralysis. At appropriate doses, the CNS actions include the production of tremors and, ultimately, convulsions. Nicotine has been reported also to have significant effects on spinal reflexes.

Considering the effects of nicotine on spinal reflex activity and locomotor behavior, it would not be surprising if some of these actions would contribute to the inhibition of nociceptive reactions. Schweitzer and Wright [1938] reported that nicotine inhibited spinal reflexes by a direct action on the spinal cord. Curtis et al. [1957] proposed that the depression was due to excitation of Renshaw cells that are normally activated by cholinergic motor axon collaterals which exert a potent inhibitory action on motor neurons. Pradhan et al. [1967], on the other hand, have demonstrated that nicotine acting centrally (i.e. following i.c.v. injections) is also capable of inhibiting both monosynaptic as well as polysynaptic spinal reflexes. Nicotine has also been shown to produce potent effects on general locomotor output of rodents [Clarke and Kumar, 1983]. In nontolerant rats, nicotine (0.1–0.4 mg/kg) depressed locomotor activity and induced ataxia in the first 20 min, but increased activity subsequently. It should be noted that the doses of nicotine that have been used to induce antinociception have almost always been higher than those reported by these investigators to depress locomotor output.

Mattila et al. [1965; 1968], of course, have observed a positive correlation between sedative and antinociceptive effects of nicotine but did not relate the motoric deficits to antinociception. Mansner [1972] also reported an inverse relationship between locomotor output and the antinociceptive effects of nicotine on the hot-plate test. Sahley and Bernston [1979] have been two of the few investigators concerned with the confounding effects of nicotine in measuring its antinociceptive properties. They concluded that it was unlikely that the elevated tail-flick latencies observed following injections of nicotine were due simply to a motor impairment of the tail, since the animals also failed to vocalize, escape, or exhibit other aversive reactions to thermal stimulation. The authors also asserted that it was likewise unlikely that the effects were due to general behavioral depression since, following the administration of nicotine, the animals still remained responsive to visual, auditory, and somatic stimuli. In fact, as noted above, nicotine was found to enhance reactions to tail-pinch in these studies.

Conclusions

It seems apparent that alterations in cholinergic function can lead to changes in nociceptive behavior. What is not certain at this time, however, is whether these changes reflect modifications in the perception of nociceptive stimuli or whether they are merely related to the inhibition of motoric reactions to nociceptive input. Electrophysiological procedures might reveal which of these alternatives is more tenable.

There are several other issues that need clarification with regard to the role of the cholinergic system in modulating nociception or nociceptive behavior. In man it is not entirely clear whether the reported antinociceptive effects of cholinergic stimulation (through cholinesterase inhibition) are determined through peripheral or central mechanisms. As noted above, neostigmine as well as physostigmine have been shown to have antinociceptive properties in man under certain conditions and experimental settings, suggesting an involvement of peripheral cholinergic systems. On the other hand, at least some of the antinociceptive effects associated with cholinergic activation in man must be mediated through the CNS, since Sitaram et al. [1977] have also demonstrated antinociceptive properties for physostigmine after blockade of peripheral muscarinic receptors with methyl atropine. It is possible that both peripheral and central cholinergic systems are involved in modulating nociceptive behaviors in man. The precise mechanisms underlying the antinociceptive effects of peripheral cholinergic activation are not at all clear at this time. It is entirely possible that these actions may not be related to the inhibition of pain perception per se.

In laboratory animals the antinociceptive effects of cholinergic activation, whether with cholinomimetics or anticholinesterases, appear to be mediated predominantly through central mechanisms, since peripherally-acting compounds generally have not been found to be effective. If enhanced cholinergic activity in the CNS does, in fact, produce analgesia in rodents (and possibly man), what is the precise neuroanatomical substrate involved in this action? Does cholinergic activation inhibit the transmission of nociceptive information by enhancing the activity of descending inhibitory pathways to the spinal cord [Fields and Buxbaum, 1978], or does it inhibit the perception or integration of nociceptive stimuli at higher levels of the neuraxis? At least part of the antinociceptive effects of cholinergic activation appear to be mediated through descending inhibitory mechanisms. Inhibition of the tail-flick reaction (a spinal reflex) has been observed following i.c.v. injections of neostigmine [Pedigo et al., 1975] or acetylcholine [Pedigo and Dewey, 1981a, b]. The descending inhibitory systems activated by enhanced cholinergic activity may originate from the parabrachial region of the pons, since Katayama et al. [1984a, b] have shown that injections of carbachol into this region were also effective in attenuating the tail-flick reflex in cats. While these central effects seem to be mediated through the muscarinic cholinergic receptor (see above), it appears that nicotinic receptor activation directly at the level of the spinal cord is also effective in attenuating the tail-flick reflex. It is not clear, however, whether the effects of nicotinic cholinergic stimulation in the cord produce a decrease in pain transmission and perception, or whether the effects of nicotinic compounds on nociceptive reflex activity are mediated through Renshaw cells.

While muscarinic cholinergic activation appears to inhibit nociceptive reactions, some studies have also reported antinociceptive effects following the administration of muscarinic antagonists. It is unlikely that both of these pharmacological effects (muscarinic activation and blockade) have physiological relevance. Of the handful of studies that have reported antinociceptive effects following muscarinic blockade, three utilized the shock-titration paradigm with monkeys as subjects, while two used the hot-plate test with rodents. Van Eick and Bock [1971] have proposed that anticholinergics may simply alter the nature of the response of rats on the hot-plate and not actually decrease the perceived intensity of noxious stimulation. The antinociceptive actions of anticholinergics in the laboratory primate may also not be related to the effects of these compounds on the protopathic component of pain. Pert and Maxey [1975] have suggested that the increases in shock threshold following administration of scopolamine may be related to attentional deficits.

Catecholaminergic Modulation

Interest in the involvement of brain catecholamines in antinociceptive mechanisms stems from early studies that attempted to relate the analgesic effects of opiates to either an inhibition or activation of these neurohumoral systems. An involvement of brain catecholamines in the action of opiates was initially suggested by Vogt [1954], who found a decrease in hypothalamic norepinephrine following the administration of morphine. This finding was confirmed by a number of subsequent reports from other laboratories [see Way and Shen, 1971, for review]. Such decreases in brain norepinephrine were initially suggested to be the result of enhanced release that exceeds the rate of synthesis of this catecholamine. Numerous subsequent studies have attempted to define the role brain catecholamines in nociception and opiate analgesia by evaluating the antinociceptive effects of compounds that affect catecholaminergic neurotransmission and by characterizing their interactive effects with opiate analgesics.

Effects of Sympathomimetics and Catecholamine Agonists on Pain Thresholds

There is relatively good evidence to suggest that increasing catecholaminergic tone produces an attenuation of nociceptive reactions. Sympathomimetics like amphetamine, which enhance relase and/or inhibit reuptake of catecholamines, have been found to increase nociceptive thresholds in a number of experimental situations [Görlitz and Frey, 1972; Major and Pleuvry, 1971; Houser and Houser, 1973; Vedernikov, 1969; Vedernikov and Afrikanov, 1969; Pleuvry and Tobias, 1971]. Pleuvry [1971] has also reported cross-tolerance between the antinociceptive effects of amphetamine and morphine. 4-Alpha-Dimethyl-m-tyramine, a potent releaser of central catecholamine stores, has also been shown to have some antinociceptive properties as measured by shock-elicited vocalization [Paalzow and Paalzow, 1976a]. It is not apparent whether the reported effects of sympathomimetics are mediated by the release of central or peripheral catecholamines. Intraventricularly administered norepinephrine, as well as dopamine and epinephrine, have been found to increase hot-plate latencies [Handley and Spencer, 1969]. The same group, however, failed to extend their observations to the paw-pressure test [Sparks and Spencer, 1969]. Intraventricular dopamine and norepinephrine have also failed to elicit antinociceptive effects in other studies [Sparks and Spencer, 1971; Calcutt et al., 1971]. However, it is interesting to note that Contreras and Tamayo [1966] have found increased nociceptive thresholds following systematically administered epinephrine and dopamine. Heller et al. [1978] have also reported increased pain thresholds following systemic epinephrine, norepinephrine, or isoproterenol. These findings appear to suggest that at least a portion of the antinociceptive effects of sympathomimetics may be mediated through peripheral mechanisms.

Dopaminergic Activation

Apomorphine, a dopamine agonist, has been found to increase nociceptive thresholds in the tail-pressure test [Vedernikov and Afrikanov, 1969; Vedernikov, 1969] and inhibit vocalization after-discharges to electrical shock [Paalzow and Paalzow, 1975]. The latter effect was blocked by pimozide, a dopaminergic blocker. Dunai-Kovacs and Szekely [1977], on the other hand, failed to find that apomorphine had appreciable effects on writhing reactions but that it did appear to suppress the reactions of rats in the Randall-Selitto paw pressure test. Rogers [1977] has reported that injections of dopamine directly into the amygdala decreased shock-elicited jump thresholds in the rat, while injections into the caudate-putamen were ineffective.

There is evidence that dopaminergic agonists may exert an effect on nociceptive reactions at the level of the spinal cord. Pharmacological as well as anatomical evidence suggest the presence of dopaminergic mechanisms in the spinal cord. Although the levels of dopamine in the cord are relatively low, they do appear to be derived from specific descending dopamine projections. Spinal levels of dopamine and norepinephrine, for example, can be differentially altered by lesioning techniques [Commissiong et al. 1979; Mouchet et al., 1982]. In addition, retrograde tracing techniques specific for dopamine have revealed that dopamine terminals in the spinal cord probably have their origin in the diencephalon [Björklund und Skagerberg, 1979; Blessing and Chalmers, 1979; Hokfelt et al., 1979; Skagerburg et al., 1982]. Finally, the presence of dopamine receptors in the cord also suggests that this may be one important site of action of exogenously administered dopamine agonists [Demenge et al., 1980, 1981].

A number of studies have evaluated the effects of apomorphine on the tail-flick reflex at the spinal level. Under ordinary conditions, apomorphine does not appear to alter tail-flick reactions following intrathecal or systemic injections, even at relatively high doses [Jensen and Smith, 1983; Jensen and Yaksh, 1984]. In rats anesthetized lightly with sodium pentobarbitone, on the other hand, apomorphine administered either intravenously or intrathecally has been found to elevate tail-flick latencies [Barasi et al., 1984, 1985]. Injections into the lateral ventricle did not alter tail-flick latencies as consistently or to the same degree. Intrathecal injections of apomorphine also attenuated the activity of dorsal horn neuron driven by noxious stimuli. Jensen and his colleagues have examined the effects of apomorphine on the tail-flick reaction following intrathecal injections. As noted above, intrathecal injections of apomorphine appear to have little effect on the tail-flick reaction in rats with an intact neuraxis [Jensen and Smith, 1983; Jensen and Yaksh, 1984]. On the other hand, Jensen and Smith [1982] reported that apomorphine administered intrathecally to spinal rats increased tail-flick latencies, an effect that was antagonized by *cis*-flupenthixol and (+)butaclamol. Methysergide and (−)butaclamol were not effective antagonists, while trans-flupenthixal was effective only at high doses. In a subsequent study [Jensen and Smith, 1983], the same authors also reported similar effects following intrathecal administration of dopamine as well as apomorphine. Again, the effects of apomorphine were antagonized by *cis*-flupenthixol and (+)butaclamol but not by their stereo isomers. Phenoxybenzamine, propranolol, methysergide, and naloxone were ineffective. While intrathecal apomorphine appeared to suppress tail-flick reactions in spinal rats to noxious stimulation, it facilitated the tail-withdrawal response to tactile, non-noxious stimulation. The authors concluded trom these findings that dopamine- and apomorphine-induced elevations in tail-flick latencies were due to an inhibition of spinal nociceptive neurotransmission. The finding that apomorphine was only effective in spinal rats suggests that descending neuronal pathways might tonically inhibit spinal dopaminergic mechanisms. The possibility that these pathways are serotonergic or noradrenergic receives support from the finding that intrathecal apomorphine was also able to elevate tail-flick latencies in intact rats pretreated with either the serotonergic antagonist methysergide or the noradrenergic antagonist phentolamine. Apomorphine was also effective following pretreatment with reserpine, *p*-chlorophenylalanine or a norepinephrine synthesis inhibitor [Jensen and Smith, 1983b]. While apomorphine does not appear to alter tail-flick reactions under normal circumstances in rodents with an intact neuraxis, it has been found to increase hot-plate latencies and inhibit writhing reactions following intrathecal application [Jensen and Yaksh, 1984]. This effect appears to be specific to the spinal cord since injections into the ventricles did not alter the hot-plate reactions. The effects on the hot-plate were probably not due to motoric deficits, since the animals exhibited normal ambulatory behavior and their non-nociceptive reflexes appeared intact. The authors suggest that the differential effects of apomorphine on the tail-flick and hot-plate reactions may be due to effects on different second-order spinal pathways mediating these behaviors.

While dopaminergic activation of the spinal cord appears to alter spinal reflex activity under certain conditions and attenuate some nociceptive reactions, there is little evidence to suggest that this class of compounds are important analgesics in man. Only one study has indicated that apomorphine may have some clinical use in treating pain. Miley et al. [1978] reported that apomorphine may be beneficial in relieving symptoms of thalamic pain in man. The relationship of this observation to dopaminergic modulation of nociception is not clear, however.

Noradrenergic Activation

While the role of dopamine in modulating nociceptive transmission is equivocal, there is considerable evidence to suggest an important function for norepinephrine, especially at the level of the spinal cord.

One adrenergic agonist that has been found to attenuate nociceptive reactions almost universally is clonidine, a compound that acts selectively on alpha-adrenergic receptors. Paalzow [1974] originally found that clonidine increased shock thresholds for vocalization in rats and mice following systemic administration, while Schmitt et al. [1974a] characterized the antinociceptive effects for clonidine as well as a number of other alpha-adrenergic stimulants following intraventricular and systemic injections. They concluded that the failure of certain alpha-agonists to produce an antinociceptive effect could be related to their inability to penetrate into the brain. Lipman and Spencer [1979] confirmed this possibility by demonstrating that while the peripherally-acting 2,6 diethyl analog of clonidine did not alter the tail-withdrawal reaction in mice following systemic injections, it was effective in this test following intraventricular injections. In a subsequent study, Schmitt et al. [1974b] characterized the pharmalogical effects of xylazine, another alpha-adrenergic agonist, by examining the efficacy of various adrenergic blocking agents in antagonizing its analgesic effects. Yohimbine and piperoxan, two presynaptic alpha-adrenergic blockers, were found to be the most efficacious. Propranolol (a beta-adrenergic blocker) and phenoxybenzamine (a postsynaptic alpha-blocker) were relatively ineffective. Fielding et al. [1978] were also not able to antagonize the antinociceptive effects of clonidine in the tail-flick test with phenoxybenzamine. These findings suggest that the activation of postsynaptic alpha-adrenergic receptors is not critical for the antinociceptive effects of systemically administered clonidine. Chance [1983] and Spaulding et al.

[1979b] have also reported that systemically administered clonidine inhibits the tail flick reflex while Hynes and Henderson [1984], as well as Bentley et al. [1977], have found an inhibition of writhing reactions following administration of clonidine.

The effects of clonidine, as well as other alpha-receptor agonists, on the writhing reaction appear to be mediated through both peripheral and central sites of action with different pharmacological profiles. Bentley et al. [1977], for example, reported that clonidine and oxymetazoline inhibited acetic acid-induced writhing following either systemic or intracisternal injections. The systemic effects appeared to involve presynaptic alpha-receptors, since they were reversed readily by piperoxane. The central effects, however, were refractory to this antagonist. The authors postulated that the central effects may have been mediated through postsynaptic alpha-receptors. This seems unlikely since neither Schmitt et al. [1974b] nor Fielding et al. [1978] were able to antagonize the antinociceptive effects of clonidine with postsynaptically acting alpha-receptor blockers using other nociceptive tests. Paalzow and Paalzow [1976b] have also found clonidine-induced antinociception to be reversed by the presynaptic alpha-adrenergic blocker, yohimbine. Chlorpromazine, atropine, PCPA, reserpine, and phenoxybenzamine, on the other hand, were all found to enhance the antinociceptive effects of clonidine. The finding that phenoxybenzamine enhanced the antinociceptive effects of clonidine may not be entirely inconsistent with the actions of clonidine. It is possible that clonidine may actually inhibit noradrenergic neurotransmission by acting preferentially on noradrenergic presynaptic or 'auto' receptors. In fact, there is evidence that inhibition of norepinephrine release in the peripheral adrenergic system by clonidine is mediated by activation of a presynaptic or prejunctional alpha-receptor [Starke and Montel, 1973; Starke and Altman, 1973]. It has also been proposed that the inhibition of noradrenergic neurons in the locus coeruleus by clonidine is determined by the activation of autoreceptors on or near the norepinephrine cell bodies [Svensson et al., 1975]. Thus, one of the actions of clonidine appears to be the inhibition of norepinephrine release. While this action appears to be at variance with the role of spinal norepinephrine in suppressing nociceptive transmission (see below), it is consistent with the finding that lesions of the locus coeruleus elevate nociceptive thresholds in rats [Sasa et al., 1977; Bodnar et al., 1978]. This, of course, would imply that the main site of action of clonidine is in the brain, a supposition that still appears to be controversial. Skingle et al. [1982] have proposed that the main focus of action of clonidine in inhibiting a variety of nociceptive reactions is the brain rather than the spinal cord. This conclusion was based on the observation that much higher doses of clonidine are needed to produce antinociception following intrathecal than following intraventricular injections. Spaulding et al. [1979a], on the other hand, concluded that the major effects of clonidine are on the spinal cord. They found that while morphine ED_{50} was increased 4 to 7-fold following spinal transsection (indicating involvement of descending inhibitory mechanisms), the effects of clonidine on the tail-flick reaction were unaltered.

There is considerable evidence to suggest that norepinephrine is, in fact, an important modulator of nociceptive information in the spinal cord. For example, iontophoretically applied norepinephrine has been found to inhibit the discharge of dorsal horn neurons that are activated by noxious stimuli [Engberg and Ryall, 1966; Belcher et al., 1978; Headley et al., 1978]. The release of endogenous spinal cord norepinephrine by electrical stimulation of brainstem noradrenergic nuclei also has been found to produce a similar inhibitory effect on dorsal horn cells [McCreery et al., 1979; Hodge et al., 1981], as well as inhibit nociceptive reactions [Segal and Sandberg, 1977; Akaike et al., 1978]. The most direct evidence implicating spinal cord norepinephrine in the suppression of nociception, however, comes from studies which have evaluated the antinociceptive effects of intrathecally applied adrenergic compounds.

Kuraishi et al. [1979b] were the first to formally demonstrate that intrathecally administered norepinephrine had the capacity to inhibit nociceptive reactions in the rat. This effect was mediated through the postsynaptic alpha-adrenergic receptor, since phenoxybenzamine reduced it considerably. Since intrathecal phenoxybenzamine by itself had no effect, it appears that the alpha-adrenergic system in the cord is not tonically active. Reddy and Yaksh [1980] also reported significant dose-dependent antinociceptive effects in rats and cats following intrathecal injections of norepinephrine. This again appeared to be mediated through the alpha-adrenergic receptor since phenylephrine but not isoproterenol was effective. In addition, the effects of norepinephrine were antagonized by phentolamine (an alpha-blocker) but not propranolol (a beta-blocker). Yaksh and Reddy [1981] have reported that alpha-adrenergic agonists also produce antinociceptive effects in the primate following intrathecal application. Howe et al. [1983] have attempted to ascertain whether alpha-1- or alpha-2-adrenergic receptors are selectively involved in mediating the antinociceptive effects of adrenergic activation. The results revealed that activation of either one of two separate populations of spinal alpha-adrenergic receptors inhibited nociceptive reactions. Since 6-hydroxy dopamine (6-OHDA) pretreatment did not appear to antagonize the antinociceptive effects of intrathecal alpha-agonists, it appears that the actions of these compounds are mediated through receptors that are postsynaptic to noradrenergic terminals in the cord. Considering these findings, it is not clear precisely how systemically administered alpha-adrenergic agonists produce antinociception, since they are not antagonized by post synaptically acting alpha-antagonists. For a more complete review of the role of spinal noradrenergic systems in modulating nociception at the spinal level, see Yaksh [1985].

Most, if not all, of the noradrenergic innervation of the spinal cord appears to originate from supraspinal structures, since drastic reduction in the fluorescent labeling of this catecholamine is found caudal to spinal transections [Dahlstrom and Fuxe, 1965]. In the rat [Westlund et al., 1981, 1982], it appears that the majority of noradrenergic input to the spinal cord originates from a variety of pontine nuclei (e.g. locus coeruleus, nucleus subcoeruleus, medial and lateral parabrachial nuclei, Kölliker-Fuse nucleus, and a region dorsal and lateral to the superior olivary nucleus). In the cat, however, the principal source of noradrenergic input to the cord comes from the Kölliker-Fuse nucleus alone [Stevens et al., 1982]. Very few catecholamine-containing cells projecting to the cord were found in the locus coeruleus, subcoeruleus, or parabrachial nuclei. If spinal norepinephrine, originating from pontine structures, plays a role in modulating nociception, then destruction or stimulation of these structures should alter nociceptive reactions. In addition, if endogenous norepinephrine exerts a tonic inhibitory effect on spinal cord nociceptive transmission, then lesions of the pontine noradrenergic nuclei should decrease nociceptive thresholds. Bodnar et al. [1977], however, found that electrolytic lesions of the locus coeruleus and adjacent regions, including the dorsal noradrenergic bundle, actually increased jump-thresholds in rats in response to electrical shock. Sasa et al. [1976] reported similar findings employing paw pressure as a nociceptive stimulus. The lesion effects in the latter study appeared to be somewhat transitory, since the reaction thresholds returned to baseline levels 27 days following surgery. Sasa et al. [1976] interpreted these findings to suggest that *decreasing* the noradrenergic tone in brain produces a decrease in sensitivity to painful stimuli. It is not entirely clear how motoric deficits contributed to these effects. Bodnar et al. [1977], for example, reported that 4 of 5 lesioned rats appeared ill following surgery (e.g. failed to groom and emitted bloody urine and semen) and 3 of the 5 lesioned rats died within 2 weeks of surgery. Kastrowski et al. [1978], on the other hand, reported that bilateral lesions of the locus coeruleus decreased tail-compression-induced vocalization thresholds in the rat. Finally, Hammon and Proudfit [1980] failed to find any alteration of hot-plate or tail-flick reactions following locus coeruleus lesions with monosodium-*L*-glutamate, which presumably lesions cell bodies while sparing axons of passage. Considered together, these findings certainly do not suggest that descending noradrenergic systems are tonically active in suppressing nociceptive transmission in the spinal cord. It is also unlikely that ascending noradrenergic pathways are involved in tonic suppression of nociception, since Price and Fibiger [1975] failed to find alterations in tail-flick latencies following 6-OHDA lesions of either the dorsal or ventral noradrenergic bundle in the rat mesencephalon. While destruction of the locus coeruleus in rats generally has failed to decrease nociceptive thresholds, stimulation of this structure does appear to produce antinociceptive effects. Segal and Sandberg [1979], for example, found that electrical stimulation of the locus coeruleus increased nociceptive thresholds of rats on the hot-plate. Hodge et al. [1981] have reported that electrical stimulation of the locus coeruleus in cats reduced the activity of dorsal horn cells that responded to noxious input. Neither study, however, provided appropriate pharmacological controls to ascertain whether the effects elicited by electrical stimulation were actually mediated by the release of norepinephrine. It is quite likely, in fact, that the effects reported by Hodge et al. [1981] were not mediated by descending noradrenergic pathways, since Stevens et al. [1982] failed to find a major descending noradrenergic input to the spinal cord of the cat originating from the locus coeruleus.

Effects of Catecholamine Blockers and Depleters on Nociceptive Thresholds

Since catecholamines appear to be involved in modulating nociception, then their depletion in brain should also alter nociceptive thresholds. While intraventricularly administered 6-OHDA, a neurotoxin that selectively destroys catecholamine neurons, does increase irritability in rats [Kastrzewa and Jacobowitz, 1971], it does not seem to have pronounced effects on pain thresholds following intraventricular injections [Ayhan, 1972; Elchisak and Rosecrans, 1973; Samanin and Bernasconi, 1972]. Only a few studies have reported a decrease in nociceptive thresholds following intraventricular injection of 6-OHDA. Bläsig et al. [1973] found a decrease in vocalization thresholds while Slater and Blundell [1978a] found a decrease in hot-plate and tail-flick latencies in mice but not in rats. Nakamura et al. [1973b], on the other hand, observed a transient increase in nociceptive thresholds following intrastriatal injections of 6-OHDA. The effect of catecholamine depletion with 6-OHDA appears to be more consistent following intrathecal application. Both Howe and Yaksh [1982] and Howe et al. [1983] reported that intrathecal 6-OHDA reduced nociceptive thresholds in the rodent. These findings, while at variance with the lesion studies cited above, were taken to imply that descending noradrenergic systems exert a tonic inhibitory influence on spinal mechanisms mediating nociception.

The majority of studies that have depleted catecholamines with alpha-methyl-tyrosine (a tyrosine hydroxylase inhibitor that prevents the biosynthesis of catecholamines) have found no effect on pain thresholds in a variety of situations [Verri et al., 1967; Fennessy and Lee, 1970; Sethy et al., 1970; Major and Pleuvry, 1971; Buxbaum et al., 1973; Reinhold et al., 1973]. Houser and Houser [1973], on the other hand, did observe increased shock thresholds in squirrel monkeys following alpha-methyl-tyrosine (AMPT). Paradoxically, however, amphetamine, which increase catecholaminergic tone, also had a similar effect in the same study. The relative lack of effect of drugs that deplete catecholamines on nociceptive thresholds suggests that catecholamines are not tonically active in suppressing nociceptive transmission.

Reserpine, which depletes the brain of both catecholamines and 5-HT, has been reported to increase nociceptive thresholds in rats in some situations. Ross and Ashford [1967] found that reserpine produced slight antinociceptive effects after 24 or 48 h in the hot-plate test but was ineffective in the tail-clip test. Garcia Leme and Rocha e Silva [1961] demonstrated antinociceptive effects for reserpine in the hot-plate test which reached asymptote in 48 to 72 h. Weller et al. [1968], however, were unable to find antinociceptive effects for reserpine in a battery of four tests. Grossman et al. [1973] have demonstrated that reserpine increases the tail-flick reaction times in rats and, at the same time, depresses the polysynaptic reflex response to sural nerve stimulation. Heinz and Jurna [1979] have concluded that the effects of reserpine as well as haloperidol on the tail-flick reflex are mediated through the nigrostriatal system. They found that reserpine and haloperidol injected systematically increased the tail-flick latencies in intact rats but not in pre-nigrally decerebrate or spinal rats. In addition, the effects of both drugs were antagonized by injections of either apomorphine or naloxone into the caudate nucleus. They concluded that the nigrostriatal feedback system is critically involved in the antinociceptive effects of dopamine blockers. Thus, while dopaminergic blockade inhibits nociceptive reflex activity, this action may not be linked to a decrease in pain perception per se. In fact, in a clinical study, it was found that reserpine actually increased rather than decreased pain in patients suffering from chronic pain [Sternbach et al., 1976].

At any rate, since a number of brain monoamines are modified by reserpine, it would be difficult to interpret the neurochemical mechanisms of action of this drug regardless of what effects it has on pain perception. Phenothiazines (e.g. chlorpromazine) and butyrophenones (e.g. haloperidol), which block, at least in part, dopaminergic neurotransmission, also have been reported to increase nociceptive thresholds in rodents. Weller et al. [1968] found that chlorpromazine as well as haloperidol were fully active analgesics in the hot-plate, tail-pinch, and tail-shock tests, but not in the tail immersion test. Chlorpromazine as well as other neuroleptics also have been found to increase nociceptive thresholds in the hot-plate test by other investigators [Garcie Leme and Rocha e Silva, 1961; Jansen et al., 1959; Ramabadran and Jacob, 1978], but not in the tail-flick test (Tulunay et al., 1975], which agrees with the findings of Weller et al. [1968] but not Grossman et al. [1973] nor Boissier et al. [1960], who found haloperidol effective in both the tail-flick and hot-plate tests. McGilliard and Takemori [1979] also evaluated the effects of neuroleptics on the tail-flick test. They found that pimozide but not haloperidol increased tail-flick latencies. The relative efficacy of pimozide in this test was assumed to be due to its greater ability to block dopamine receptors. Analgesic effects of haloperidol have also been documented in the clinical situation [Maltbie et al., 1979]. It is not clear, however, whether haloperidol has true analgesic properties in man or merely alters a patient's complaint of pain.

Adrenergic blocking agents have also been examined for antinociceptive properties. Neither alpha- nor beta-adrenergic blockers have been found to increase tail-flick withdrawal reaction latencies in rats [Heller et al., 1978; Tulunay et al., 1975]. Cicero et al. [1974a], however, have reported that phenoxybenzamine increased hot-plate latencies, while beta-blockers were ineffective. Bentley et al. [1977] have also reported that intraventricular injections of piperoxane (a presynaptic alpha-receptor blocker) inhibited writhing reactions in mice. Interestingly, phenoxybenzamine-induced analgesia has been found to be antagonized by naloxone, suggesting the involvement of an opiate mechanism in its antinociceptive effect [Elliott et al., 1976].

Opiate and Catecholamine Interactions

At the spinal level it appears certain that opiate-induced analgesia involves, at least in part, the release of norepinephrine. Shiomi and Takagi [1974] originally proposed that the analgesic effects of opiates in the brainstem are mediated by the activation of descending noradrenergic pathways. In a series of studies, described elsewhere in this volume, Takagi and his colleagues further developed this notion. Similarly, Yaksh [1979] found that the antinociceptive effects of morphine following injections into the periaqueductal gray matter were antagonized by subarachnoid administration of phentolamine. These findings emphasize the importance of the spinal noradrenergic system in mediating the analgesic actions of morphine. If norepinephrine is critically involved in opiate analgesia, then catecholaminergic drugs should alter this action of opiates.

Reserpine. A number of early studies utilizing reserpine or tetrabenazine, which deplete the brain of both catecholamines (norepinephrine and dopamine), and 5-hydroxytryptamine (5-HT), demonstrated that the analgesic effects of morphine were clearly attenuated in rodents which had been pretreated with these compounds [Schneider, 1954; Sigg et al., 1958; Medakovic and Banic, 1964; Takagi et al., 1964; Rudzik and Mennear, 1965; Contreras and Tamayo, 1966; Verri et al., 1967, 1968; Vedernikov and Afrikanov, 1969; Fennessy and Lee, 1970; Contreras et al., 1973]. The obvious conclusion from such findings is that the analgesic effects of opiates are somehow dependent on the integrity of the catecholaminergic or 5-HT systems. Other studies, however, reported that reserpinization can also potentiate morphine analgesia [Tripod and Gross, 1957; Garcia Leme and Rocha e Silva, 1961; Dandiya and Meman, 1963].

These discrepancies have been partially resolved by the demonstration that the effects of reserpine on morphine-induced antinociception are task- and time-dependent. For example, Ross and Ashford [1967] found that reserpine antagonized the effect of morphine in the tail-clip task, but potentiated it in the hot-plate paradigm. It is interesting to note that studies reporting potentiation of morphine by reserpine utilized the hot-plate task to assess nociception [Tripod and Gross, 1957; Garcia Leme and Rocha e Silva, 1961; Dandiya and Meman, 1963]. An additional variable appears to be the injection interval between reserpine and morphine. Apparently, the initial action of reserpine (release of catecholamines and 5-HT) can potentiate morphine, whereas the later effect (brain monoamine depletion) antagonizes morphine [Harris, 1970].

Since reserpine depletes both brain catecholamines and 5-HT, a number of investigators have attempted more critical analyses by utilizing drugs and other manipulations which are more specific to a given neurohumoral system.

Depletion of Catecholamines. A number of investigators have examined the analgesic potency of morphine in the rodent following the inhibition of catecholamine synthesis with AMPT, a tyrosine hydroxylase inhibitor. The majority of such studies have reported that pretreatment with AMPT antagonizes morphine analgesia [Verri et al., 1967; Sethy et al., 1970; Görlitz and Frey, 1972; Contreras et al., 1973; Reinhold et al., 1973], although several others have reported either no effect [Sethy et al., 1970; Fennessy and Lee, 1970] or potentiation [Major and Pleuvry, 1971; Calcutt et al., 1971]. In addition, it is interesting to note that Houser and Houser [1973] and Houser and Van Hart [1973] have reported that AMPT by itself raises the aversive threshold for electrical shock in both rats and monkeys, suggesting antinociceptive properties for this compound in at least volitional nociceptive behavior.

Findings from studies which have used 6-OHDA, a compound that destroys catecholaminergic nerve terminals by auto-oxidation after uptake, also suggest the involvement of CNS catecholamine systems in opiate analgesia. Intracerebral and intraventricular injections of 6-OHDA in rodents have been found to antagonize the antinociceptive actions of opiates concomitant with reductions in brain norepinephrine and dopamine [Fennessy and Lee, 1970; Friedler et al., 1972, Bläsig et al., 1973; Slater, 1974; Slater and Blundell, 1978a, b]. Several studies in rats, on the other hand, have reported potentiation following such pretreatment [Ayhan, 1972; Samanin and Bernasconi, 1972; Bhargawa et al.,

1972]. Nakamura et al. [1973c] have also reported that intraventricular 6-OHDA restored the analgetic actions of morphine in morphine-tolerant rats, suggesting that the development of tolerance to morphine analgesia may, in some way, be due to changes in central catecholamine neurons.

Slater and Blundell [1978] have suggested that some of the inconsistencies regarding the interaction between 6-OHDA and morphine may be due to species differences. These investigators have found a consistent antagonism of morphine analgesia by intraventricular 6-OHDA in mice which they ascribed to an effect on central norepinephrine. Rats, however, were found to be unaffected by 6-OHDA pretreatment unless also pretreated with pargyline, which enhanced the depletion of dopamine from brain and potentiated morphine analgesia. They proposed that, in the rat, dopamine appears to be more critical (in some unspecified way) for opiate analgesia than norepinephrine. Elchisak and Rosecrans [1973], on the other hand, have found that pretreatment with des-methylimipramine before 6-OHDA lesions (which results in preferential depletion of dopamine) produces an even greater antagonism of morphine analgesia in rats than 6-OHDA alone. The reasons behind these clearly contradictory findings are not readily apparent.

Catcholamine Agonists and Potentiators

There are a variety of other miscellaneous compounds affecting catecholaminergic activity that have been reported to influence morphine analgesia. For example, monoamine oxidase inhibitors (MAOI), which potentiate catecholamine functions, have been reported to enhance the analgesic actions of opiate agonists following acute administration [Defalque, 1965; Gupta and Kulkarni, 1966; Jounela and Mattila, 1966; Paalzow and Paalzow, 1971; Iwamoto et al., 1976]. Opposite effects, however, have been reported after chronic administration by Iwamoto et al. [1976]. These investigators have proposed that the interactive effects between MAOI and morphine are due to norepinephrine and not dopamine or 5-HT.

Sympathomimetics, which increase the availability of norepinephrine and dopamine at the synapse (e.g. cocaine and amphetamine), are catecholaminergic compounds which have been reported consistently to enhance morphine analgesia. Ivy and co-workers were the first to report the potentiation of morphine analgesia by amphetamine in both man [Ivy et al., 1944], dog [Goetzl et al., 1944b] and mouse [Goetzl et al.,

1944a]. These findings have been confirmed subsequently by a number of investigators in a variety of situation and species [Nickerson, 1950; Saxena and Gupta, 1957; Sigg et al., 1958; Witkin et al., 1961; Evans and Bergner, 1964]. The antinociceptive effects of codeine are modified by amphetamine as well. Ahmed et al. [1970] reported that high doses (4–5 mg/kg) of amphetamine potentiated the antinociceptive actions of codeine in the rat tail-clip test, while low doses produced a slight antagonism. Nott [1968] has also reported potentiation of morphine analgesia by cocaine. More recently, Webb et al. [1978] and Forrest et al. [1977] have confirmed the potentiation of opiate analgesia by amphetamine for both experimental and clinical pain. Sprague and Takemori [1978] have shown that the potentiation of opiate analgesia by amphetamine is not due simply to alterations in the brain disposition of morphine by this sympathomimetic. The combination of sympathomimetics with opiates may be an interesting and beneficial new drug combination with less side effects than opiate alone [Forrest et al., 1977].

Unfortunately, most of the above pharmacological manipulations affect both dopamine and norepinephrine and some have profound effects on other neurohumors as well. This lack of pharmacological specificity would be a basis for some of the inconsistencies reported, especially if dopamine and norepinephrine (or other affected neurohumors) had divergent functions in nociception or morphine-induced analgesia, of if either catecholamine was preferentially critical in mediating a given nociceptive response. This problem has been recently highlighted by Price and Fibiger [1975], who demonstrated divergent effects of ascending dopamine and norepinephrine pathways in morphine analgesia. These investigators reported that 6-OHDA lesions of the substantia nigra (origin of the nigrostriatal dopamine pathways) abolished the antinociceptive actions of morphine in the tail-flick test, while lesions to the dorsal norepinephrine bundle potentiated morphine analgesia. Ventral norepinephrine bundle lesions were found to have no effect.

Do Norepinephrine and Dopamine Play a Differential Role in Morphine Analgesia?

Norepinephrine. Attempts have been made to separate the relative roles of norepinephrine and dopamine in opiate analgesia with more selective pharmacological manipulations. In several studies, selective reductions in brain norepinephrine have been found to decrease the analgesic efficacy of morphine. Dopamine-beta-hydroxylase (DBH) inhibitors, which prevent the conversion of dopamine into norepinephrine, have generally been found to decrease opiate analgesia [Reinhold et al., 1973; Paalzow et al., 1974]. Slater and Blundell [1978b] have also reported that 6-hydroxydopa (which specifically depletes norepinephrine) reduced morphine analgesia in mice. Similar effects, however, were not seen in the rat. Cicero [1974], on the other hand, have found that DBH inhibitors had no effect on the analgesic effects of morphine, while Watanabe et al. [1969] have actually reported a potentiation. To add further confusion to the issue, intraventricular injections of norepinephrine seem to decrease the effects of a number of opiate agonists and partial agonists in the tail-immersion test [Sewell and Spencer, 1975]. Several investigators have evaluated the interactive effects between alpha- and beta-adrenergic blocking agents and opiates. Surprisingly, the majority of such studies have found no significant effect of either alpha- or beta-blockers and opiate analgesia [Major and Pleuvry, 1971; Gorlitz and Frey, 1972; Sprague and Takemori, 1978; Slater and Blundell, 1978a]. Cicero et al. [1974a] have reported that phenoxybenzamine (an alpha-adrenergic blocker), produced analgesic effects in the hot-plate test and also potentiated morphine analgesia. Interestingly, the same laboratory [Cicero et al., 1974b] has reported that phenoxybenzamine has an affinity for the opiate receptor (producing 50% inhibiton of [^{3}H]naloxone binding at 9.2×10^{-7}). Phenoxybenzamine analgesia has also been reported to be reversed by naloxone, indicating the involvement of opiate mechanisms [Elliott et al., 1976]. Yaksh [1979], on the other hand, has reported that intrathecal phenoxybenzamine is capable of blocking the analgesic effects of morphine following periaqueductal gray injections. This finding is more consistent with the notion that spinopetal norepinephrine systems may be involved in opiate analgesia, especially since alpha-adrenergic receptors appear to mediate spinal functions.

If spinopetal noradrenergic systems are involved in opiate analgesia, then destruction of brainstem nuclei containing the noradrenergic cell bodies giving rise to these systems should decrease the antinociceptive effects of opiates. The results of studies that have evaluated the effects of locus coeruleus lesions on morphine analgesia are relatively consistent. Bilateral electrolytic lesions of the locus coeruleus have been found to decrease the analgesic effects of morphine in studies by both Kostowiki et al. [1978] and Sasa et al. [1976]. While locus coeruleus lesions result in a destruction of both ascending and descending noradrenergic pathways originating from this structure, it is unlikely that ascending pathways contribute to the interactive effects reported above. Price and Fibiger [1977] have reported that destruction of the dorsal noradrenergic pathway with 6-OHDA actually enhanced the analgesic effects of morphine, while destruction of the ventral noradrenergic pathway had no significant effect on the actions of morphine. Nakamura et al. [1973] also reported that 6-OHDA induced depletion of norepinephrine in the hypothalamus increased morphine analgesia.

Since electrolytic lesions destroy fibers of passage as well as cell bodies, it is not clear whether the effects of locus coeruleus lesions described above are due solely to destruction of cell bodies or to other fibers coursing through this structure. Hammond and Proudfit [1980] utilized monosodium-*L*-glutamate to selectively destroy cell bodies in the locus coeruleus and adjacent structures. They found that injections of this substance into the locus coeruleus as well as adjacent nuclei attenuated the analgesic effects of morphine as measured with the hot-plate and tail-flick tests. The authors, however, failed to find any correlation between attenuation of morphine-induced analgesia and decreases in the norepinephrine content of the cortex, spinal cord, and medial brain stem. Furthermore, since lesions confined primarily to the nucleus parabrachialis ventralis were as effective as lesions that destroyed both the locus coeruleus and parabrachial region, the authors concluded that the parabrachial region, and not the locus coeruleus, was most likely involved in mediating the analgesic actions of morphine.

Pert et al. [1979] have reported that 6-OHDA lesions of the A-1 catecholamine nuclei, which decreased the norepinephrine content of the spinal cord dorsal horn by 64%, also attenuated the antinociceptive effects of morphine. Descending noradrenergic pathways, however, do not appear to originate from the A-1 noradrenergic cell bodies. It is possible that 6-OHDA injections into this medullary region destroyed descending noradrenergic pathways originating from more rostral pontine catecholamine nuclei. Alternatively, it is also possible that the 6-OHDA destroyed descending adrenalin-containing pathways originating from this brain region [Ross et al., 1981], which may be partially involved in mediating the analgesic effects of morphine.

While lesions of the locus coeruleus and adjacent structures appear to reduce the analgesic effects of morphine, Segal and Sandberg [1977] have reported that electrical stimulation of this structure potentiates morphine-induced analgesia. Sasa et al. [1975], on the other hand, found that morphine selectively reduced the inhibitory effects of locus coeruleus stimulation on the spinal trigeminal nucleus and concluded that morphine may act on the presynaptic terminals of locus coeruleus neurons to block norepinephrine release.

The results of lesion studies appear to suggest that opiates induce at least part of their analgesic effects by activating neurons in the locus coeruleus or adjacent structures. Findings from studies employing electrophysiological techniques, however, have not supported this prediction. Korf et al. [1974] have reported that systematically administered morphine decreases the spontaneous as well as pain-induced activation of locus coeruleus neurons. Bird and Kuhar [1977] reported similar findings following the iontophoretic application of morphine onto locus coeruleus cells. It is difficult to reconcile these findings with the notion that opiate-induced analgesia involves the activation of noradrenergic pathways originating from the locus coeruleus.

Dopamine. Considerable attention has been focused on the involvement of dopamine in opiate analgesia. Dopamine agonists and compounds which increase dopamine availability have been reported almost unequivocally to antagonize opiate analgesia in a number of species and a variety of tasks [Vedernikov, 1969; Vander-Wende and Spoerlein, 1973; Tulunay et al., 1975; Sprague and Takemori, 1978]. Some of these interactive effects may be mediated through the amygdala. Rogers [1977] reported that dopamine injections into the amygdala decreased shock-elicited jump thresholds in the rat and significantly attenuated the elevation in threshold following systemically administered morphine. Conversely, dopamine antagonists (e.g. haloperidol, spiroperidol, pimozide) have been found to enhance morphine analgesia [Vander-Wende and Spoerlin, 1973; Eidelberg and Erspamer, 1975; Tulunay et al., 1975; Sprague and Takemori, 1978; McGilliard and Takemori, 1979]. Phenothiazines (e.g. chlorpromazine), which block dopamine receptors, have also been reported to potentiate opiate analgesia [Courvoisier et al., 1953; Sadove et al., 1954; Schneider, 1954; Wirth, 1954]. However, since phenothiazines have a number of other effects in the CNS, it is not simple to ascribe their interactive effects with opiates to an action on dopamine receptors alone. In contrast to the above findings, Calcutt et al. [1971] have reported that intraventricular dopamine potentiates morphine analgesia and Görlitz et al. [1972] have found that morphine analgesia is antagonized by both chlorpromazine and haloperidol.

One of the interactions between dopaminergic agents and opiates probably occurs in the nigrostriatal dopamine system. Nakamura et al. [1973b] have found that 6-OHDA lesions of the caudate-putamen decreased morphine analgesia which was restored by *L*-DOPA, a precursor for dopamine. In the same study, A-10 lesions (origin of the mesolimbic dopamine system) were not found to affect morphine analgesia. Grossman et al. [1973] have found that bilateral lesions of the substantia nigra (origin of the nigrostriatal dopamine pathways) also decreased morphine analgesia in the tail-flick test. This effect, however, was interpreted to be due to a complex interactive effect of the lesions and morphine on alpha- and gamma-motoneuron activity in the spinal cord. Jurna et al. [1978] examined the effects of various manipulations to the substantia nigra on the reflex discharges of alpha- and gamma-motoneurons, as well as the reaction time of the tail-flick response. These investigators found that stimulation of the substantia nigra enhanced the alpha-motoneuron discharges, inhibited the tail-flick reaction, and potentiated the effects of morphine on the same measure. It is interesting to note that in rats with prenigral decerebration, morphine failed to prolong the reaction time, suggesting the involvement of structures rostral to the nigra in morphine's action on this reflex. Morphine has also been observed to facilitate monosynaptically activated alpha-motoneurons. This effect was ascribed to an impairment of dopaminergic impulse transmission in the nigrostriatal system. Jurna et al. [1978] also found that blockade of the nigra with microinjections of GABA or procaine likewise inhibited the tail-flick reaction in rats. This effect was also abolished by prenigral decerebration. These findings were interpreted to imply that antinociception induced by inhibition of the nigra had to involve the nigrostriatal feedback system, which is activated following inhibition of the nigrostriatal pathways. While these findings are quite intriguing, it is not clear at this time whether such interactions of opiates with the nigrostriatal dopamine system influence the motor neurons in the ventral horn or the sensory neurons in the dorsal horn.

Summary and Conclusions

There is certainly compelling evidence in support of a critical role for catecholamines in modulating nociception as well as mediating opiate-induced analgesia. Findings from studies employing pharmacological agents which alter catecholamine function generally appear to support the notion that increases in catecholaminergic function result in increased nociceptive thresholds and enhanced opiate analgesia. Norepinephrine rather than dopamine appears to be the critical neurohumor involved. Only a few studies have found that systemic injections of dopamine agonists increase nociceptive thresholds. Since dopamine agonists induce enhanced locomotor output and stereotypy, it is highly likely that competing behaviors elicited by these drugs might interfere with the expression of nociceptive reactions. The effects of dopamine agonists on spinal nociceptive reflex activity are also somewhat puzzling, since inhibition of such reflexes is only seen in anesthetized or spinal animals. It has been argued that the inability of apomorphine or intrathecal dopamine to alter the tail-flick reflex in rats reflects the operation of descending, tonically active, inhibitory influences on spinal dopamine mechanisms. Whether this explanation is tenable remains to be determined. At any rate, there is little clinical evidence in support of an important role for dopamine in modulating nociception.

Norepinephrine, on the other hand, does appear to alter nociception as well as nociceptive reactions. Neuroanatomical, neurophysiological, and behavioral findings all seem to indicate that the spinal cord is an important focus for the antinociceptive actions of norepinephrine as well as pharmacological agents that increase its activity. The antinociceptive effects of norepinephrine in the spinal cord appear to be related to its ability to inhibit the activity of dorsal horn cells concerned with the transmission of nociceptive information to the brain. This effect appears to be mediated through the alpha-adrenergic receptor located postsynaptically to noradrenergic terminals, since postsynaptic alpha-blockers antagonize the antinociceptive effects of norepinephrine in the cord. Systemically administered alpha-agonists (e.g. clonidine) have also been found to increase nociceptive thresholds consistently. It is not certain, however, whether the site of action of these compounds is the spinal cord or the brain. A spinal site of action would be difficult to reconcile with the findings that presynaptic but not postsynaptic alpha-blockers antagonize clonidine-induced analgesia. The actions of clonidine on presynaptic receptors located on noradrenergic cells is thought to decrease their firing rate, resulting in an inhibition of norepinephrine release. If the antinociceptive effects of clonidine are indeed mediated through presynaptic norepinephrine receptors in brain, it is not clear what neutral mechanisms are involved. The participation of descending noradrenergic pathways has to be ruled out, since clonidine, acting presynaptically, should inhibit norepinephrine release in the cord. On the other hand, it is also unlikely that ascending noradrenergic pathways are involved, since lesions of the ascending dorsal and ventral noradrenergic bundles have not been found to alter nociceptive reactions. The precise mechanism of action of other sympathomimetic compounds is not known, but likely involves, at least in part, the release of norepinephrine in the spinal cord.

It seems apparent that norepinephrine must not exert an important tonic inhibitory influence on nociception, since neither blockade nor destruction of brain catecholamines has been found to decrease nociceptive thresholds in the majority of studies. In fact, a number of reports have actually described antinociceptive effects following the administration of drugs that block either dopamine or norepinephrine function. The antinociceptive effects of alpha-noradrenergic blockade are especially puzzling in light of the possibility that norepinephrine suppresses nociceptive activity through alpha receptors in the spinal cord. It is possible that the antinociceptive effects of catecholaminergic blockade are also not related to inhibition of pain, but to other factors. Heinz and Jurna [1979], for example, have noted that haloperidol-induced suppression of the tail-flick reaction in rats may be mediated through the nigrostriatal system. This may consequently translate into an inhibition of spinal cord motor neurons producing an attenuation of the nociceptive reflex.

The proposal that some of the antinociceptive effects of opiates may be mediated through descending noradrenergic pathways has considerable merit. There are substantial data to support the notion that the actions of morphine in the mesencephalon and pons serve to enhance the release of norepinephrine in the spinal cord. One finding that has always been difficult to reconcile with the biochemical and behavioral data, however, is that opiates administered either systemically or isophoretically seem to decrease the activity of noradrenergic neurons in the locus coeruleus. One possibility is that the effects of opiates on descending noradrenergic systems are mediated through other pontine nuclei (e.g. subcoeruleus or parabrachial regions). While the activation of descending noradrenergic pathways (whatever their origin) by morphine appears to be critical for the analgesic effects of this substance, there is little evidence for a similar participation of ascending noradrenergic pathways. Price and Fibiger [1975] have, in fact, reported that lesions of the dorsal noradrenergic bundle potentiate the antinociceptive effects of morphine, while lesions of the ventral bundle had no effect on morphine-induced analgesia. Therefore, it seems that the opiate-potentiating effects of sympathomimetics and other compounds that increase noradrenergic activity must be mediated through the spinal cord mechanisms and not the brain.

References

Abaza, A.; Gregoire, M.: Potentiation of analgesic effects of opiates by combination with prostigmine. Presse med *60:* 331–333 (1952).

Aceto, M.D.; Awaya, H.; Martin, B.R.; May, E.: Antinociceptive action of nicotine and its methiodide derivates in mice and rats. Br. J. Pharmacol. *79:* 869–876 (1983).

Aceto, M.D.; Bagley, R.S.; Dewey, W.L.; Tsu-Ching, F.; Martin, B.R.: Spinal cord as a major site for nicotine's antinociceptive action in the rat. Neuropharmacology in press, 1986.

Aceto, M.D.; Martin, B.R.; Tripathi, H.L.; May, E.L.; Jacobson, A.E.; Antinociceptive effects of the optically pure stereoisomers of nicotine. Pharmacologist *22:* 302 (1980).

Ahmed, S.S.; Abraham, G.J.S.; Ansari, M.Y.: Dose dependent modification of codeine analgesia by *d*-amphetamine in albino rats. Archs. int. Pharmacodyn. *184:* 240–244 (1970).

Akaike, A.; Shibata, T.; Satoh, M.; Takagi, H.: Analgesia induced by microinjection of morphine into, and electrical stimulation of, the nucleus reticularis gigantocellularis of rat medulla oblongata. Neuropharmacology *17:* 775–778 (1978).

Amir, S.; Amit, Z.: Endogenous opioid ligands may mediate stress-induced changes in the effective properties of pain-related behavior in rats. Live Sci. *23:* 1143–1152 (1978).

Andrews, H.L.: The effect of morphine and prostigmine methylsulfate on measurement of pain threshold. J. Am. Med. Ass. *120:* 525–527 (1942).

Armitage, A.K.; Hall, G.H.: Further evidence relating to the mode of action of nicotine in the central nervous system. Nature, Lond. *214:* 977–979 (1967).

Armitage, A.K.; Milton, A.S.; Morrison, C.F.: Effects of nicotine and some nicotine-like compounds injected into the cerebral ventricles of the cat. Br. J. Pharmac. Chemother. *27:* 33–45 (1966).

Ayhan, I.H.: Effect of 6-hydroxydopamine on morphine analgesia. Psychopharmacologia *25:* 183–188 (1972).

Barasi, S.; Bensreti, M.M.; Duggal, K.N.; Gonzalez, J.P.; Sewell, R.D.E.: Antinociceptive activity of intravenous and intrathecal dopamine agonists in naive and supersensitive rats. Br. J. Pharmacol. *84:* 166P (1985).

Barasi, S.; Duggal, K.N.; Russel, D.: Effects of apomorphine on the tail-flick latency in the anaesthetized rat. J. Physiol. *353:* 105P (1984).

Baxter, B.L.: Induction of both emotional behavior and a novel form of REM sleep by chemical stimulation applied to cat mesencephalon. Expl. Neurol. *23:* 220–229 (1969).

Belcher, G.; Ryall, R.W.; Schaffner, R.: The differential effects of 5-hydroxytryptamine, noradrenaline and raphe stimulation on nociceptive and nonnociceptive dorsal horn interneurons in the cat. Brain Res. *151:* 307–321 (1978).

Bentley, G.A.; Copeland, I.W.; Starr, J.: The actions of some alpha-adrenoreceptor agonists and antagonists in an antinociceptive test in mice. Clin. exp. Pharmacol. Physiol. *4:* 405–419 (1977).

Bernheim, F.; Bernheim, M.L.C.: Action of drugs on the choline esterase of the brain. J. Pharmac. exp. Ther. *57:* 427–436 (1936).

Bhargawa, H.N.; Afifi, A.-H.; Way, E.L.: Effect of chemical sympathectomy on morphine antinociception and tolerance development in the rat. Biochem. Pharmac. *22:* 2769–2772 (1973).

Bhargawa, H.N.; Way, E.L.: Acetylcholinesterase inhibition and morphine effects in morphine tolerant and dependent mice. J. Pharmac. exp. Ther. *183:* 32–40 (1972).

Bird, S.J.; Kuhar, M.J.: Iontophoretic application of opiates to the locus coeruleus. Brain Res. *122:* 523–533 (1977).

Bite, A.; György, L.: Studies on the potentiation of the antinociceptive effect of oxotremorine by tetrabenazine in mice. Psychopharmacologia *34:* 63–70 (1974).

Björklund, A.; Skagerberg, G.: Evidence for a major spinal cord protection from the diencephalic All dopamine cell group in the rat using transmitter specific fluorescent retrograde tracing. Brain Res. *177:* 170–175 (1979).

Bläsig, J.; Reinhold, K.; Herz, A.: Effect of 6-hydroxydopamine,5,6-dihydroxytryptamine and raphe lesions on the antinociceptive actions of morphine in rats. Psychopharmacologia *31:* 111–119 (1973).

Blessing, W.W.; Chalmers, J.P.: Direct projection of catecholamine (presumably dopamine)-containing neurons from hypothalamus to spinal cord. Neurosci. Lett. *11:* 35–40 (1979).

Blinn, G.; Heinz, G.; Jurna, I.: Effects of substantia nigra stimulation on suralis-evoked spinal reflex activity: comparison with the effects of morphine and stimulation in the periaqueductal gray matter. Neuropharmacology *19:* 75–85 (1980).

Blockus, L.E.; Everett, G.M.: Tremor producing drug, 1,4-dipyrrolidino-2-butyne (tremorine). Fed. Proc. *16:* 283 (1957).

Bodnar, R.J.; Ackermann, R.F.; Kelly, D.D.; Glusman, M.: Elevations in nociceptive thresholds following locus coeruleus lesions. Brain. Res. Bull. *3:* 125–130 (1978).

Boissier, J.R.; Pagny, J.; Mouille, P.; Forest, J.: Quelques actions pharmacologiques du haloperidol. Acta neurol. belg. *60:* 39–53 (1960).

Bunney, B.S.; Aghajanian, G.K.: Evidence for drug actions on both pre- and postsynaptic catecholamine receptors in the CNS; in Usdin, Bunney, Pre- and postsynaptic receptors, pp. 89–120 (Marcel Dekker, New York 1975).

Buxbaum, D.N.; Yarbrough, G.G.; Carter, M.E.: Biogenic amines and narcotic effects. I. Modification of morphine-induced analgesia and motor activity after alteration of cerebral amine levels. J. Pharmac. exp. Ther. *185:* 317–327 (1973).

Calcutt, C.R.; Doggett, N.S.; Spender, P.S.J.: Modification of the anti-nociceptive activity of morphine by centrally administered ouabain and dopamine. Psychopharmacologia *21:* 111–117 (1971).

Canepa, F.G.; Pauling, P.; Sörum, H.: Structure of acetylcholine and other substrates of cholinergic systems. Nature, Lond. *210:* 907–909 (1966).

Chance, W.T.: Clonidine analgesia. Tolerance and cross-tolerance to autoanalgesia. Life Sci. *33:* 2241–2246 (1983).

Chen, G.: The anti-tremorine effect of some drugs as determined by Haffner's method of testing analgesia in mice. J. Pharmac. exp. Ther. *124:* 73–76 (1958).

Cho, A.K.; Haslett, W.L.; Jenden, D.J.: The peripheral actions of oxotremorine, a metabolite of tremorine. J. Pharmac. exp. Ther. *138:* 219–257 (1962).

Christensen, E.M.; Gross, E.G.; Analgesic effects in human subjects of morphine, meperidine and methadon. J. Am. Med. Ass. *137:* 594–599 (1948).

Cicero, T.J.; Effects of alpha-adrenergic blocking agents on narcotic-induced analgesia. Archs. int. Pharmacodyn. Thér. *208:* 5–13 (1974).

Cicero, T.J.; Meyer, E.R.; Smithloff, B.R.: Alpha adrenergic blocking agents: antinociceptive activity and enhancement of morphine-induced analgesia. J. Pharmac. exp. Ther. *189:* 72–82 (1974a).

Cicero, T.J.; Wilcox, G.E.; Meyer, E.R.: Effect of alpha-adrenergic blockers on naloxone binding in brain. Biochem. Pharmac. *23:* 2349–2352 (1974b).

Clarke, P.B.S.; Kumar, R.: The effects of nicotine on locomotor activity in nontolerant and tolerant rats. Br. J. Pharmacol. *78:* 329–337 (1983).

Clarke, P.B.S.; Pert, C.B.; Pert, A.: Autoradiographic distribution of nicotine receptors in rat brain. Brain Res. *323:* 390–395 (1984).

Clarke, P.B.S.; Schwartz, R.D.; Paul, S.M.; Pert, C.B.; Pert, A.: Nicotinic binding in rat brain: autoradiographic comparison of [^{3}H]acetylcholine, [^{3}H]nicotine, and [^{125}I]alpha-bungarotoxin. J. Neurosci. *5:* 1307–1315 (1985).

Cocolas, G.H.; Robinson, E.L.; Dewey, W.L.; Spaulding, T.C.: Preparation and activity of beta-substituted acetylcholine iodides. J. pharm. Sci. *60:* 1749–1752 (1971).

Commissiong, J.W.; Gentlemann, S.; Neff, N.H.: Spinal cord dopaminergic neurons. Evidence for an incrossed nigrospinal pathway. Neuropharmacology *18:* 565–568 (1979).

Contreras, E.; Quijada, L.; Tamayo, L.: A comparative study of reserpine and *p*-chlorophenylalanine on morphine analgesia in mice. Psychopharmacologia *28:* 319–324 (1973).

Contreras, E.; Tamayo, L.: Effects of drugs acting in relation to sympathetic functions on the analgesic action of morphine. Archs. int. Pharmacodyn. Thér. *160:* 312–320 (1966).

Courvoisier, S.; Fournel, J.; Ducrat, R.; Kalsky, M.; Koetschet, P.: Propriétés pharmacodynamiques du chlorhydrate de chloro-3(dimethylamino-3′ propyl)-10 phenothiazine. Archs. int. Pharmacodyn. Thér. *92:* 305–361 (1953).

Cox, B.; Tha, S.J.: The antinociceptive activities of oxotremorine physostigmine and dyflos. J. Pharm. Pharmac. *24:* 547–551 (1972).

Curtis, D.R.; Eccles, J.C.; Eccles, R.M.: Pharmacological studies on spinal reflexes. J. Physiol. *136:* 420–434 (1957).

Dahlström, A.; Fuxe, K.: Evidence for the existence of monoamine-containing neurons in the central nervous system. II. Experimentally induced changes in the intraneuronal amine levels of the bulbospinal neuron systems. Acta. physiol. scand. *64:* suppl. 247, pp. 1–36 (1965).

Dahlström, B.; Paalzow, G.; Paalzow, L.: A pharmacokinetic approach to morphine analgesia and its relation to regional turnover of rat brain catecholamines. Life Sci. *17:* 11–16 (1975).

Dandiya, P.C.; Meman, M.K.: Studies on central nervous system depressants. III. Influence of some tranquillizing agents on morphine analgesia. Archs. int. Pharmacodyn. Thér. *141:* 223–232 (1963).

Davis, L.; Pollock, L.J.; Stone, T.T.: Visceral pain. Surgery Gynec. Obstet. *55:* 418–426 (1932).

Dayton, H.E.; Garrett, R.L.: Production of analgesia by cholinergic drugs. Proc. Soc. exp. Biol. Med. *142:* 1011–1013 (1973).

Defalque, R.J.: Potentiation of morphine by intraperitoneal injections of iproniazid in rabbits. Anesth. Analg. *44:* 190 (1965).

de Jongh, D.K.: Remarks on the mechanism of analgesic action of morphine. Acta physiol. pharmacol. neer *3:* 164–172 (1954).

Demenge, P.; Feuerstein, C.; Mouchet, P.; Guerin, B.: Stereospecific binding of ^{3}H-haloperidol in rat dorsal spinal cord. Eur. J. Pharmacol. *66:* 117–120 (1980).

Demenge, P.; Mouchet, P.; Guerin, B.; Feuerstein, C.: Identification and distribution of neuroleptic binding sites in the rat spinal cord. J. Neurochem. *37:* 53–59 (1981).

Denisenko, P.P.: Pharmacological blocking of central cholinoreactive systems and the possibilities of its therapeutic application; Koelle, Douglas, Carlson, Trika, Pharmacology of cholinergic and adrenergic transmission; pp. 147–152 MacMillan, New York (1965).

Dewey, W.L.; Chou-Pham, T.T.; Cocolas, G.L.: The antinociceptive effects of acetylcholine analogs and isomers and their antagonism by isomers of narcotic antagonists; Kosterlitz, Opiates and endogenous peptides, pp. 185–190 Elsevier North Holland/Biomedical Press, Amsterdam (1976).

Dewey, W.L.; Cocolas, G.; Daves, E.; Harris, L.S.: Stereospecificity of intraventricularly administered acetylmethylcholine antinociception. Life Sci. *17:* 9–10 (1975).

Dundee, J.W.; Nicholl, R.M.; Moore, J.: Alterations in response to somatic pain associated with anaesthesia. VIII. The effects of atropine and hyoscine. Br. J. Anaesth. *33:* 565–571 (1961).

Dunai-Kovács, Z.; Szèkely, J.I.: Effect of apomorphine on the antinociceptive activity of morphine. Psychopharmacology *53:* 65–72 (1977).

Eidelberg, E.; Erspamer, R.: Dopaminergic mechanisms of opiate actions in brain. J. Pharmac. exp. Ther. *192:* 50–57 (1975).

Elchisak, M.A.; Rosecrans, J.A.: Effect of central catecholamine depletions by 6-hydroxydopamine on morphine antinociception in rats. Res. Commun. chem. Pathol. Pharmacol. *6:* 349–352 (1973).

Elliott, H.W.; Spiehler, V.; Navarro, G.: Effect of naloxone on antinociceptive activity of phenoxybenzamine. Life Sci. *19:* 1637–1644 (1976).

Engberg I.; Ryall, R.W.: The inhibitory action of noradrenaline and other monoamines on spinal neurons. J. Physiol. Lond. *185:* 298–322 (1966).

Essman, W.B.: Nicotine-related neurochemical changes: Some implications for motivational mechanisms and differences; in Dunn, Smoking behavior: Motives and incentives (Winston, Washington 1973).

Evans, W.O.; Bergner, D.P.: A comparison of the analgesic potencies of morphine, pentazocine and a mixture of methamphetamine and pentacozine in the rat. J. New Drugs *4:* 82–85 (1964).

Fennessy, M.R.; Lee, J.R.: Modification of morphine analgesia by drugs affecting adrenergic and tryptaminergic mechanisms. J. Pharm. Pharmac. *22:* 930–935 (1970).

Fielding, S.J.; Wilker, M.; Hynes, M.; Szewczak, M.; Novick, W.; Lal, H.A.: Comparison of clonidine with morphine for antinociceptive and antiwithdrawal actions. J. Pharmac. exp. Ther. *207:* 899–905 (1978).

Fields, H.L.; Buxbaum, A.I.: Brainstem control of spinal pain-transmission neurons. A. Rev. Physiol. *40:* 217–248 (1978).

Floodmark, S.; Wrammer, T.: The analgesic action of morphine, eserine and prostigmine. Acta physiol. scand. *9:* 88–96 (1945).

Forrest, W.H., Jr.; Brown, B.W.; Colin, R.B.; Defalque, R.; Gold, M; Gordon, E.H.; James, K.E.; Katz, J.; Mahler, D.L.; Schraff, P.; Teutsch, G.: Dextroamphetamine with morphine for the treatment of postoperative pain. New Engl. J. Med. *296:* 712–715 (1977).

Frederickson, R.C.A.; Burgis, V.; Edwards, J.D.: Hyperalgesia induced by naloxone follows diurnal rhythm in responsivity to painful stimuli. Science *198:* 756–758 (1977).

Friedler, G.; Bhargava, H.H.; Quock, R.; Way, E.L.: The effect of 6-hydroxydopamine on morphine tolerance and physical dependence. J. Pharmac. exp. Ther. *183:* 49–55 (1972).

Garcia-Leme, L.J.; Rocha e Silva, M.: Analgesic action of chlorpromazine and reserpine in relation to that of morphine. J. Pharm. Pharmac. *13:* 734–742 (1961).

George, R.; Haslett, W.L.; Jenden, D.J.: A cholinergic mechanism in the brainstem reticular formation. Induction of paradoxical sleep. Int. J. Neuropharmacol. *3:* 541–552 (1964).

Goetzl, F.R.; Burrill, D.Y.; Ivy, S.C.: The analgesic effect of morphine alone and in combination with dextroamphetamine. Proc. Soc. exp. Biol. Med. *55:* 248–250 (1944a).

Goetzl, F.R.; Burrill, D.Y.; Ivy, S.C.: The analgesic effects of sympathomimetics and the combination of morphine with amphetamine. Proc. int. Med. Chic. *15:* 87–88 (1944b).

Görlitz, B.-D.; Frey, H.H.: Central monoamines and antinociceptive drug action. Eur. J. Pharmacol. *20:* 171–180 (1972).

Grevert, D.; Goldstein, A.: Some effects of naloxone on behavior in the mouse. Psychopharmacology *53:* 111–113 (1977).

Grossmann, W.; Jurna, I.; Nell, T.; Theres, C.: The dependence of the antinociceptive effect of morphine and other analgesic agents on spinal motor activity after central monoamine depletion. Eur. J. Pharmacol. *24:* 67–77 (1973).

Gupta, S.K.; Kulkarni, H.J.: Modification of morphine analgesia in rats by MAO inhibitors. J. Indian med. Ass. *46:* 197 (1966).

Hammond, D.L.; Broudfit, H.K.: Effects of locus coeruleus lesions on morphine-induced antinociception. Brain Res. *188:* 79–91 (1980).

Handley, S.L.; Spencer, P.S.J.: Analgesic activity after intracerebral injection in the mouse. Br. J. Pharmacol. *35:* 361–362P (1969).

Harris, L.S.: Central neurohumoral systems involved with narcotic agonists and antagonists. Fed. Proc. *29:* 28–32 (1970).

Harris, L.S.; Demen, W.L.; Haines, J.F.; Kennedy, J.S.; Pars, H.: Narcotic-antagonists analgesics interaction with cholinergic systems. J. Pharmac. exp. Ther. *169:* 17–22 (1969).

Headley, P.M.; Duggan, A.W.; Griersmith, B.T.: Selective reduction by noradrenergic and 5-hydroxytryptamine of nociceptive responses of cat dorsal horn neurons. Brain Res. *145:* 185–189 (1978).

Heinz, G.; Jurna, I.: The anti-nociceptive effect of reserpine and haloperidol mediated by the nigrostriatal system. Antagonism by naloxone. Arch. Pharmacol. *306:* 97–100 (1979).

Heller, B.; Saavedra, J.M.; Fischer, E.: Influence of adrenergic blocking agents upon morphine and catecholamine analgesic effect. Experientia *24:* 804–805 (1978).

Hernandez-Peón, R.; Chavez-Ibarra, G.; Morgane, P.J.; Timo-Laria, C.: Limbic cholinergic pathways involved in sleep and emotional behavior. Expl. Neurol. *8:* 93–111 (1963).

Herz, A.: Actions of arecoline on the central nervous system. Naunyn-Schmeideberg's Arch. exp. Path. Pharmak. *242:* 414–420 (1962).

Herz, A.; Yacoub, F.: Hemmung nociceptiver und bedingter Reaktion durch Cholinomimetica im Vergleich mit der Wirkung anderer zentral angreifender Substanzen. Psychopharmacologia *5:* 115–125 (1964).

Hodge, C.J.; Apkarian, A.V.; Stevens, R.; Vogelsang, G.; Wisnicki, H.J.: Locus coeruleus modulation of dorsal horn unit responses to cutaneous stimulation. Brain Res. *204:* 415–420 (1981).

Hökfelt, T.; Phillipson, O.; Goldstein, M.: Evidence for a dopaminergic pathway in the rat descending from the All cell group to the spinal cord. Acta physiol. scand. *107:* 393–395 (1979).

Houser, V.P.: Modulation of the aversive quality of shock through a central inhibitory cholinergic system in the rat. Pharmacol. Biochem. Behav. *4:* 561–568 (1976).

Houser, V.P.; Hart, D.A. van: The effects of scopolamine and pilocarpine upon the aversive threshold of the rat. Pharmacol. Biochem. Behav. *1:* 427–431 (1973).

Houser, V.P.; Hart, D.A. van: Modulation of cholinergic activity and aversive threshold in the rat. Pharmacol. Biochem. Behav. *2:* 631–637 (1974).

Houser, V.P.; Houser, F.L.: The alteration of aversive thresholds with cholinergic and adrenergic agents. Pharmacol. Biochem. Behav. *1:* 433–444 (1973).

Howe, J.R.; Wang, J.-W.; Yaksh, T.L.: Selective antagonism of the antinociceptive effect of intrathecally applied alpha-adrenergic agonists by intrathecal prazasin and intrathecal yohimbine. J. Pharmac. exp. Ther. *224:* 552–558 (1983).

Howe, J.R.; Yaksh, T.L.: Changes in sensitivity to intrathecal norepinephrine and serotonin after 6-hydroxydopamine (6-OHDA), 5,6-dihydroxytryptamine (5,6 DHT) on repeated monoamine administration. J. Pharmac. exp. Ther. *220:* 311–321 (1982).

Howes, J.F.; Harris, L.S.; Dewey, W.L., Voyda, C.S.: Brain acetylcholine lends an inhibition of the tail-flick reflex in mice. J. Pharmac. exp. Ther. *169:* 23–28 (1969).

Hynes, M.D.; Henderson, J.K.: Modification of alpha-adrenoceptor agonist antinociceptive activity by nisoxetine. A selective inhibition of noradrenergic uptake. Pharmacol. Biochem. Behav. *20:* 463–466 (1984).

Ireson, J.D.: A comparison of the antinociceptive actions of cholinomimetic and morphine-like drugs. Br. J. Pharmacol. *40:* 92–101 (1970).

Ivy, A.C.; Goetzl, F.R.; Burrill, D.Y.: Morphine-dextro-amphetamine analgesia. War Med. *6:* 67–71 (1944).

Iwamoto, E.T.; Ho, I.K.; Way, E.L.: Effect of pargyline on morphine tolerance and physical dependence development in mice. Eur. J. Pharmacol. *38:* 261–268 (1976).

Jacob, J.J.C.; Ramabadran, K.: Enhancement of a nociceptive reaction by opioid antagonists in mice. Br. J. Pharmacol. *64:* 91–98 (1978).

Jacob, J.J.; Tremblay, E.C.; Colombel, M.C.: Enhancement of nociceptive reactions by naloxone in mice and rats. Psychopharmacologia *37:* 217–223 (1974).

Jansen, P.A.J.; Van de Westeringh, C.; Jageneau, A.H.M.; Demaen, P.J.A.; Hermans, K.F.; Van Daele, G.H.P.; Schellekens, H.L.; Van der Eijchen, C.H.M; Niemeyeers, C.J.E.: Chemistry and pharmacology of CNS depressants related to 4-(4-hydroxy-4-phenyl-piperidino) butyrophenone. I. Synthesis and screening. J. med. Pharmac. Chem. *1:* 281–297 (1959).

Jensen, T.S.; Smith, D.F.: Dopaminergic effects on tail-flick response in spinal rats. Eur. J. Pharmacol. *79:* 129–133 (1982).

Jensen, T.S.; Smith, D.F.: Stimulation of spinal dopaminergic receptors. Differential effects on tail reflexes in rats. Neuropharmacology *22:* 477–483 (1983a).

Jensen, T.S.; Smith, D.F.: Role of 5-HT and NA in spinal dopaminergic analgesia. Eur. J. Pharmacol. *86:* 65–70 (1983b).

Jensen, T.S.; Yaksh, T.L.: Effects of intrathecal dopamine agonist, apomorphine, on thermal and chemical evoked noxious responses in rats. Brain Res. *296:* 285–293 (1984).

Jounela, A.J.; Mattila, M.J.: Modification by phenelzine of morphine and pethidine analgesia in mice. Ann. Med. exp. Biol. Fenn. *46:* 66 (1966).

Jurna, I.; Heinz, G.: Anti-nociceptive effect of morphine, opioid analgesics and haloperidol injected into the caudate nucleus of the rat. Arch. Pharmacol. *309:* 145–151 (1979).

Jurna, I.; Heinz, G.; Blinn, G.; Nell, T.: The effect of substantia nigra stimulation and morphine on alpha-motoneurones and the tail-flick response. Eur. J. Pharmacol. *51:* 239–250 (1978).

Kaakkola, S.; Ahtee, L.: Effects of muscarinic cholinergic drugs on morphine-induced catalepsy, antinociception and changes in brain dopamine metabolism. Psychopharmacology *52:* 7–15 (1977).

Kamat, U.G.; Pradhan, R.J.; Sheth, U.K.: Potentiation of a non-narcotic analgesic, dipyrone, by cholinomimetic drugs. Psychopharmacologia *23:* 180–186 (1972).

Kamerling, S.G.; Wettstein, J.G.; Sloan, J.W.; Su, T.-P.; Martin, W.R.: Interaction between nicotine and endogenous opioid mechanisms in the unanesthetized dog. Pharmacol. Biochem. Behav. *17:* 733–740 (1982).

Kastrzewa, R.M.; Jacobowitz, D.M.: Pharmacological actions of 6-hydroxydopamine. Pharmac. Rev. *26:* 199–288 (1974).

Katayama, Y.; DeWitt, D.S.; Becker, D.P.; Hayes, R.L.: Behavioral evidence for a cholinoceptive pontine inhibitory area. Descending control of spinal motor output and sensory input. Brain Res. *296:* 241–262 (1984a).

Katayama, Y.; Watkins, L.R.; Becker, D.P.; Hayes, R.L.: Non-opiate analgesia induced by carbachol microinjection into the pontine parabrachial region of the cat. Brain Res. *296:* 263–283 (1984b).

Katayama, Y.; Watkins, L.R.; Becker, D.P.; Hayes, R.L.: Evidence for involvement of cholinoceptive cells of the parabrachial region in environmentally induced nociceptive suppression in the cat. Brain Res. *299:* 348–353 (1984c).

Keranen, G.M.; Zaratzian, V.L.; Coleman, R.: Studies on 1,4-dipyriolidine-2-butyne (tremorine) in mice. Toxicol. appl. Pharmacol. *3:* 481–492 (1961).

Knapp, D.E.; Domini, E.F.: Action of nicotine on the ascending reticular activating system. Int. J. Neuropharmacol. *1:* 333–351 (1962).

Knoll, J.; Komlas, E.: The analgesic effect of atropine and its synergism with morphine and prostigmine. Acta physiol. hung. *2:* 57–63 (1951).

Koehn, G.L.; Hendersson, G.; Karczmar, A.G.: Di-isopropyl phosphofluoridate-induced antinociception. Possible role of endogenous opioids. Eur. J. Pharmacol. *61:* 167–173 (1980).

Koehn, G.L.; Karczmar, A.G.: Effect of diisopropyl phosphofluoridation analgesia and motor behavior in the rat. Prog. Neuropsychopharmacol. *2:* 169–177 (1978).

Komlas, E.; Komlas-Szarz, V.: The role of the liver in synergism between analgesic agents and parasympathomimetic drugs. Acta physiol. hung. *6:* 443–451 (1954).

Komlas, E.; Porszarz, J.; Knoll, J.: Morphine-prostigmine synergism. Acta physiol. hung. *1:* 77–83 (1950).

Korf, J.; Bunney, B.S.; Aghajanian, G.K.: Noradrenergic neurons. Morphine inhibition of spontaneous activity. Eur. J. Pharmacol. *25:* 165–169 (1974).

Kostowiki, W.; Jerlicz, M.; Bidzinski, A.; Hauptmann, M.: Reduced analgesic effects of morphine after bilateral lesions of the locus coeruleus in rats. Pol. J. Pharmacol. Pharm. *30:* 49–53 (1978).

Kuraishi, Y.; Fukui, K.; Shiami, H.; Akaike, A.; Takagi, H.: Microinjection of opioids into the nucleus reticularis giganto cellularis of the rat; analgesia and increase in the normetanephrine level in the spinal cord. Biochem. Pharmac. *27:* 2756–2758 (1978).

Kuraishi, Y.; Harada, Y.; Satoh, M.; Takagi, H.: Antagonism by phenoxybenzamine of the analgesic effects of morphine injected into the nucleus reticularis giganto cellularis of the rat. Neuropharmacology *18:* 107–110 (1979a).

Kuraishi, Y.; Harada, Y.; Takagi, H.: Noradrenaline regulation of pain-transmission in the spinal cord mediated by alpha-adenoceptors. Brain Res. *174:* 333–336 (1979b).

Kurchinsky, K.; Rapte, H.; Meseke, R.; Cremer, H.; Sotag, K.-H.: Specific action of narcotics on reflex activation of rat alpha-motoneurones. Arch. Pharmacol. *296:* 249–253 (1977).

Lenke, D.: Narkosepatenzierende und analgetische Wirkung von 1,4-Dipyrralidino-2-butin. Arch. exp. Path. Pharmak. *234:* 35–45 (1958).

Leslie, G.B.: The effect of anti-Parkinsonian drugs on oxotremorine-induced analgesia in mice. J. Pharm. Pharmac. *21:* 248–250 (1969).

Lipman, J.J.; Spencer, P.S.J.: Further evidence for a central site of action for the antinociceptive effect of clonidine-like drugs. Neuropharmacology *18:* 731–733 (1979).

Lipman, J.J.; Spencer, P.S.J.: A comparison of muscarinic cholinergic involvement in the antinociceptive effects of morphine and clonidine in the mouse. Eur. J. Pharmacol. *64:* 249–258 (1980).

Major, C.T.; Pleuvry, B.J.: Effects of alpha-methyl-para-tyrosine, para-chlorophenylalanine, *L*-beta(3-4-dihydroxyphenyl) alanine, 5-hydroxytryptophan and diethyldithiocarbamate on the analgesic activity of morphine and methylamphetamine in the mouse. Br. J. Pharmacol. *42:* 512–521 (1971).

Malec, D.; Langwinski, R.: Central action of narcotic analgesics. VIII. The effect of dopaminergic stimulans on the action of analgesics in rats. Pol. J. Pharmacol. Pharm. *33:* 273:282 (1982).

Maltbie, A.A.; Cavenar, J.O.; Sullivan, J.L.; Hammett, E.; Zung, W.W.K.: Analgesia and haloperidol: a hypothesis. J. clin. Psychiat. *40:* 323–326 (1979).

Mansner, R.: Relation between some central effects of nicotine and its brain levels in the mouse. Ann. Med. exp. Biol. Fenn. *50:* 205–212 (1972).

Mattila, M.J.; Ahtee, L.; Vartiainen, A.: Some central and peripheral effects of nornicotine derivatives; in: von Euler, Tobacco alkaloids and related compounds, pp. 321–331 (Pergamon Press, Oxford, 1965).

Mattila, M.J.; Ahtee, L.; Saarnivaara, L.: The analgesic and sedative effects of nicotine in white mice, rabbits and golden hamsters. Ann. Med. exp. Biol. Fenn. *46:* 78–84 (1968).

McCreery, D.B.; Bloedel, J.R.; Hames, E.G.: Effects of stimulating in raphe nuclei and reticular formation on response of spinothalamic neurons to mechanical stimuli. J. Neurophysiol. *42:* 166–182 (1979).

McGilliard, K.L.; Takemori, A.E.: The effect of dopaminergic modifiers on morphine-induced analgesia and respiratory depression. Eur. J. Pharmacol. *54:* 61–68 (1979).

Medakovic, M.; Banic, B.: The action of reserpine and alpha-methyl-m-tyrosine on the analgesic effect of morphine in rats and mice. J. Pharm. Pharmac. *16:* 198–206 (1964).

Metys, J.; Wagner, N.; Metysova, J.; Herz, A.: Studies on the central antinociceptive action of cholinomimetic agents. Int. J. Neuropharmacol. *8:* 413–425 (1969).

Migdal, W.; Frumin, M.J.: Amnesic and analgesic effects of centrally acting anticholinergics. Fed. Proc. *22:* 188 (1963).

Miley, D.P.; Abrams, A.A.; Atkinson, J.H.; Janowsky, D.S.: Successful treatment of thalamic pain with apomorphine. Am. J. Psychiat. *135:* 1230–1232 (1978).

Misantone, L.J.: Effects of damage to the monoamine axonal constituents of the medial forebrain bundle on reactivity to foot shock and ingestive behavior in the rat. Expl. Neurol. *50:* 448–464 (1976).

Mitler, M.M.; Dement, W.C.: Cataleptic-like behavior in cats after micro-injections of carbachol in pontine reticular formation. Brain Res. *68:* 335–343 (1974).

Mouchet, D.; Guerin, B.; Feuerstein, C.: Dissociate destruction of noradrenaline and dopamine descending projections in the thoracic cord of the rat. Life Sci. *30:* 373–382 (1982).

Mudgill, L.; Friedhoff, A.J.; Tobey, J.: Effect of intraventricular administrations of epinephrine, norepinephrine, dopamine, acetylcholine and physostigmine on morphine analgesia in mice. Archs. int. Pharmacodyn. Thér. *210:* 85–91 (1974).

Nakamura, K.; Kuntzman, R.; Maggio, A.; Conney, A.H.: Effect of 6-hydroxydopamine on catecholamine concentrations and behaviors in the morphine-tolerant rat. J. Pharm. Pharmac. *24:* 484–487 (1972).

Nakamura, K.; Kuntzman, R.; Maggio, A.C.; Angulis, V.; Conney, A.H.: Influence of 6-hydroxydopamine on the effect of morphine on the tail-flick latency. Psychopharmacologia *31:* 177–189 (1973a).

Nakamura, K.; Kuntzmann, R.; Maggio, A.C.; Conney, A.H.: Decrease in morphine's analgesic action and increase in its cataleptic action by 6-hydroxydopamine injected bilaterally into caudate and putamen areas; partial restoration by *L*-DOPA plus decarboxylase inhibition. Neuropharmacology *12:* 1153–1160 (1973b).

Nakamura, K.; Kuntzman, R.; Maggio, A.; Conney, A.H.: Restoration of morphine analgesia in morphine-tolerant rats after the intraventricular administration of 6-hydroxydopamine. J. Pharm. Pharmac. *25:* 584–587 (1973c).

Nanaka, R.; Moraji, T.: Quantitative autoradiography of muscarinic cholinergic receptors in the rat brain. Brain Res. *296:* 295–303 (1984).

Nickerson, M.: Analgesic effects of nitrous oxide and meperidine alone and combined with amphetamine. West. Soc. clin. Res. *April:* 541 (1950).

Nott, M.W.: Potentiation of morphine analgesia by cocaine in mice. Eur. J. Pharmacol. *5:* 93–99 (1968).

Paalzow, L.: Analgesia produced by clonidine in mice and rats. J. Pharm. Pharmac. *26:* 361–363 (1974).

Paalzow, L.; Paalzow, G.: Studies on the relationship between morphine analgesia and the brain catecholamines in mice. Acta. pharmac. tox. *30:* 104 (1971).

Paalzow, G.; Paalzow, L.: Enhancement of apomorphine-induced inhibition of vocalization afterdischarge response by thephylline. Life Sci. *17:* 1145–1151 (1975).

Paalzow, G.; Paalzow, L.: Evaluation of the antinociceptive effects of 4-alpha-dimethyl-m-tyramine (H77/77) in the rat. Eur. J. Pharmacol. *36:* 321–329 (1976a).

Paalzow, G.; Paalzow, L.: Clonidine antinociceptive activity: effects of drugs influencing central monoaminergic and cholinergic mechanisms in the rat. Arch. Pharmacol. *292:* 119–126 (1976b).

Paalzow, G.; Paalzow, L.; Stalby, B.: Pentazocine analgesia and regional rat brain catecholamines. Eur. J. Pharmacol. *27:* 78–88 (1974).

Pedigo, N.W.; Dewey, W.L.: Comparison of the antinociceptive activity of intraventricularly administered acetylcholine to narcotic antinociception. Neurosci. Lett. *26:* 85–90 (1981a).

Pedigo, N.W.; Dewey, W.L.: Acetylcholine induced antinociception: comparisons to opiate analgesia; in Pepeu, Ladinsky, Cholinergic mechanisms, pp. 795–807 (Plenum Press, New York 1981b).

Pedigo, N.W.; Dewey, W.L.; Harris, L.S.: Determination and characterization of the antinociceptive activity of intraventricularly administered acetylcholine in mice. J. Pharmac. exp. Ther. *193:* 845–852 (1975).

Pellandra, C.L.: La geneserine-morphine adjuvant de l'anesthesia générale. Lyon med. *151:* 653–655 (1933).

Pert, A.: The cholinergic system and nociception in the primate: Interactions with morphine. Psychopharmacologia *44:* 131–137 (1975).

Pert, A.; Massari, J.V.; Tizabi, Y.; O'Donahue, T.L.; Jacobowitz, D.: Effects of 6-hydroxydopamine and 5,7-dihydroxytryptamine brainstem lesions on morphine analgesia in the rat; in Way, Endogenous and exogenous opiate agonists and antagonists, pp. 151–154 (Pergamon Press, New York 1979).

Pert, A.; Maxey, G.: Asymmetrical cross-tolerance between morphine and scopolamine induced antinociception in the primate. Differential sites of action. Psychopharmacologia *44:* 139–145 (1975).

Phan, D.V.; Dóda, M.; Bite, A.; György, L.: Antinociceptive activity of nicotine. Acta physiol. acad. hung *44:* 85–93 (1973).

Pleuvry, B.J.: Cross tolerance between methylamphetamine and morphine in the mouse. J. Pharm. Pharmac. *23:* 969–970 (1971).

Pleuvry, B.J.; Tobias, M.A.: Comparison of the antinociceptive activities of physostigmine, oxatremarine and morphine in the mouse. Br. J. Pharmacol. *43:* 706–714 (1971).

Porszasz, J.; Knoll, J.; Komlas, E.: Effect of parasympathomimetic drugs on analgesia. Acta physiol. hung. *2:* 469–474 (1951).

Pradhan, S.N.; Bhattacharya, I.C.; Atkinson, K.S.: The effects of intraventricular administration of nicotine on blood pressure and some somatic reflexes. Ann. N.Y. Acad. Sci. *142:* 50–66 (1967).

Price, M.T.C.; Fibiger, H.C.: Ascending catecholamine systems and morphine analgesia. Brain Res. *99:* 189–193 (1975).

Ramabadram, K.; Jacob, J.J.C: Facilitatory effect of naloxone and involvement of specific ligand-opiate receptor system in the antinociceptive effects of non-opioid drugs. Archs. int. Pharmacodyn. Thér. *236:* 27–42 (1978).

Ramabadran, K.; Jacob, J.C.C.: Stereospecific effects of opiate antagonists on superficial and deep nociception and on motor activity suggest involvement of endorphins on different opioid receptors. Life Sci. *24:* 1959–1970 (1979).

Reddy, S.V.R.; Yaksh, T.L.: Spinal noradrenergenic terminal system mediates antinociception. Brain Res. *189:* 391–401 (1980).

Reinhold, K.; Bläsig, J.; Herz, A.: Changes in brain concentration of biogenic amines and the antinociceptive effect of morphine in rats. Arch. Pharmacol. *278:* 69–80 (1973).

Rogers, R.J.: Attenuation of morphine analogues in rats by intra-amygdaloid injection of dopamine. Brain Res. *130:* 156–162 (1977).

Rogers, R.J.: Effects of nicotine, mecamylamine, and hexamethonium on shock-induced fighting, pain reactivity, and locomotor behaviour in rats. Psychopharmacology *66:* 93–98 (1979).

Romano, J.A.; Shih, T.-M.: The effect of adrenalectomy and dexamethasone on the antinociceptive effects of physostigmine. Psychopharmacology *84:* 426–430 (1984).

Rosecrans, J.A.; Elchisak, M.A.; Harry, G.J.: Morphine and methadone-induced antinociception in rats permanently depleted of brain dopamine. Archs. int. Pharmacodyn. Thér. *229:* 287–300 (1977).

Ross, C.A.; Armstrong, D.M.; Ruggiero, D.A.; Pickels, V.M.; Joh, T.H.; Reis, D.J.: Adrenaline neurons in the rostral ventrolateral medulla innervate thoracic spinal cord. A combined immunocytochemical and retrograde transport demonstration. Neurosci. Lett. *25:* 257–262 (1981).

Ross, J.W.; Ashford, A.: The effect of reserpine and alpha-methyldopa on the analgesic action of morphine in the mouse. J. Pharm. Pharmac. *19:* 709–713 (1967).

Rotter, A.; Birdsall, N.J.M.; Field, P.M.; Rossman, G.: Muscarinic receptors in the central nervous system of the rat. II. Distribution of binding of [^{3}H]propylbenzilcholine mustard in the midbrain and hindbrain. Brain Res. Rev. *1:* 167–183 (1979).

Rudzik, A.D.; Mennear, J.H.: Antagonism of analgeticx by amine-depleting agents. J. Pharm. Pharmac. *17:* 326–327 (1965).

Sadove, M.S.; Levin, M.J.; Rose, R.F.; Schwartz, L.; Witt, F.W.: Chlorpromazine and narcotics in the management of pain or malignant lesions. J. Am. Med. Ass. *155:* 626–628 (1954).

Sahley, T.L.; Berntson, G.G.: Antinociceptive effects of central and systemic administration of nicotine in the rat. Psychopharmacology *65:* 279–283 (1979).

Samanin, R.; Bernasconi, S.: Effects of intraventricularly injected 6-OH dopamine or midbrain raphe lesion on morphine analgesia in rats. Psychopharmacologia *25:* 175–182 (1972).

Sasa, M.; Munekiyo, K.; Osumi, Y.; Takaori, S.: Attenuation of morphine analgesia in rats with lesions of the locus coeruleus and dorsal raphe nucleus. Eur. J. Pharmacol. *42:* 53–62 (1977).

Sasa, M.; Munekiyo, K.; Takaori, S.: Morphine interference with noradrenaline-mediated inhibition from the locus coeruleus. Life Sci. *17:* 1373–1380 (1975).

Saxena, P.N.: Mechanism of cholinergic potentiation of morphine analgesia. Indian J. med. Res. *46:* 653–658 (1958).

Saxena, P.N.; Gupta, G.P.: Analgesic potentiating effects of ephedrine and methamphetamine. J. Indian med. Ass. *4:* 1553–1554 (1957).

Schneider, J.A.: Reserpine antagonism of morphine analgesia in mice. Proc. Soc. exp. Biol. Med. *87:* 614–615 (1954).

Schmitt, H.; Le Douarec, J.-C.; Petillat, N.: Antinociceptive effects of some alpha-sympathomimetic agents. Neuropharmacology *13:* 289–294 (1974a).

Schmitt, H.; Le Douarec, J.-C.; Petillat, N.: Antagonism of the antinociceptive action of xylazine, an alpha-sympathomimetic agent, by adrenoceptor and cholinoceptor blocking agents. Neuropharmacology *13:* 295–303 (1974b).

Schweitzer, A.; Wright, S.: Action of nicotine on the spinal cord. J. Physiol. *94:* 136–147 (1938).

Segal, M.; Sandberg, D.: Analgesia produced by electrical stimulation of catecholamine nuclei in rat brain. Brain Res. *123:* 369–372 (1977).

Sethy, V.H.; Naik, S.R.; Sheth, U.K.: The effects of drugs influencing amine synthesis on the analgesic action of tremorine. Psychopharmacologia *19:* 73–80 (1971).

Sethy, V.H.; Pradhan, R.J.; Mandrekan, S.S.; Sheth; V.K.: Role of brain amines in the analgesic action of meperidine hydrochloride. Psychopharmacologia *17:* 320–326 (1970).

Sewell, R.D.E.; Spencer, P.S.J.: Anti-nociceptive activity of narcotic agonists and partial agonists in mice given biogenic amines by intracerebroventricular injection. Psychopharmacologia *42:* 67–71 (1975).

Shiomi, H.; Takagi, H.: Morphine analgesia and bulbospinal noradrenergic system: increase in the concentration of normetanephrine in the spinal cord of rat caused by analgesia. Br. J. Pharmacol. *52:* 519–526 (1974).

Sigg, E.B.; Caprio, G.; Schneider, J.A.: Synergism of amines and antagonism of reserpine to morphine analgesia. Proc. Soc. exp. Biol. Med. *97:* 97–100 (1958).

Sitaram, N.; Buchsbaum, M.S.; Gillin, J.C.: Physostigmine analgesia and somatosensory evolked responses in man. Eur. J. Pharmacol. *42:* 285–290 (1977).

Skagerberg, G.; Björklund, A.; Lindvall, L.O.; Schmidt, R.H.: Origin and termination of the diencephalic dopamine system in the rat. Brain Res. Bull. *9:* 237–244 (1982).

Skingle, M.; Hayes, A.G.; Tyers, M.B.: Antinociceptive activity of clonidine in the mouse, rat and dog. Life Sci. *31:* 1123–1132 (1982).

Slater, P.: Effect of 6-hydroxydopamine on some actions of tremorine and oxotremorine. Eur. J. Pharmacol. *25:* 130–137 (1974).

Slater, P.; Blundell, C.: The effect of 6-hydroxydopamine on the antinociceptive action of morphine. Eur. J. Pharmacol. *48:* 237–247 (1978a).

Slater, P.; Blundell, C.: The effects of a permanent and selective depletion of brain catecholamines on the antinociceptive action of morphine. Arch. Pharmacol. *305:* 227–232 (1978b).

Slaughter, D.: Neostigmine and opiate analgesia. Archs. int. Pharmacodyn. Thér. *83:* 143–148 (1950).

Slaughter, D.; Gross, E.G.: Some new aspects of morphine action. Effect on intestine and blood pressure; toxicity studies. J. Pharmac. exp. Ther. *68:* 96–102 (1939).

Slaughter, D.; Munsell, D.W.: Some new aspects of morphine action effects on pain. J. Pharmac. exp. Ther. *68:* 104–112 (1940).

Slaughter, D.; Parsons, J.C.; Munal, H.D.: New clinical aspects of the analgesic action of morphine. J. Am. Med. Ass. *115:* 2058–2060 (1940).

Smith, D.L.; D'Amour, M.C.; D'Amour, F.E.: The analgesic properties of certain drugs and drug combinations. J. Pharmac. exp. Ther. *77:* 184–194 (1943).

Sparks, C.G.; Spencer, P.S.J.: Antinociceptive activity of morphine after injection of biogenic amines in the cerebral ventricles of the conscious rat. Br. J. Pharmacol. *35:* 362P-363P (1969).

Sparks, C.G.; Spencer, P.S.J.: Antinociceptive activity of morphine after injection of biogenic amines in the cerebral ventricles of the conscious rat. Br. J. Pharmacol. *42:* 230–242 (1971).

Spaulding, T.C.; Venafro, J.J.; Ma, M.G.; Fielding, S.: The dissociation of the antinociceptive effect of clonidine from supraspinal structures. *18:* 103–105 (1979a).

Spaulding, T.C.; Fielding, S.; Venafro, J.J.; Lal, H.: Antinociceptive activity of clonidine and its potentiation of morphine analgesia. Eur. J. Pharmacol. *58:* 14–25 (1979b).

Sprague, G.L.; Takamori, A.E.: Effect of methamphetamine on the development of acute morphine tolerance and dependence in mice. Life Sci. *23:* 2491–2498 (1978).

Starke, K.; Altman, K.P.: Inhibition of adrenergic neurotransmission by clonidine. An action on prejunctional alpha-receptors. Neuropharmacology *12:* 339–347 (1973).

Starke, K.; Montel, H.: Involvement of alpha-receptors in clonidine-induced inhibition of transmitter release from central monoamine neurones. Neuropharmacology *12:* 1073–1080 (1973).

Sternbach, R.A.; Janowsky, D.S.; Huey, L.Y.; Segal, D.S.: Effects of altering brain serotonin activity on human chronic pain. In Bonica, Albe-Ferraud, Advances in pain research and therapy, vol. 1, pp. 601–606 (Raven Press, New York, 1976).

Stevens, R.T.; Hodge, C.J.; Apkarian, A.V.: Kolliker-Fuse nucleus: the principal source of pontine catecholaminergic cells projecting to the lumbar spinal cord of cat. Brain Res. *239:* 589–594 (1982).

Svensson, T.H.; Bunney, B.S.; Aghajanian, G.K.: Inhibition of both noradrenergic and serotonergic neurons in brain by the alpha-adrenergic agonist clonidine. Brain Res. *92:* 291–306 (1975).

Takagi, H.; Takashima, T.; Kumura, K.: Antagonism of the analgetic effect of morphine in mice by tetrabenazine and reserpine. Archs. int. Pharmacodyn. Thér. *149:* 484–492 (1964).

Takemori, A.E.; Tulunay, F.C.; Yano, I.: Differential effects on morphine analgesia and naloxone antagonism by biogenic amine modifiers. Life Sci. *17:* 21–28 (1975).

Tang, A.H.; Yim, G.K.W.: The effects of arecoline on spinal reflexes of the cat. Int. J. Neuropharmacol. *4:* 309–316 (1965).

Tripathi, H.L.; Martin, B.R.; Aceto, M.D.: Nicotine-induced antinociception in rats and mice. Correlation with nicotine brain levels. J. Pharmac. exp. Ther. *221:* 91–96 (1982).

Tripathi, H.L.; Vocci, F.J.; Dewey, W.L.: Effect of antinociceptive doses of oxotremorine on mouse brain acetylcholine turnover rate. Life Sci. *27:* 2055–2062 (1980).

Tripod, J.; Gross, F.: Different effects of central depressant drugs on the analgetic and central stimulating effects of morphine. Helv. physiol. pharmacol. Acta *15:* 105–116 (1957).

Tulunay, F.C.; Sparber, S.B.; Takemori, A.E.: The effect of dopaminergic stimulation and blockade on the nociceptive and antinociceptive responses of mice. Eur. J. Pharmacol. *33:* 65–70 (1975).

Tulunay, F.C.; Takemori, A.E.: Further studies on the alteration of analgesic receptor-antagonists interaction induced by morphine. J. Pharmac. exp. Ther. *190:* 401–407 (1974).

Vander-Wende, C.; Spoerlein, M.T.: Role of dopaminergic receptors in morphine analgesia and tolerance. Res. Commun. chem. Pathol. Pharmacol. *3:* 35–43 (1973).

Van Dongen, P.A.M.; Broekkamp, C.L.E.; Cools, A.R.: Atonia after carbachol microinjections near the locus coeruleus in cats. Pharmacol. Biochem. Behav. *8:* 527–532 (1978).

Van Eick, A.J.; Bock, J.: Comparison of analgesic, cholinomimetic, anticholinergic and sympathomimetic drugs by means of the hot-plate test. Archs. int. Pharmacodyn. Thér. *189:* 384–387 (1971).

Vedernikov, Yu.D.: Interaction of amphetamine, apomorphine, disulfiram with morphine and the role played by catecholamines in morphine analgesic action. Archs. int. Pharmacodyn. Thér. *182:* 59–64 (1969).

Vedernikov, Yu.D.; Afrikanov, I.I.: On the role of a central adrenergic mechanism in morphine analgesic action. J. Pharm. Pharmac. *21:* 845–847 (1969).

Velasco, M.; Velasco, F.; Romo, R.; Estrada-Villanueva, F.: Carbachol 'push-pull' perfusion in the reticular formation: effect on the contiguous multiple-unit activity and other sleep-waking parameters in cats. Expl. Neurol. *72:* 318–331 (1981).

Verri, R.A.; Graeff, F.G.; Corrado, A.D.: Antagonism of morphine analgesia by reserpine and alpha-methyltyrosine and the role played by catecholamines in morphine analgesic action. J. Pharm. Pharmac. *19:* 264–265 (1967).

Verri, R.A.; Graeff, F.G.; Corrado, A.D.: Effect of reserpine and alphamethyl-tyrosine on morphine analgesia. Int. J. Neuropharmacol. *7:* 283–292 (1968).

Vogt, M.: The concentration of sympathin in different parts of the central nervous system under normal conditions and after the administration of drugs. J. Physiol. *123:* 451–481 (1954).

Watanabe, K.; Matsui, Y.; Iwata, H.: Enhancement of the analgesic effects of morphine by sodium diethyldithiocarbamate in rats. Experientia *9:* 950–951 (1969).

Wamsley, J.K.; Lewis, M.S.; Young, S.W. III; Kuhar, M.: Autoradiographic localization of muscarinic cholinergic receptors in the rat brain stem. J. Neurosci. *1:* 176–191 (1981).

Way, E.L.; Shen, F.-H.: Catecholamines and 5-hydroxytryptamine; in Clouet, Narcotic Drugs: biochemical pharmacology, pp. 229–253 (Plenum Press, New York, 1971).

Webb, S.S.; Smith, G.M.; Evans, W.O.; Webb, N.C.: Toward the development of a potent, nonsedating, oral analgesic. Psychopharmacology *60:* 25–28 (1978).

Weller, C.P.; Ibrahim, I.; Sulman, F.G.: Analgesic profile of tranquillizers in multiple screening in mice. Archs. int. Pharmacodyn. Thér. *176:* 176–192 (1968).

Westlund, K.N.; Bowker, R.M.; Ziegler, M.G.; Coulter, J.D.: Origins of spinal noradrenergic pathways demonstrated by retrograde transport of antibody to dopamina-beta-hydroxylase. Neurosci. Lett. *25:* 243–249 (1981).

Westlund, K.N.; Bowker, R.M.; Ziegler, M.G.; Coulter, J.D.: Descending noradrenergic projections and their spinal terminations. Prog. Brain Res. *57:* 219–238 (1982).

Wirth, W.: Versuche zur kombinierten Wirkung von Megaphen mit stark wirksamen Analgeticis. Arch. exp. path. Pharmak. *22:* 75–76 (1954).

Witkin, L.B.; Heubner, C.F.; Goldi, F.; O'Keefe, E.; Spitaletta, P.; Plummer, A.J.: Pharmacology of z-amino-indane hydrochloride (Su-8629). A potent non-narcotic analgesic. J. Pharmac. exp. Ther. *133:* 400–408 (1961).

Yaksh, T.: Direct evidence that spinal serotonin and noradrenaline terminals mediate the spinal nociceptive effects of morphine in the periaqueductal gray. Brain Res. *160:* 180–185 (1979).

Yaksh, T.L.: Pharmacology of spinal adrenergic systems which modulate spinal nociceptive processing. Pharmacol. Biochem. Behav. *22:* 845–858 (1985).

Yaksh, T.L.; Hammond, D.L.; Tyce, G.M.: Functional aspects of bulbospinal monoaminergic projections in modulating processing of somatosensory information. Fed. Proc. *40:* 2786–2794 (1981).

Yaksh, T.L.; Reddy, S.V.R.: Studies on the analgetic effects of intrathecal opiates, alpha-adrenergic agonists and baclofen: their pharmacology in the primate. Anesthesiology *54:* 451–467 (1981).

Young, D.C.; Ploeg, R.A.V.; Featherstone, R.M.; Gross, E.G.: The interrelationship among the central, peripheral and anticholinesterase effects of some morphinan derivates. J. Pharmac. exp. Ther. *114:* 33–42 (1956).

Zetler, G.: Cataleptic state and hypothermia in mice, caused by central cholinergic stimulation and antagonized by anticholinergic and antidepressant drugs. Int. J. Neuropharmacol. *7:* 325–335 (1986).

Dr. Agu Pert, Bldg. 10, Room 3N212, Biological Psychiatry Branch,
NIMH, 9000 Rockville Pike, Bethesda, MD 20892 (USA)

Pain Headache, vol. 9, p. 64–100 (Karger, Basel 1987)

Descending Serotoninergic Systems

J.M. Besson, A. Chaouch

Unité de Recherches de Neurophysiologie Pharmacologique,
Inserm U-161, Paris, France

One of the functional roles of the raphe-spinal system consists of the control of the activity of neurons in the dorsal horn of the spinal cord. In the early 1960s, Lundberg's group [see refs. in Lundberg, 1982] demonstrated that the tonic inhibition exerted by the brain stem on flexion reflex pathways originated in the raphe nuclei of the lower brain stem and the adjacent reticular formation. These descending influences passed through the dorsolateral funiculus and were partly mediated by serotoninergic mechanisms [Engberg et al., 1968a, b].

Following these initial studies devoted to the control of flexion reflex afferents, many studies have considered influences exerted by the brain stem on the activity of dorsal horn neurons, notably on the responses due to unmyelinated nociceptive afferent fiber inputs [Besson, 1980; Willis, 1982]. However, it is the behavioral studies on the analgesic effects of deep brain stimulation (stimulation-produced analgesia; SPA) which have mainly provided evidence implicating descending serotoninergic systems in these effects. These studies prompted a variety of electrophysiological and neuroanatomical studies which, without doubt, underlined the important role played by the descending serotoninergic systems in the modulation of the spinal transmission of nociceptive messages.

In this chapter on descending serotoninergic systems we will successively cover the behavioral aspects of SPA, the electrophysiological mechanisms and the anatomical substrates which could underlie these effects, and the involvement of serotonin in these phenomena.

Analgesia Induced by Stimulation of the Periaqueductal Gray Matter and Raphe Nuclei

Since the initial observation by Reynolds [1969] which demonstrated that electrical stimulation of the periaqueductal gray matter (PAG) in the rat could induce powerful antinociceptive effects, this type of analgesia has been extensively studied. The numerous behavioral, electrophysiological and neurochemical data obtained in various animal species (rat, cat and monkey) have emphasized the predominant role of the brain stem in the control of nociception [Basbaum and Fields, 1978; Besson et al., 1981; Liebeskind et al., 1976; Mayer, 1979; Mayer and Price, 1976; Willis, 1982]. Schematically, the 'key' areas can be defined as the PAG at the mesencephalic level and the nucleus raphe magnus region at the bulbar level.

However, although there is general agreement on the antinociceptive effects induced by stimulation of these two zones, the data related to the PAG in the rat, cat and humans are still controversial, particularly with regard to the exact location of the effective sites.

In the Cat

On the basis of the extensive mapping studies [Oliveras et al., 1974, 1979] which include 300 sites of stimulation spread over the brain stem between the mesencephalon and the medulla, it appears that two main areas play a major role in SPA (fig. 1).

At the level of the PAG, the points from which analgesia was obtained are mainly located in the ventral area in or close to the dorsal raphe nucleus (DRN), rich in serotoninergic cell bodies. Frequently, stimulation of sites adjacent to this region or stimulation in dorsal PAG elicited aversive reactions such as directed attack, fear, escape, apparent pain, and prominent motor disturbances.

The location of the effective stimulation, in or close to the DRN, suggests the involvement of serotoninergic mechanisms in SPA, and for this reason we have investigated the effects elicited by the stimulation of other raphe nuclei rich in serotonin-containing neurones. The exploration of more posterior areas of the mesencephalon and medulla confirmed the importance of the raphe nuclei in analgesia, as the effective stimulation zones corresponded successively to the nucleus centralis superior followed by the more ventral raphe centralis inferior, including the nuclei raphe pontis and raphe magnus.

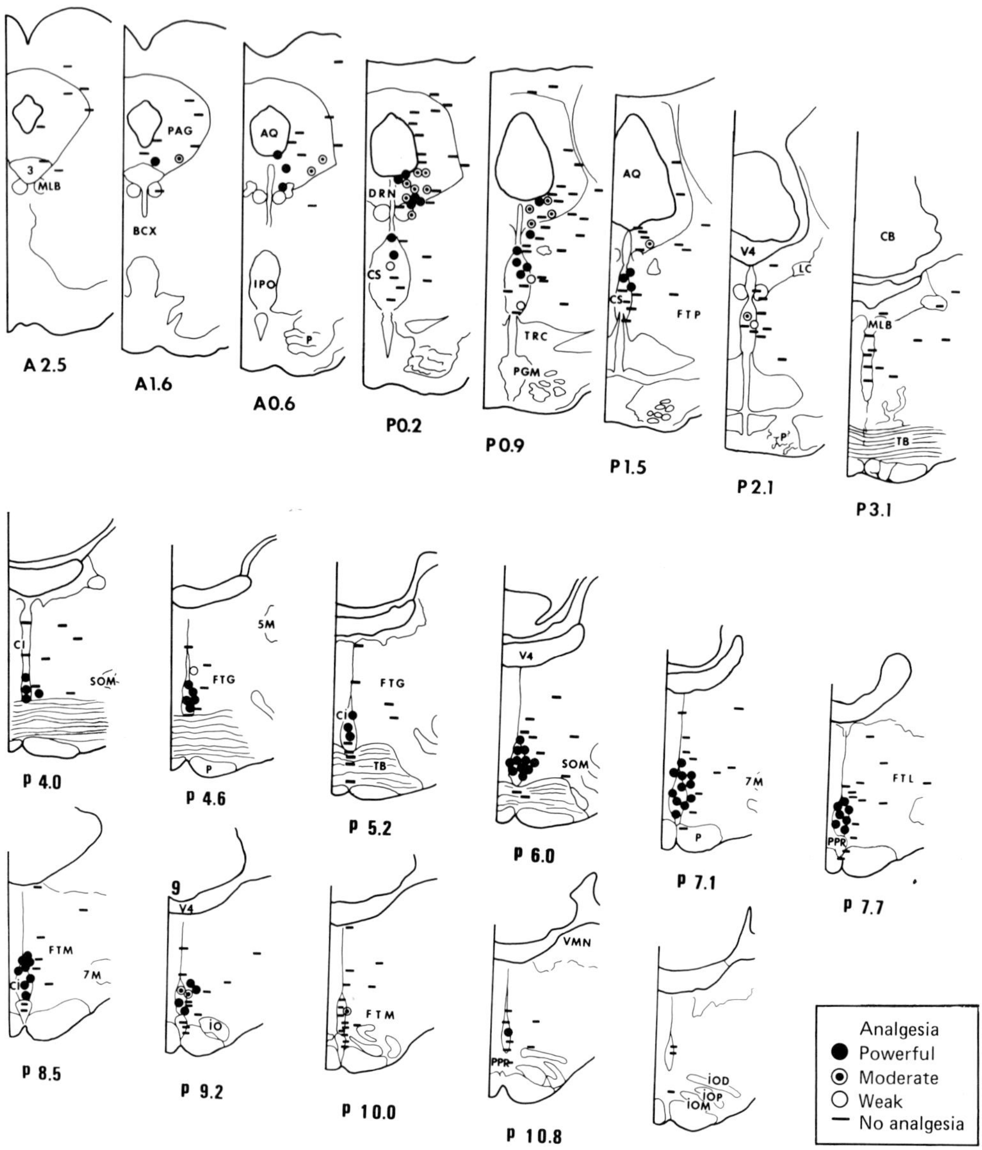

3
MLB
A 2.5
PAG
BCX
A1.6
AQ
IPO
P
A0.6
DRN
CS
P0.2
TRC
PGM
P 0.9
AQ
CS
FTP
P 1.5
V4
LC
P
P 2.1
CB
MLB
TB
P 3.1
CI
SOM
P 4.0
5M
FTG
P
P 4.6
FTG
CI
TB
P 5.2
V4
SOM
P 6.0
7M
P
P 7.1
FTL
PPR
P 7.7
FTM
CI
7M
P 8.5
9
V4
IO
P 9.2
FTM
P 10.0
VMN
PPR
P 10.8
IOD
IOP
IOM
P 11.6
Analgesia
Powerful
Moderate
Weak
No analgesia

1

Without doubt, the most marked analgesic effects were obtained from sites located in the nucleus raphe magnus (NRM). Central stimulation totally blocked, in almost every case, the behavioral responses to intense peripheral pinch. Furthermore, the threshold of the jaw-opening reflex evoked by tooth pulp stimulation was considerably elevated by 100- or 200-fold in some animals, and the nociceptive manifestations induced by trains of stimulation delivered in the tooth pulp were also strongly affected by the central stimulation [Oliveras et al., 1977].

The analgesic effect cannot be attributed to more general stimulation-evoked deficits in sensory, emotional, or motor functions, since during central stimulation the animals appeared alert, calm, able to walk normally and to be attentive to other innocuous stimuli (touch, visual, auditory). It must be underlined that very often these animals seemed to be hyperreactive to light tactile stimulation such as touch or air jet, yet were totally unresponsive to strong pinches applied to the same area.

Furthermore, recordings of the jaw opening reflex elicited by either tooth pulp or central trigeminal stimulation in the chronic cat demonstrate that the suppression of the reflex by the raphe nuclei stimulation was not due to direct effects on motoneurons (fig. 2).

Considering the peripheral extent of the analgesia which, in most cases, included widespread areas of the body, DRN stimulation, in few cases, only produced clear effects on posterior regions of the body, with normal reactions being elicited from painful stimulation of the anterior areas. Generally, the onset of the analgesia was rapid, with the effects being obvious a few seconds after the onset of central stimulation. For brief periods of stimulation (20 to 30 s) and when gauging behavioral reactions to strong pinch, analgesia rarely outlasted the period of brain stimulation by more than between a few seconds to 1 min. In contrast, we observed larger poststimulation effects (up to 60 min) with the jaw-opening reflex following a 5 min raphe magnus stimulation [Oliveras et

---

Fig. 1. Sites for electrical analgesia throughout the mesencephalon and medulla of the cat. The analgesia was gauged by the behavioral responses of the animal to pinch. The most efficacious sites are located in the nuclei raphe dorsalis, central superior (superior part) and centralis inferior (principally raphe magnus). AQ = Aqueduct; BCX = brachium conjunctivum onjunctivum; CB = cerebellum; CI = nucleus centralis inferior; CS = nucleus centralis superior; DRN = dorsal raphe nucleus; FIG = gigantocellularis tegmental field; FIL = lateral tegmental field; IO = inferior olive; FTM = magnocellular tegmental field; PAG = periaqueductal gray matter; PPR = postpyramidal raphe nucleus; TB = trapezoid body. Modified from Oliveras et al. [1974, 1979].

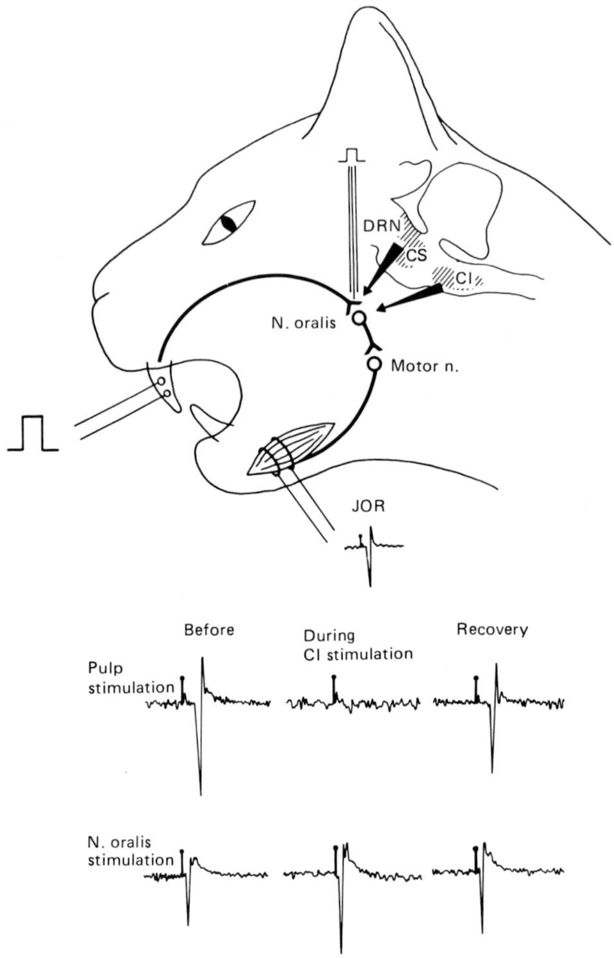

Fig. 2. Nucleus raphe centralis inferior stimulation effects upon the jaw-opening reflex (JOR) of peripheral tooth pulp and central trigeminal origins in the cat. As shown in the upper diagram, the JOR is a disynaptic reflex involving a first relay in the pars oralis of the spinal trigeminal nucleus (N. Oralis) and a second one in the motor trigeminal nucleus (Motor n). In this experiment, the JOR was triggered by an electrical stimulation of the upper canine tooth pulp and of the N. Oralis. The JOR was recorded at the level of the digastric, a jaw opener muscle. The lower part of the figure shows that the electrical stimulation of the nucleus raphe centralis inferior (CI) completely abolishes the JOR of pulpar origin; in contrast, the same central stimulation slightly facilitates the JOR of central trigeminal origin.

al., 1977]. Poststimulation effects of 5 min to 1 h have also been reported in the rat after PAG stimulation [Mayer et al.; 1981; Mayer and Liebeskind, 1974].

In the Rat

In this species, powerful analgesic effects were also obtained by stimulation of the NRM in which the effective stimulation sites equally include the nucleus paragigantocellularis [Akaike et al., 1978; Hammond and Yaksh, 1984; Oleson et al., 1978; Oliveras et al., 1978; Proudfit and Anderson, 1975; Zorman et al., 1981, 1982]. This area, NRM and nucleus paragigantocellularis corresponds to the group of serotoninergic cell bodies B3 first described by Dalhstrom and Fuxe [1965]. The analgesia induced from this bulbopontine area is generally pure, i.e. not associated with side effects.

At the level of the PAG, there are important interspecies differences between cat and rat. The numerous studies performed in the rat clearly indicate that analgesic effects can be induced by stimulation of areas widely spread over the whole of the PAG (fig. 3a). In addition, with the exception of the study of Basbaum et al. [1977], very few ineffective stimulation sites have been described; in this latter study, ineffective areas were generally found in the dorsal and lateral PAG. By contrast, as already mentioned, in the cat (fig. 3b), antinociceptive effects have only been reported during the stimulation of the ventral part of the PAG [Dubuisson and Melzack, 1977; Gebhart and Toleikis, 1978; Liebeskind et al., 1973; Oliveras et al., 1974, 1979], whilst stimulation of the dorsal and lateral parts produced other behavioral effects, notably aversive reactions [Liebeskind et al., 1973; Oliveras et al., 1974].

These dramatic behavioral responses have not been clearly described in the rat by the main groups studying 'electrical analgesia'. These discrepancies between cat and rat seem surprising [see discussion of Kiser et al., 1978] since it is well known that stimulation of the dorsal PAG, in the rat, can also trigger behavioral effects which have been used as a model for the study of aversive reactions [Kiser et al., 1978; Olds and Frey, 1971; Schmitt et al., 1977].

We postulated that analgesia from PAG in the rat could be due to the fact that several of the above-mentioned studies were performed on restrained animals, a condition which could mask the possible side effects induced by the stimulation. In recent studies, we described the complex nature of the behavioral reactions observed during PAG stimu-

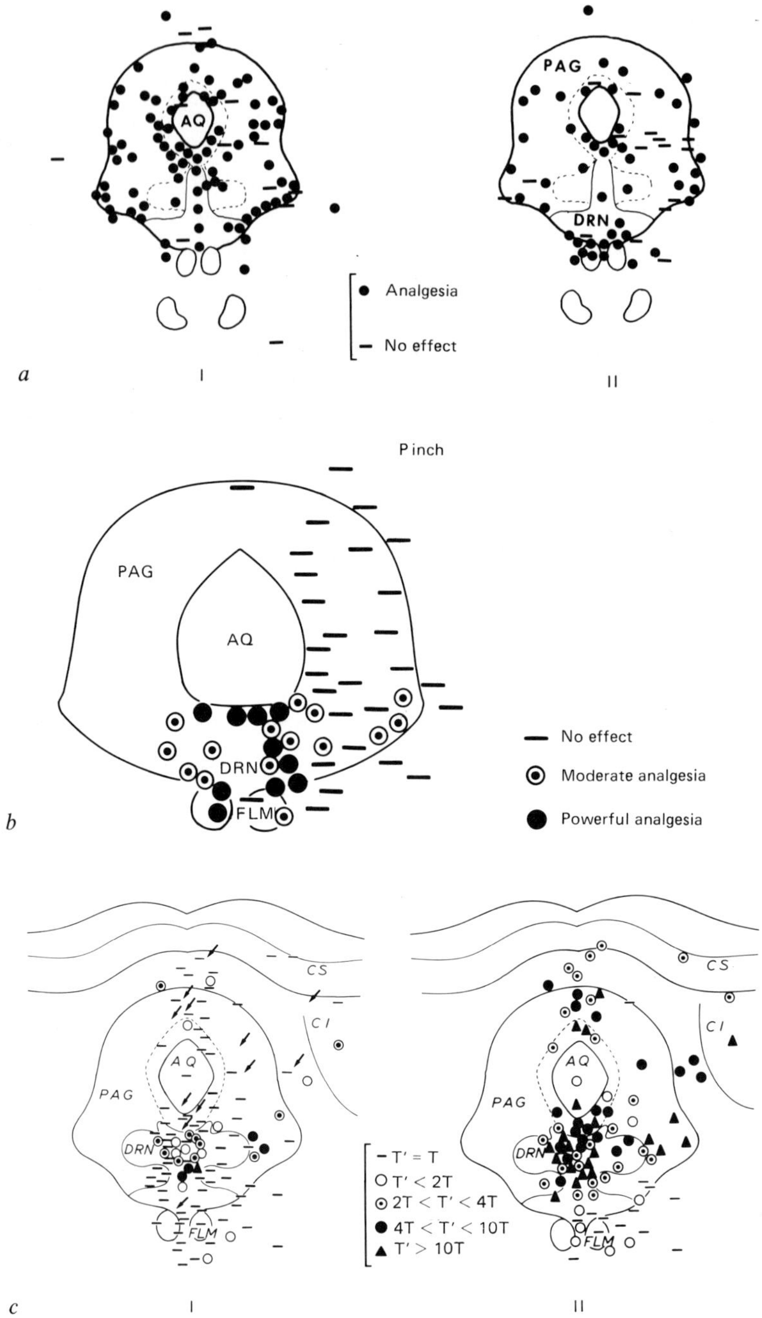
a
AQ
PAG
DRN
Analgesia
No effect
I
II
b
Pinch
PAG
AQ
DRN
FLM
No effect
Moderate analgesia
Powerful analgesia
c
CS
CI
AQ
PAG
DRN
FLM
T' = T
T' < 2T
2T < T' < 4T
4T < T' < 10T
T' > 10T
I
II

lation in the freely moving rat [Fardin et al., 1984a, b] (fig. 3c). It has been demonstrated that: (a) At intensities of stimulation which did not induce side effects, it was possible to define two 'pure analgesic regions', both located in the ventral PAG; the dorsomedial part of the DRN and a ventrolateral zone. (b) When the intensity of central stimulation (applied to the various parts of the PAG) was increased, some stereotypic 'behavioral responses' were observed; they varied with the stimulation site: motor effects (gnawing, rotation or tremor) in the ventral PAG and aversive effects (flight, jumping and on occasions, distress vocalizations) in the dorsal, dorsolateral PAG and a ventral region surrounding the cerebral aqueduct. Under these conditions, the nociceptive thresholds were increased, sometimes to a very great extent.

These results demonstrating the major role of the ventral PAG in SPA in the rat are in good agreement with previous observations in the cat [Dubuisson and Melzack, 1977; Gebhart and Toleikis, 1978; Oliveras et al., 1974, 1979]. In contrast, they question some of the data reported in the literature in the rat and raise the problem of the possible involvement of 'stress-produced analgesia' in most of the analgesic effects elicited by PAG stimulation.

The various side effects of PAG stimulation reported in the rat and cat are particularly important, since such deep brain stimulation is used for the treatment of severe chronic pain in humans. There are many discrepancies between the various neurological groups involved in this field [Adams, 1976; Boivie and Meyerson, 1982; Gybels, 1979; Hosobuchi et al., 1977; Richardson and Akil, 1977], and observations in animals could partly explain some of these contradictory results. Indeed, Fardin et al. [1984a, b] illustrate the relatively limited extent of 'pure analgesic areas',

Fig. 3. Effective stimulation sites for stimulation produced antinociceptive effects in the periaqueductal gray matter. *a.* In the rat: data taken from Akil and Liebeskind [1975], Akil and Mayer [1972], Giesler and Liebeskind [1976], Lewis and Gebhart [1977], Mayer and Liebeskind [1974], Morrow and Casey [1976], Rhodes and Liebeskind [1978] and Yaksh et al. [1976] for tail flick (I), and from Basbaum et al. [1976], Lewis and Gebhart [1977], Mayer and Liebeskind [1974], Mayer et al. [1971], Reynolds [1969], Rhodes and Liebeskind [1978] and Yaksh et al. [1976] for pinch (II), although some were restricted to the study of the ventral part of the PAG. *b.* In the cat [modified from Oliveras et al., 1974, 1979]. *c.* In the rat, from the study of Fardin et al. [1984], performed on freely moving rats in the absence (I) or in the presence (II) of additional behavioral reactions. Analgesia was assessed by alteration in the vocalization threshold induced by electrical tail shocks. T = Initial threshold; T′ = threshold during central stimulation; the arrow indicates a decrease in threshold.

and, consequently, one could presume that these zones would be difficult to locate in humans. Although analgesic effects have been reported in primates [Goodman and Holocombe, 1976; Oleson et al., 1980], unfortunately no systematic mapping study in this species has been made. Hence, there is no guide to clinical investigations in humans [Boivie and Meyerson, 1982]. The functional heterogeneity of the PAG exemplified by the behavioral studies should be considered in the analysis of the electrophysiological and pharmacological studies of the effects of stimulation of this region.

Neural Correlates of SPA

It is generally admitted that analgesia is an active phenomenon that results, at least partly, from the activation of descending inhibitory systems, which block the transmission of nociceptive messages at the spinal level. This idea is supported by the suppression of certain nociceptive reflexes in the rat and in the cat by central stimulation [Mayer et al., 1971; Oliveras et al., 1974]. A more direct argument has been put forward by our groups. Stimulation of the dorsal PAG induces a powerful inhibition on some dorsal horn neurones involved in nociception [Liebeskind et al., 1973; Oliveras et al., 1974]. It must be pointed out that, in some of these experiments, we had previously checked that some sites of stimulation produced strong analgesia in freely moving animals. Numerous subsequent studies [Besson et al., 1982; Willis, 1984] confirmed that PAG stimulation inhibits to a marked extent the responses of dorsal horn neurons to natural (pinch, heat) nociceptive stimuli and electrical stimuli, activating C fibres. It is of interest to note that these inhibitory effects have been observed on both nociceptive-specific and nociceptive-nonspecific (driven by both light and intense cutaneous stimuli) neurons, some of which are at the origin of the spinothalamic tract [Besson et al., 1982; Willis, 1984].

Similarly, stimulation of the NRM powerfully inhibits the activity of cells of the dorsal horn (fig. 4) and trigeminal nuclei, including those at the origin of the main ascending tracts (spinothalamic; trigeminothalamic, and spinoreticular) involved in the transmission of nociceptive messages [Besson et al., 1982; Willis, 1984]. The first studies, carried out in the cat, showed that the inhibitory effects observed on nociceptive-nonspecific neurons were preferential for the responses induced by

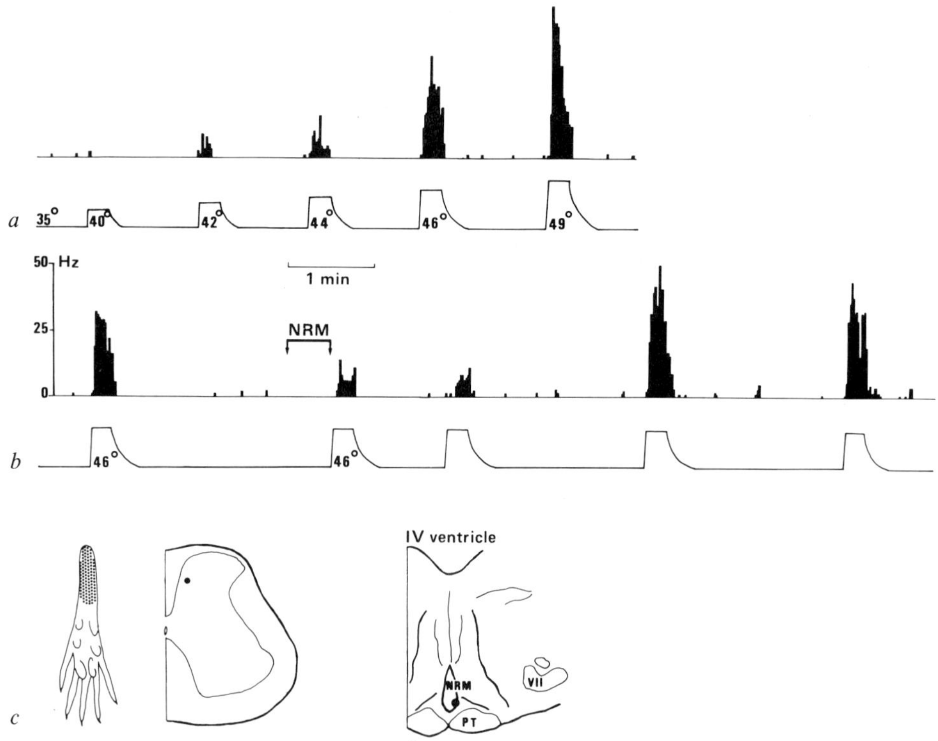

Fig. 4. Effects of nucleus raphe magnus stimulation on responses of dorsal horn neurons to radiant heat. *a.* Control: typical response of a class 2 dorsal horn unit to radiant heat application. Note that responses increase progressively with the increases in skin temperature. *b.* Effects of NRM stimulation (30 s duration, 100 μA alternating current at 50 Hz, trains of 100 ms at 0.3 cycles/s) applied between arrows immediately before the onset of the radiant heat application. *c.* Cutaneous receptive field and spinal location of the neuron: location of the stimulating electrode at the level of the nucleus raphe magnus [from Rivot et al., 1980].

nociceptive stimulation [Duggan and Griersmith, 1979; Guilbaud et al., 1977]. In addition, some authors [Fields et al., 1977; Guilbaud et al., 1977] found that NRM stimulation did not alter the responses of neurons of the dorsal horn exclusively driven by light tactile or by proprioceptive non-nociceptive stimulation. However, the selectivity of action of the NRM on responses induced by nociceptive stimulation is highly controversial, since some authors reported clear inhibitory effects on the responses of nociceptive-nonspecific neurons to light mechanical stimulation [Ammons et al., 1984; Chapman et al., 1985; Gerhart et al., 1981;

Willis et al., 1977]. These conflicting results could be due to methodological factors; however, in the rat [Rivot et al., 1980] and in the monkey [Gerhart et al., 1981], NRM stimulation produces a particularly powerful inhibition on the responses due to C fiber activation. Thus, in these experiments the descending inhibitory effects are preferential on responses evoked by thin afferent fibres.

Direct anatomical spinal projections from the PAG do not seem particularly dense, at least with regard to the lumbar cord and more caudal levels. By contrast, it has been long recognized that the NRM sends massive projections to the cord [Basbaum and Fields, 1984], and so this nucleus could relay the effects produced from the mesencephalon.

Indeed, direct PAG-NRM connections have been clearly demonstrated by anatomical investigations [Basbaum and Fields, 1984] and electrophysiological studies confirm these results and demonstrate that electrical or chemical stimulation of the PAG modulates the activity of NRM neurons including raphe-spinal cells [Behbehani and Fields, 1979; Fields and Anderson, 1978; Lovick et al., 1978; Mason et al., 1985; Shah and Dostrovsky, 1980; Vanegas et al., 1984; Willis et al., 1984]. The most commonly described effect was excitation, although in some cases inhibitory actions were reported. In addition, the antinociceptive action of PAG is, at least in part, exerted through the NRM [Behbehani and Fields, 1979; Gebhart et al., 1983; Prieto et al., 1983; Sandkuhler et al., 1984; Morton et al., 1984]. However, in spite of these PAG-NRM connections, one cannot exclude the involvement of other relays in the descending inhibitory controls from the PAG. The PAG also projects onto the bulbopontine reticular formation [Abols and Basbaum, 1981; Chung et al., 1983; Gallagher and Pert, 1978; Mantyh, 1983], areas of which, i.e. nucleus gigantocellularis, are at the origins of bulbospinal pathways [Walberg, 1982). These anatomical data are in agreement with the fact that the inhibition of spinal nociceptive transmission induced from the PAG is strongly reduced only when both NRM and lateral reticular formation are simultaneously blocked by microinjection of lidocaine [Gebhart et al., 1983; Sandkuhler and Gebhart, 1984; Mohrland et al., 1982]. Similarly, the inhibitory effects of PAG stimulation on dorsal horn neurons have been reported to be only slightly reduced by midline medullary electrocoagulation, whilst bilateral lesions restricted to the more caudal lateral reticular nucleus abolished the inhibitory effects [Morton et al., 1984]. It therefore seems necessary to reconsider in detail the relative roles of these different components.

Direct projections from the NRM onto the spinal cord have been known since the work of Brodal et al. [1960]. In addition, Dahlstrom and Fuxe [1965] have shown that serotoninergic fibres, probably unmyelinated, descend from the raphe nuclei towards the spinal cord. Later studies [Basbaum, et al., 1976, 1978; Basbaum and Fields, 1979; Bobillier et al., 1976; Kneisley et al., 1978; Leichnetz et al., 1978; Martin et al., 1978; Ruda et al., 1981; Tohyama et al., 1979; Watkins et al., 1981; Zemlan and Pfaff, 1979] have demonstrated the numerical importance of these projections, since in the cat following an HRP injection at L_6 over 50% of the neurons found in NRM region were labelled. The serotoninergic nature of a great number of these raphe-spinal fibers has been demonstrated. The descending fibers from raphe magnus pass in the dorsolateral funiculus (DLF) and terminate at different levels of the spinal grey matter [Basbaum et al., 1978; Basbaum and Fields, 1979]. Spinal serotoninergic terminals have been demonstrated by the use of anatomical techniques [Glazer and Basbaum, 1984; Hoffert et al., 1983; Lamotte and de Lanerolle, 1983; Light et al., 1983; Maxwell et al., 1983; Miletic et al., 1984; Nishikawa et al., 1983; Ruda et al., 1982; Oliveras et al., 1977].

Is There a Clear Relationship between Behavioral Analgesia and Electrophysiological Effects?

A major difficulty in correlating electrophysiological and behavioral data arises from the fact that the inhibition of dorsal horn neuronal activity can also be obtained from stimulation sites, particularly the adjacent mesencephalic and bulbar reticular structures, where stimulation does not induce analgesia in freely moving animals. In brief, it is easy to inhibit dorsal horn neurons from different brain stem areas, but it is much more difficult to locate stimulation sites which induce 'pure' analgesic effects. For example, at the mesencephalic level, it must be pointed out that analgesia associated with various types of side effects (motor, aversive, etc.) could be also obtained from the whole PAG and the adjacent reticular formation [Fardin et al., 1984a, b]. It is then necessary in the behavioral studies to dissociate pure analgesic from secondary analgesic effects due, for example, to stress which could be subserved by partially different mechanisms. It should be recalled that NRM lesions reduce the analgesia obtained by stimulation of the ventral PAG [Prieto et al., 1983] but does not influence that produced from dorsal regions where the stimulation is often aversive [Lovick, 1985]. This underlines the caution with which one should interpret the results of behavioral studies carried out in restrained or slightly anesthetized animals.

To approach this problem under suitable conditions, it would be necessary to develop experiments in chronic animals where the evaluation of the analgesic effects and the recording of dorsal horn neurons could be simultaneously carried out.

In spite of these difficulties, the involvement of descending control systems in stimulation producing analgesia is strongly suggested by the experiments of Basbaum et al. [1977] showing that analgesia induced by stimulation of the medioventral PAG is almost totally suppressed by lesions of the dorsolateral funiculus which contains numerous fibers originating from the NRM. Interestingly, Basbaum and co-workers [Basbaum et al., 1976; Fields et al., 1977] as well as Willis et al. [1977] have demonstrated that such a lesion of the dorsolateral funiculus is able to suppress the inhibitory effects of raphe magnus on the activities of dorsal horn neurons. However, the functional importance of the DLF with respect to nociception and pain [Basbaum et al., 1978; Martin et al., 1978; Watkins et al., 1980] leads to a brief consideration of the nature of the fibers in this tract. In addition to propriospinal fibers [Chung and Coggeshal, 1983], it is known that this tract contains ascending fibers [Apkarian et al., 1985; McMahon and Wall, 1983], amongst which are spinocervical tract fibers. Interestingly, it has also been shown that dorsal horn neurons send mesencephalic and thalamic projections in this quadrant of the cord [Jones et al., 1985; McMahon and Wall, 1985; Zemlan et al., 1978]. In addition, descending fibers passing in this pathway have various origins [Basbaum and Fields, 1979; Björklund and Skagerberg, 1982; Kuypers and Huisman, 1982; Watkins et al., 1981] such as the ventromedial medulla, locus coeruleus, paralemniscal reticular formation, mesencephalic central gray, red nucleus, hypothalamic paraventricular nucleus, etc. The multiple origins of fibers comprising the DLF stress the care which should be taken in interpreting the results of behavioral studies based on lesions of this tract. The role attributed to these descending fibers is of importance, since their terminations mainly involve the superficial laminae of the dorsal horn where nociceptive primary afferents terminate. However, one cannot totally exclude the participation of descending fibers travelling in other zones of the spinal white matter which could equally modulate the activity of nociceptive neurons in the intermediate and ventral regions of the spinal gray matter.

It is interesting that the effect of DLF lesions on the analgesia induced by stimulation of the PAG tends to wear off after several weeks [Basbaum et al., 1977]; this would agree with the electrophysiological studies demonstrating that some of the inhibitory effects exerted from the PAG are mediated by fibers passing outside the DLF.

In addition, even though a section of the DLF reduces the analgesia elicited from the brain stem [Basbaum et al., 1976, 1977], one cannot exclude that the stimulation modulates the transmission and integration of nociceptive messages at supraspinal levels where the 'analgesic zones' could have different influences from the nonanalgesic areas. Interestingly, such an approach has been used [Morrow and Casey, 1976, 1983] to evaluate the effects of mesencephalic stimulation on the activity of medullary reticular formation neurons. It was reported that inhibition of these neurons in the awake animal was only obtained from stimulation sites able to evoke analgesic effects.

In summary, taking into account electrophysiological, anatomical, and behavioral data, it appears that the ventral PAG and the raphe spinal system issuing from the NRM seem to be crucial areas for SPA. In addition, it is conceivable that the activity of this system or these two regions could be modulated by other areas of the central nervous system as both the PAG and NRM are connected to such areas as hypothalamus, periventricular gray matter, cortex, thalamus and cuneiformis area [Brodal et al., 1960; Conrad et al., 1974; Grofova et al., 1978; Sakai et al., 1977]. These multiple connections suggest that the inhibition of the responses of dorsal horn neurons to noxious stimuli by septal [Carstens et al., 1982], hypothalamic [Carstens, 1982] and cortical [Yezierski et al., 1983] stimulation could be mediated by the PAG or the NRM.

Involvement of 5-HT in Descending Controls

The pharmacology of descending systems is extraordinarily complex and involves 5-HT, norepinephrine, endogenous opiates and certainly various other putative substances, as revealed by recent immunocytochemical techniques. This review only considers 5-HT.

From a general point of view, the role of descending serotonin pathways in the modulation of spinal transmission has been investigated by many authors. Evidence has been obtained by Lundberg's group that tonic descending inhibition on the transmission of flexor reflex afferents in the decerebrate cat is partly sustained by serotoninergic mechanisms. This tonic inhibition is reduced but not abolished after electrolytic lesions of the raphe nuclei [Engberg et al., 1968]. Moreover, the decerebrate tonic inhibitory control is partially released by the administration of 5-HT antagonists. These results are in agreement with a report demonstrating the inhibitory effect of 5-hydroxytryptophan (5-HTP) on transmission from the flexor reflex afferents. The remaining controls which still persisted after raphe nuclei destruction are probably mediated by the dorsal reticulospinal tract [Engberg et al., 1968], which conducts at 30 m/s and so is unlikely to be serotonergic, since it is thought [Dahlström and Fuxe, 1965] that 5-HT fibers are unmyelinated.

More recent studies have considered the involvement of 5-HT in SPA and its participation in the inhibitory effects induced by brain stem stimulation on nociceptive dorsal horn neurons.

From the behavioral point of view, the localization of effective stimulation sites for SPA, in or close to nuclei where 5-HT-containing neurons had been demonstrated, are indispensably in favor of the participation of this monoamine. Interestingly, in the rat, the administration of the serotonin synthesis inhibitor, p-CPA, blocks the analgesic effects of stimulation only from electrodes near the DRN; the analgesia reappears after administration of 5-HT [Akil and Mayer, 1972]. A decreased affectiveness of SPA evoked from the PAG has also been reported after administration of lysergic acid diethylamide (LSD) [Hayes et al., 1977], which blocks the activity of serotoninergic neurons. In addition, tolerance to PAG stimulation was initially reported in the rat [Mayer and Hayes, 1979], and we found [Oliveras et al., 1978] in the cat that the analgesia induced by electrical stimulation of the feline nucleus raphe magnus progressively disappeared when repeated prolonged stimulation was applied. This phenomenon seems to be sustained by serotoninergic mechanisms, since the effectiveness of central stimulation was clearly restored after administration of the serotonin precursor, 5-hydroxytryptophan (5-HTP) (fig. 5). The time course of the effects of 5-HTP on SPA was comparable to the time course of 5-HT synthesis after administration of the precursor. The recovery of the efficacy of central stimulation after 5-HTP indicated that tolerance was not due to tissue damage caused by intense periods of stimulation. It is also of considerable interest that tolerance to PAG stimulation in humans is reduced by the oral administration of *L*-tryptophan [Hosobuchi, 1978].

The results from studies observing the effect of different drugs

a

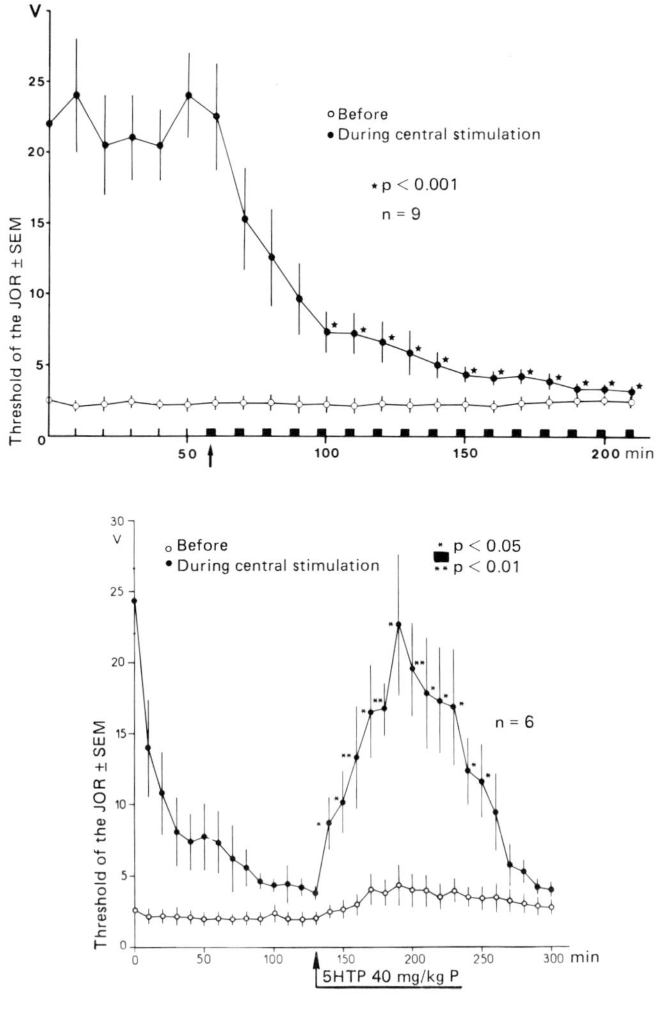

b

Fig. 5. Tolerance to central stimulation of the nucleus raphe magnus. *a.* Threshold of the jaw opening reflex (JOR) triggered by tooth pulp stimulation before and during raphe magnus stimulation. In the first part of the experimental procedure (left), the duration of the central stimulation sequences was 10 s and the large increase of the JOR during central stimulation was constantly obtained. Then the central stimulation was delivered during 3 min each 10 min. In this case note that central stimulation became less and less effective. *b.* In this experimental procedure (6 cats), the central stimulation was applied for 3 min during 10 min. Note the progressive disappearance of the efficacy of central stimulation and its restoration after administration of 5-HTP. Modified from Oliveras et al. [1978].

administered systemically are difficult to interpret with regard to the controls exerted by the brain stem, since the effects could be due to an action at levels other than those under question. The results of these studies are also still controversal [Messing and Lyttle, 1977; Roberts, 1984]. Thus, after administration of parachlorophenylalanine (pCPA), a serotonin synthesis inhibitor, the sensitivity of the animal to nociceptive stimulation can be either exaggerated [Berge et al., 1984; Dennis and Melzack, 1980; Fibiger et al., 1972; Gorlitz and Frey, 1972; Tenen, 1968] or unchanged [Akil and Liebeskind, 1975; Long et al., 1984; Vogt, 1974]. These different effects could result from the use of different nociceptive tests [Berge et al., 1983; Dennis and Melzack, 1980]. On the other hand, the effects of pCPA cannot be explained only by a reduction in 5-HT in the CNS. The consequences of depletion of 5-HT in the nociceptive afferents should be considered, as should the recently reported changes in the levels of substance P in the CNS [Barden et al., 1983; Savard et al., 1984]. Similar problems arise from the analysis of results from studies with systemic 5-HT antagonists [Berge, 1982; Berge et al., 1983; Fasmer et al., 1984; Gorlitz and Frey, 1972].

Intrathecal 5-HT antagonists induce a hyperalgesia [Fasmer et al., 1984; Hammond and Yaksh, 1984; Proudfit and Hammond, 1981]. Similar effects have also been reported following lesions of 5-HT raphe spinal fibers by intrathecal 5–6 or 5–7 DHT. However, this latter hyperalgesia lasts only a short period, and in the rat the nociceptive threshold regains normal levels 2 to 3 weeks after the administration of the neurotoxin [Berge et al., 1983, 1984; Fasmer et al., 1983; Sagen et al., 1983]. This decay with time of the hyperalgesia, noted in different nociceptive tests, could explain the negative results of some studies [Howe and Yaksh, 1982; Johannessen et al., 1982; Vogt, 1974; Watkins et al., 1984]. Whatever the case, the hyperalgesia observed after the intrathecal administration of 5-HT antagonists and neurotoxins supports a tonic inhibitory action of the descending serotoninergic fibers.

From the neurochemical point of view, two studies by our group suggest the hypothesis of a release of 5-HT during SPA induced by NRM stimulation: (1) In the anesthetized rat, stimulation of the B3 area (NRM + nucleus paragigantocellularis) under conditions similar to those producing analgesia provoked a significant increase (+50%) in 5-HT synthesis in the spinal cord on the basis of an increased rate of 5-HTP accumulation after the blockade of central 5-HTP decarboxylase by benserazide. In contrast, no change was detected in 5-HT metabolism in the forebrain of stimulated rats [Bourgoin et al., 1980]. (2) In vivo studies using electrochemical detection of 5-HT [Rivot et al., 1982] have shown that NRM stimulation in the anesthetized rat produces a release of extracellular 5-HT in the dorsal horn. In these experiments, carbon fiber microelectrodes were used with the differential pulse voltammetry method for determination of indolamines. Moreover, the increase in the 5-hydroxyindolaminergic signal was significantly reduced during a second series of NRM stimulations, indicating some degree of tolerance to central stimulation. The accuracy of these observations is strengthened by the fact that the basal 5-hydroxyindolaminergic signal was strongly depressed after pretreatment of the animal with *p*-chlorophenylalanine; in addition, under these conditions, NRM stimulation was totally inefficient (fig. 6).

More recently a direct demonstration of an increase in 5-HT and its metabolism in the CSF has been described after stimulation of the ventromedial medulla [Hammond et al., 1985]. Such a demonstration of a release of 5-HT during NRM stimulation is of importance since: (1) as will be discussed later, iontophoresis of 5-HT in the vicinity of dorsal horn neurons strongly depresses their responses to noxious stimuli; (2) the analgesic effect induced by electrical stimulation of the ventromedial medulla are reduced by intrathecal 5-HT antagonists [Hammond and Yaksh, 1984; Satoh et al., 1980]; (3) the intrathecal administration of 5-HT can produce analgesia [Howe and Yaksh, 1982; Schmauss et al., 1983; Wang, 1977; Yaksh and Wilson, 1979]: these effects of serotonin were potentiated by blockade of monoaminoxidase or reuptake inhibition, whilst they were antagonized by receptor blockers such as methysergide and cyproheptadine [Yaksh and Wilson, 1979].

From the electrophysiological point of view, if we consider [Dahlstrom and Fuxe, 1965] that 5-HT fibers are unmyelinated, the involvement of this amine in the inhibitory effects induced by the brain stem is questioned by the fact that the short latency (or latency) of this inhibition is inconsistent with unmyelinated conduction velocities [Belcher et al., 1978; McCreery et al., 1979; Willis et al., 1977; Rivot et al., 1980]. In addition, electrophysiological studies using antidromic activation techniques have shown that raphe spinal fibers issuing from the NRM most often correspond to myelinated fibers [Anderson et al., 1977; Fields and Anderson, 1978; Lovick et al., 1978; West and Wolstencroft, 1977; Willcockson et al., 1983; Willis et al., 1984]. However, it is well known that the electrical threshold of myelinated fibers is lower than unmyeli-

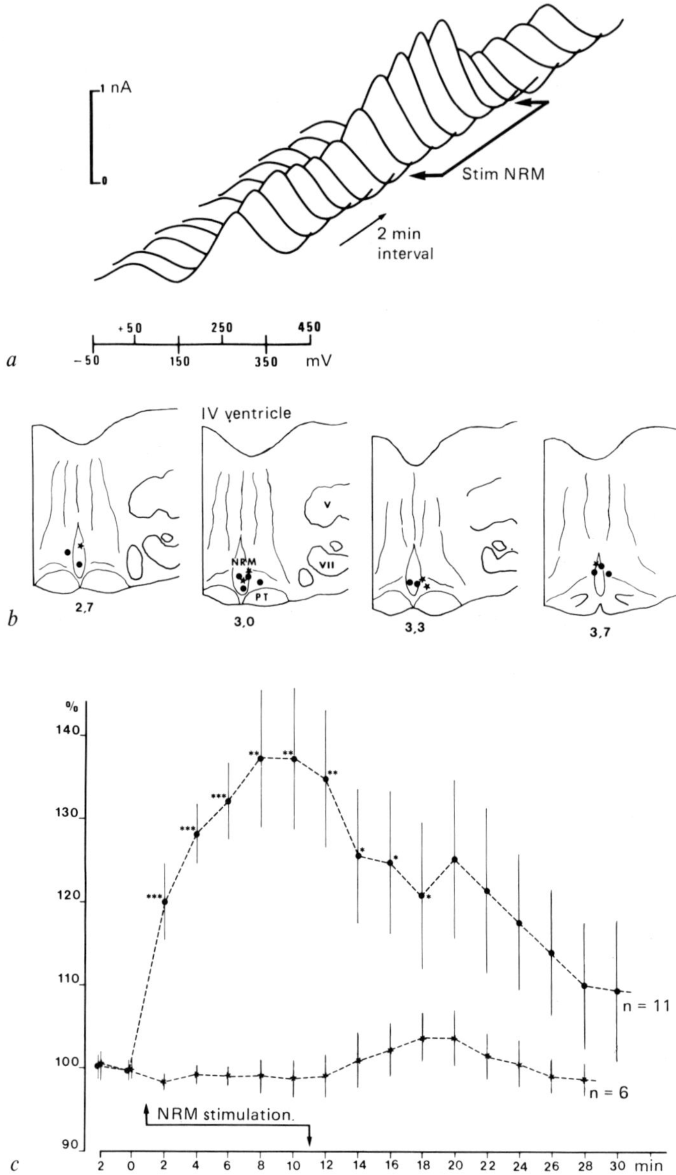
1 nA
0
Stim NRM
2 min
interval
+50 250 450
-50 150 350 mV
a
IV ventricle
V
NRM
VII
PT
2,7
3,0
3,3
3,7
b
%
140
130
120
110
100
90
n = 11
n = 6
NRM stimulation
2 0 2 4 6 8 10 12 14 16 18 20 22 24 26 28 30 min
c

nated fibers; consequently, studies using antidromic activation from the spinal cord will overestimate the number of large fibers. More recent studies in the rat have shown that nearly 25% of raphe-spinal fibers are unmyelinated (conduction velocity 0.7–1 m/s) [Wessendorf et al., 1981]. Thus, it appears that NRM stimulation-induced inhibitory effects at the spinal level could be due to the activation of both serotoninergic and nonserotoninergic raphe-spinal fibers. These former fibers represent 40 to 80% of the total number of raphe spinal neurons [Bowker et al., 1983; Skagerberg and Bjorklund, 1985]. Thus, NRM stimulation activates simultaneously both myelinated and unmyelinated pathways, which results in an overlapping in their inhibitory effects; under these conditions one can only determine the short latency of the inhibition due to large fibers, and the effects due to 5-HT fibers will follow later.

We have used another approach to evaluate the participation of 5-HT in the inhibitory effects of NRM. This study was based on the comparison of the effects of NRM on responses of dorsal horn nociceptive neurones to C fiber stimulation in normal rats and in rats pretreated with high doses of pCPA [Rivot et al., 1980]. After 300 mg/kg of p-CPA for three days there is no detectable amount of 5-HT within the spinal cord. After administration of this 5-HT synthesis inhibitor, the depressive action of NRM on dorsal horn neuronal C fiber evoked discharges was markedly reduced, since no more than 40% of the neurones were exclusively inhibited, compared to 93% in control rats. In addition, after pCPA, NRM stimulation has been found to induce strong and long duration (several hundred milliseconds) excitatory effects (fig. 7), which we have never observed in normal animals. These results clearly reveal the role of 5-HT in descending inhibitions from the NRM. However, it must

---

Fig. 6. Electrochemical detection of changes in 5-hydroxyindoles in the dorsal horn of the spinal cord of the cat during stimulation of the nucleus raphe magnus. *a.* An individual example of the time course of the electrochemical signal before, during, and after nucleus raphe magnus stimulation (Stim NRM). The interval between each observation is 2 min. Note the stability of the measurements and the increase of the peak during NRM stimulation. *b.* Stimulation sites within the medulla plotted onto frontal planes at 2.7, 3.0, 3.3 and 3.7 mm caudal. Same symbols as in *c.* V = nucleus of the fifth nerve; VII = nucleus of the seventh nerve; PT = pyramidal tract; NRM = nucleus raphe magnus *c.* Mean curves illustrating the effect on the electrochemical signal of NRM stimulation in control rats (solid circles, n = 11) and in pCPA pretreated animals (stars, n = 6). The values are expressed as mean ± SEM of percentage of control (*$p < 0.05$; **$p < 0.01$; *** $p < 0.0001$). Modified from Rivot et al. [1982].

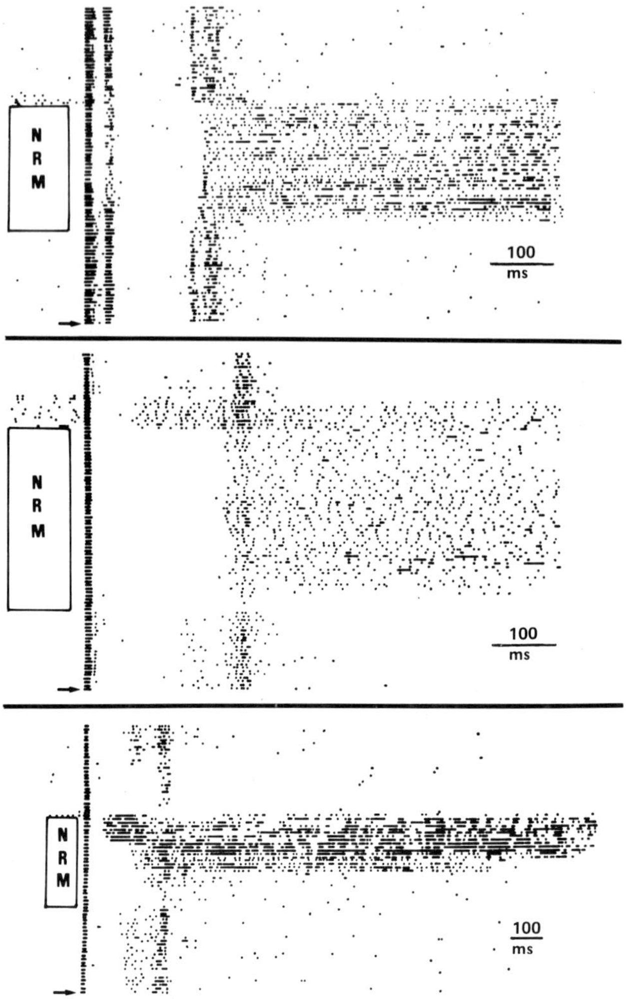

a b c

Fig. 7. Three examples *(a, b* and *c)* of effects of NRM stimulation on responses of dorsal horn neurons to transcutaneous stimulation in the pCPA-pretreated rats. Note the pronounced excitation produced by the central stimulation. Discharges of neurons are displayed as dots running left to right. At the end of every circle (1.5 s in this case), a new sweep is started, displaced above the preceding one by a synchronizing pulse from the stimulator. Peripheral stimulation (arrow) consisted of a single rectangular shock (1.5 ms duration, threshold for C-fiber response). Central stimulation represented by the boxed areas to the left consisted of a train of 100 ms duration of bipolar pulses of 0.5 ms delivered at 300 Hz. From Rivot et al. [1980].

be pointed out that some inhibitory effects could still be evoked in 5-HT depleted rats, suggesting the additional implication of nonserotoninergic raphe-spinal systems. This assertion is in good agreement with the previously mentioned results showing high conduction velocities for raphe-spinal pathways.

Although speculative, the long latency excitatory effects observed after pCPA could result from a spinal release of substance P, which is also contained in some 5-HT raphe-spinal neurons [Bowker et al., 1981, 1983; Hokfelt et al., 1978] and which has been shown to excite dorsal horn and trigeminal nociceptive neurons. In the animal pretreated with pCPA, the excitatory effect of SP is no longer masked by the inhibitory effect of 5-HT, although obviously the involvement of other excitatory substances cannot be excluded. Interestingly, no excitatory effects were detected after lesions of raphe spinal 5-HT descending fibers by intrathecal administration of 5–7 DHT [Chaouch and Besson, 1985].

Other pharmacological investigations have shown that the inhibitory action of the PAG, unlike that evoked from the mesencephalic lateral reticular formation, is reduced by metergoline [Carstens et al., 1981; Yezierski et al., 1982] or LSD [Guilbaud et al., 1973]. Recently, it has been shown that 5-HT antagonists reduce the inhibitory effects from the PAG more reliably than those induced from the NRM [Yezierski et al., 1982]. The absence or slight effect of 5-HT antagonists on inhibition produced from the NRM questions the specificity of action of 5-HT antagonists with regard to the inhibitory effects and/or suggests that the antagonists could equally act at supraspinal levels (i.e., the PAG-NRM connections).

On the other hand, in most cases, iontophoresis of 5-HT strongly depresses the responses of dorsal horn neurones to nociceptive stimulation, including those at the origin of the spinothalamic tract [Belcher et al., 1978; Griersmith and Duggan, 1980; Headley et al., 1978; Jordan et al., 1978, 1979; Willcockson et al., 1984]. The inhibitory effects of 5-HT could, at least in part, result from postsynaptic action, since the monoamine can depress responses induced by excitatory amino acids [Jordan et al., 1978; Willcockson et al., 1984]. This is in good agreement with several electrophysiological investigations that have described postsynaptic effects produced by NRM stimulation [Giesler et al., 1981; Shah and Dostrovsky, 1982; Belcher et al., 1978]. In addition, immunocytochemical studies have clearly revealed contacts between 5-HT terminals and spinal neurons, some of which are at the origin of ascending tracts involved in pain pathways [Hoffert et al., 1983; Lamotte and de Lanerolle, 1983; Light et al., 1983; Maxwell et al., 1983; Miletic et al., 1984; Nishikawa et al., 1983; Ruda et al., 1982].

These findings support those made using autoradiographic marking with ^{3}H-5-HT of spinal terminals [Glazer and Basbaum, 1984; Ruda and Gobel, 1980] or that seen after intracellular injection into chemically unidentified fibers descending in the DLF [Light, 1985].

The functional significance of these differential serotoninergic inputs is still poorly understood, but a role in the selective control of nociceptive messages has been put forward [Miletic et al., 1984]. The site at which 5-HT terminals modulate the activity of spinal neurons is still under dispute, and the interactions could be via both synaptic and non-synaptic contacts [Glazer and Basbaum, 1984; Maxwell et al., 1983; Ruda et al., 1982; Ruda and Gobel, 1980]. On the other hand, since the work of Proudfit and Anderson [1974] showing that NRM stimulation evokes dorsal root potentials, some of which are due to activation of 5-HT fibers, numerous studies have shown that the inhibitory effects could result at least partially from presynaptic action [Dostrovsky et al., 1981; Hentall and Fields, 1979; Hu et al., 1978; Lovick, 1983; Lovick et al., 1978; Martin et al., 1979; Proudfit et al., 1980]. NRM stimulation induces primary afferent depolarization (PAD) at the terminals of the afferent fibers. This presynaptic action is often proposed as a basis for the selective NRM inhibition of spinal nociceptive messages. However, during NRM stimulation, PAD is produced not only by A fibers, some of which are certainly nociceptive [Lovick, et al., 1978; Martin et al., 1979], but equally by non-nociceptive A fibers. Hence, the presynaptic inhibitory action of the NRM seems unselective, although one cannot exclude a depolarization of the fine afferent fibers influencing the release of mediators to a greater extent in these fibers than the larger ones. The involvement of 5-HT in the depolarization is still not proven, since 5-HT antagonists are without effect on the PAD [Lovick, 1983]. In addition morphological studies of axo-axonic synapses favoring a presynaptic action of 5-HT fibers are rare [Ruda et al., 1982; see, however, Lamotte and de Lanerolle, 1983]. However, a presynaptic action of the NRM and 5-HT fibers could be produced via interneurons of the dorsal horn (GABAergic) and/or via nonsynaptic interactions, thus explaining the morphological mismatch.

Whatever the mechanisms of action, 5-HT is able to inhibit the transmission of spinal nociceptive messages, and, as previously mentioned, its intrathecal administration can produce analgesia [Howe and Yaksh, 1982; Schmauss et al., 1983; Wang, 1977; Yaksh and Wilson, 1979].

Concluding Remarks and Physiological Significance of Descending Systems

This review has been restricted to descending serotoninergic systems, but in reality the pharmacology of these systems is considerably more complex. As an example, certain serotoninergic raphe spinal neurons also contain different peptides (substance P, TRH), and the regions from which electrical stimulation can give rise to analgesic effects contain opioid receptors and cell terminals containing high levels of endogenous opioids. In addition, as previously discussed, the importance of descending noradrenergic controls has been illustrated by several studies. Analgesic effects could well result from the activation of descending noradrenergic systems, and future work should consider in detail the interactions between different types of descending systems.

To explain the functional significance of descending systems, the hypothesis most often put forward postulates [Basbaum and Fields, 1978; Liebeskind et al., 1976; Mayer, 1979] the existence of an endogenous analgesic system: the PAG, ventromedial medulla and their spinal projections make up a negative feedback loop activated by nociceptive stimulation producing an inhibition of the spinal transmission of nociceptive messages. An argument cited in favor of this hypothesis is the activation of raphe magnus neurons by nociceptive stimulation [Basbaum and Fields, 1984]. Results reported by Le Bars et al. [1979a, b] describing diffuse noxious inhibitory controls (DNIC) could be an important step forward in understanding, at least in part, the activation of this system by natural stimulation. DNIC [Le Bars et al., 1982] affect all neuronal activities of dorsal horn nociceptive nonspecific neurons, are triggered exclusively by noxious stimuli, and are mediated by a supraspinal loop which ascends in the contralateral ventrolateral quadrant [Villanueva et al., 1986] and redescends via the ipsilateral DLF [Villanueva et al., 1985].

Electrophysiological studies related to DNIC are in good agreement with various behavioral studies in animal and man and can explain the phenomenon of counterirritation [Le Bars et al., 1984]. In this context, it is interesting to mention that brain stem descending systems could also be involved in analgesia elicited by various stressors [Lewis et al., 1984; Mayer and Watkins, 1984]; such findings suggest that, whilst pain is an important warning signal, there are certain conditions, such as emergency or stress, under which pain suppression may be yet more adaptive.

In addition to their activation by various forms of natural nociceptive stimuli, it has been suggested that the activity of these descending brain stem inhibitory controls is increased by morphine. Fields and co-workers [Fields et al., 1983a, b; Vanegas et al., 1984] described in the ventral medulla 'off cells', the activity of which stops just prior to the tail flick. Interestingly, morphine has been shown to excite these neurons. Consequently, Fields and co-workers proposed that 'off cells' are ventromedial medulla spinal neurons which tonically inhibit spinal cord nociception transmission. This hypothesis fits well with an increase in descending controls after morphine. However, it is difficult to integrate these 'off cells' into a negative feedback loop, since they are inhibited by noxious stimulation.

In addition, there are many discrepancies when considering the effects of microinjection of morphine, into the PAG or NRM, on the activity of dorsal horn neurons. Some electrophysiological studies report an inhibitory effect after morphine in the PAG [Bennett and Mayer, 1979; Gebhard et al., 1984] or raphe magnus [Du et al., 1984], while others did not observe depressive effects [Azami et al., 1982; Dickenson and Le Bars, 1983; Le Bars et al., 1980], but, by contrast, on occasions, excitations. Moreover, DNIC are depressed by low doses of morphine [Le Bars et al., 1981]. It appears that supraspinal descending inhibitory controls are reduced by morphine, at least when such controls are triggered by noxious stimuli. Taking into account these effects of morphine on DNIC, Le Bars et al., [1983] proposed a new hypothesis for the functional significance of descending controls.

In conclusion, there is no doubt about the existence and the effectiveness of descending controls and the involvement of monoaminergic and opioid mechanisms in these phenomena. We need much more information about their relative involvement, their functional significance and the effects of morphine. Regarding this latter point, one must emphasize that most of the groups working in this field have studied either the direct or indirect spinal effects of morphine, but it is well known that low doses of the drug which have a minimal depressive effect at the spinal level are able to depress strongly the responsiveness of supraspinal neurons to nociceptive stimulation.

So although great attention had been devoted to the modulation of the transmission of painful messages at the spinal level, future investigations will have to consider in more detail supraspinal mechanism involved in pain modulation.

References

Abols, I.A.; Basbaum, A.I.: Afferent connections of the rostral medulla of the cat. A neural substrate for midbrain-medullary interactions in the modulation of pain. J. comp. Neurol. *201:* 285–297 (1981).

Adams, J.E.: Naloxone reversal of analgesia produced by brain stimulation in the human. Pain *2:* 161–166 (1976).

Akaike, A.; Shibata, T.; Satoh, M.; Takagi, H.: Analgesia induced by microinjection of morphine into and electrical stimulation of the nucleus reticularis paragigantocellularis of rat medulla oblongata. Neuropharmacol. *17:* 775–778 (1978).

Akil, H.; Liebeskind, J.C.: Monoaminergic mechanisms of stimulation produced analgesia. Brain Res. *94:* 279–296 (1975).

Akil, H.; Mayer, D.J.: Antagonism of stimulation produced analgesia by pCPA, a serotonin synthesis inhibitor. Brain Res. *44:* 692–697 (1972).

Ammons, W.S.; Blair, R.W.; Foreman, R.D.: Raphe magnus inhibition of primate T1–T4 spinothalamic cells with cardiopulmonary visceral input. Pain *20:* 247–260 (1984).

Anderson, S.D.; Basbau, A.I.; Fields, H.L.: Response of medullary raphe neurons to peripheral stimulation and to systemic opiates. Brain Res. *123:* 363–368 (1977).

Apkarian, A.V.; Stevens, R.T.; Hodge, C.J.: Funicular location of ascending axons of lamina 1 cells in the cat spinal cord. Brain Res. *334:* 160–164 (1985).

Azami, J.; Llewellyn, M.B.; Roberts, M.H.T.: Spinal cord mechanisms of the analgesia produced by microinjection of morphine into nucleus raphe magnus (NRM) of the cat. J. Physiol. *326:* 47–48P (1982).

Barden, N.; Daigle, M.; Picard, V.; Di Paolo, T.: Perturbation of rat brain serotoninergic systems results in an inverse relation between substance P and serotonin concentrations measured in discrete nuclei. J. Neurochem. *41:* 834–840 (1983).

Basbaum, A.I.; Clanton, C.H.; Fields, H.L.: Opiate and stimulus produced analgesia. Functional anatomy of a medullo-spinal pathway. Proc. nat. Acad. Sci. USA *73:* 4685–4688 (1976).

Basbaum, A.K.; Clanton, C.H.; Fields, H.L.: Three bulbospinal pathways from the rostral medulla of the cat: an autoradiographic study of pain modulating systems. J. comp. Neurol. *178:* 209–224 (1978).

Basbaum, A.I.; Fields, H.L.: Endogenous pain control mechanisms: review and hypothesis. Ann. Neurol. *4:* 451–462 (1978).

Basbaum, A.I.; Fields, H.L.: The origin of descending pathways in the dorsolateral funiculus of the spinal cord of the cat and rat; further studies on the anatomy of pain modulation. J. comp. Neurol. *187:* 513–532 (1979).

Basbaum, A.I.; Fields, H.L.: Endogenous pain control system. Brain stem spinal pathways and endorphin circuitry. A. Rev. Neurosci. *7:* 309–338 (1984).

Basbaum, A.I.; Marley, N.J.E.; O'Keefe, J.: Spinal cord pathways involved in the production of analgesia by brain stimulation; in Bonica, Albe-Fessard, Advances in pain research and therapy, vol. I, pp. 511–525 (Raven Press, New York 1976).

Basbaum, A.I.; Marley, N.J.E.; O'Keefe, J.; Clanton, C.H.: Reversal of morphine and stimulation-produced analgesia by subtotal spinal cord lesion. Pain *3:* 43–56 (1977).

Behbehani, M.M.; Fields, H.L.: Evidence that an excitatory connection between the periaqueductal gray and nucleus raphe magnus mediates stimulation produced analgesia. Brain Res. *170:* 85–93 (1979).

Belcher, G.; Ryall, R.W.; Schaffner, R.: The differential effects of 5-hydroxytryptamine, noradrenaline and raphe stimulation on nociceptive and non nociceptive dorsal horn interneurons in the cat. Brain Res. *151:* 307–321 (1978).

Bennett, G.J.; Mayer, D.J.: Inhibition of spinal cord interneurons by narcotic microinjection and focal electrical stimulation in the periaqueductal central gray matter. Brain Res. *172:* 243–257 (1979).

Berge, O.G.: Effects of 5-HT receptor agonists and antagonists on a reflex response to radiant heat in normal and spinally transected rats. Pain *13:* 253–266 (1982).

Berge, O.G.; Fasmer, O.B.; Flatmark, T.; Hole, K.: Time-course of changes in nociception after 5,6-dihydroxytryptamine lesions of descending 5-HT pathways. Pharmacol. Biochem. Behav. *18:* 637–643 (1983).

Berge, O.G., Fasmer, O.B.; Hole, K.: Serotonin receptor antagonists induce hyperalgesia without preventing morphine antinociception. Pharmacol. Biochem. Behav. *19:* 873–878 (1983).

Berge, O.G.; Hole, K.; Ogren, S.O.: Attenuation of morphine-induced analgesia by *p*-chlorophenylalanine and *p*-chloroamphetamine. Test-dependent effects and evidence for brain stem 5-hydroxytryptamine involvement. Brain Res. *271:* 51–64 (1983).

Berge, O.G.; Ogren, S.O.: Selective lesions of the bulbospinal serotoninergic pathways reduce the analgesia induced by *p*-chloroamphetamine in the hot-plate test. Neurosci. Lett. *44:* 25–29 (1984).

Besson, J.M.; Guilbaud, G.; Abdelmoumene, M.; Chaouch, A.: Physiologie de la nociception. J. Physiol., Paris *78:* 7–107 (1982).

Besson, J.M.; Oliveras, J.L.; Chaouch, A.; Rivot, J.P.: Role of the raphe nuclei in stimulation producing analgesia; in Haber, Gabay, Issidorides, Alivisatos, Serotonin: current aspects of neurochemistry and function, pp. 153–176 (Plenum Press, New York 1981).

Bjorklund, A.; Skagerberg, G.: Descending monoaminergic projections to the spinal cord; in Sjolund, Bjorklund, Brain stem control of spinal mechanisms, pp. 55–88 (Elsevier/Biomedical Press, Amsterdam 1982).

Bobillier, P.; Seguin, S.; Petitjean, F.; Salvert, D.; Touret, M.; Jouvet, M.: The raphe nuclei of the cat brain stem: a topographical atlas of their afferent projections as revealed by autoradiography. Brain Res. *113:* 449–486 (1976).

Boivie, J.; Meyerson, B.A.: A correlative anatomical and clinical study of pain suppression by deep brain stimulation. Pain *13:* 113–126 (1982).

Bowker, R.M.; Steinbusch, H.V.M.; Coulter, J.D.: Serotoninergic and peptidergic projections to the spinal cord demonstrated by a combined retrograde HRP histochemical and immunocytochemical staining method. Brain Res. *211:* 412–417 (1981).

Bowker, R.M.; Westlund, K.N.; Sullivan, M.C.; Wilber, J.F.; Coulter, J.D.: Descending serotoninergic, peptidergic and cholinergic pathways from the raphe nuclei. A multiple transmitter complex. Brain Res. *288:* 33–48 (1983).

Brodal, A.; Taber, E.; Walberg, F.: The raphe nuclei of the brain stem in the cat. II. Efferent connections. J. comp. Neurol. *114:* 239–259 (1960).

Carstens, E.: Inhibition of spinal dorsal horn neuronal responses to noxious skin heating by medial hypothalamic stimulation in the cat. J. Neurophysiol. *48:* 808–822 (1982).

Carstens, E.; Fraunhoffer, M.; Zimmermann, M.: Serotoninergic mediation of descending inhibition from midbrain periaqueductal gray but not reticular formation of spinal nociceptive transmission in the cat. Pain *10:* 149–167 (1981).

Carstens, E.; MacKinnon, J.D.; Guinan, M.J.: Inhibition of spinal dorsal horn neuronal responses to noxious skin heating by medial preoptic and septal stimulation in the cat. J. Neurophysiol. *48:* 981–991 (1982).

Chaouch, A.; Besson, J.M.: Nucleus raphe magnus effect on dorsal horn neurones in the rat after intrathecal 5-7DHT. Neurosci. Lett., suppl. 22, p. 475 (1985).

Chapman, C.D.; Ammons, W.S.; Foreman, R.D.: Raphe magnus inhibition of feline T_1-T_4 spinoreticular tract cell responses to visceral and somatic inputs. J. Neurophysiol. *53:* 773–785 (1985).

Chung, J.M.; Kevetter, G.A.; Yezierski, R.P.; Haber, L.H.; Martin, R.F.; Willis, W.D.: Midbrain nuclei projecting to the medial medulla oblongate in the monkey. J. comp. Neurol. *214:* 93–102 (1983).

Chung, K.; Coggeshall, R.E.: Propriospinal fibers in the rat. J. comp. Neurol. *217:* 47–53 (1983).

Conrad, L.C.A.; Leonard, C.M.; Pfaff, D.W.: Connections of the medium and dorsal raphe nuclei in the rat. An autoradiographic and degeneration study. J. comp. Neurol. *156:* 179–206 (1974).

Dahlstrom, A.; Fuxe, K.: Evidence for the existence of monoamine neurons in the central nervous system. II. Experimentally induced changes in the intraneuronal amine levels of bulbospinal neuron systems. Acta physiol. scand. *64:* suppl. 247, pp. 1–36 (1965).

Dennis, S.G.; Melzack, R.: Pain modulation by 5-hydroxytryptaminergic agents and morphine as measured by three pain tests. Exp. Neurol. *69:* 260–270 (1980).

Dickenson, A.H.; Le Bars, D.: Morphine microinjections into periaqueductal grey matter of the rat: effects on dorsal horn neuronal responses to C-fibre activity and diffuse noxious inhibitory controls. Life Sci. *33:* suppl. 1, 549–554 (1983).

Dostrovsky, J.O.; Sessle, B.J.; Hu, J.W.: Presynaptic excitability changes produced in brain stem endings of tooth pulp afferents by raphe and other central and peripheral influences. Brain Res. *218:* 141–160 (1981).

Du, H.J.; Kitahata, L.M.; Thalhammer J.G.; Zimmermann, M.: Inhibition of nociceptive neuronal responses in the cat's spinal dorsal horn by electrical stimulation and morphine microinjection in nucleus raphe magnus. Pain *19:* 249–257 (1984).

Dubuisson, D.; Melzack, R.: Analgesic brain stimulation in the cat: effect of intraventricular serotonin, norepinephrine and dopamine. Exp. Neurol. *57:* 59–66 (1977).

Duggan, A.W.; Griersmith, B.T.: Inhibition of the spinal transmission of nociceptive information by supraspinal stimulation in the cat. Pain *6:* 149–161 (1979).

Engberg, I.; Lundberg, A.; Ryall, R.W.: Is the tonic decerebrate inhibition of reflex paths mediated by monoaminergic pathways? Acta Physiol. Scand. *72:* 123–133 (1968).

Engberg, I.; Lundberg, A.; Ryall, R.W.: Reticulospinal inhibition of interneurones. J. Physiol. (London), *194:* 225–236 (1968).

Fardin, V.; Oliveras, J.L.; Besson, J.M.: A reinvestigation of the analgesic effects induced by stimulation of the periaqueductal gray matter in the rat. I. The production of behavioral side effects together with analgesia. Brain Res. *306:* 105–123 (1984).

Fardin, V.; Oliveras, J.L.; Besson, J.M.: A reinvestigation of the analgesic effects induced by stimulation of the periaqueductal gray matter in the rat. II. Differential characteristics of the analgesia induced by ventral and dorsal PAG stimulation. Brain Res. *306:* 125–139 (1984).

Fasmer, O.B.; Berge, O.G.: Hole, K.: Methergoline elevates or reduced nociceptive thresholds in mice depending on test method and route of administration. Psychopharmacology *82:* 306–309 (1984).

Fasmer, O.B.; Berge, O.G.; Walther, B.; Hole, K.: Changes in nociception after intrathecal administration of 5-6-dihydroxytryptamine in mice. Neuropharmacology *22:* 1197–1201 (1983).

Fibiger, H.C.; Mertz, P.H.; Campbell, B.A.: The effect of para-chlorophenylalanine on aversion thresholds and reactivity to foot shock. Physiol. Behav. *8:* 259–263 (1972).

Fields, H.L.; Anderson, S.D.: Evidence that raphe-spinal neurons mediate opiate and midbrain stimulation-produced analgesia. Pain *5:* 333–349 (1978).

Fields, H.L.; Basbaum, A.I.; Clanton, C.H.; Anderson, S.D.: Nucleus raphe magnus inhibition of spinal cord dorsal horn neurons. Brain Res. *126:* 441–453 (1977).

Fields, H.L.; Bry, J.; Hentall, I.; Zorman, G.: The activity of neurons in the rostral medulla of the rat during withdrawal from noxious heat. J. Neurosci. *3:* 2545–2552 (1983a).

Fields, H.L.; Vanegas, H.; Hentall, I.D.; Zorman, G.: Evidence that disinhibition of brain stem neurones contributes to morphine analgesia. Nature, Lond. *306:* 684–686 (1983b).

Gallagher, D.W.; Pert, A.: Afferents to brain stem nuclei (brain stem raphe, nucleus reticularis pontis caudalis and nucleus gigantocellularis) in the rat as demonstrated by microiontophoretically applied horseradish peroxidase. Brain Res. *144:* 257–276 (1978).

Gebhart, G.F.; Sandkuhler, J.; Thalhammer, J.G.; Zimmermann, M.: Inhibition of spinal nociceptive information by stimulation in midbrain of the cat is blocked by lidocaine microinjected in nucleus raphe magnus and medullary reticular formation. J. Neurophysiol. *50:* 1446–1459 (1983).

Gebhart, G.F.; Sandkuhler, J.; Thalhammer, J.G.; Zimmermann, M.: Inhibition in spinal cord of nociceptive information by electrical stimulation and morphine microinjection at identical sites in midbrain of the cat. J. Neurophysiol. *51:* 75–89 (1984).

Gebhart, G.F.; Toleikis, J.R.: An evaluation of stimulation produced analgesia in the cat. Expl. Neurol. *62:* 570–579 (1978).

Gerhart, K.D.; Wilcox, T.K.; Chung, J.M.; Willis, W.D.: Inhibition of nociceptive and non-nociceptive responses of primate spinothalamic cells by stimulation in medial brain stem. J. Neurophysiol. *45:* 121–136 (1981).

Giesler, G.J. Jr.; Gerhart, K.D.; Yezierski, R.P.; Wilcox, T.K.; Willis, W.D.: Postsynaptic inhibition of primate spinothalamic neurons by stimulation in nucleus raphe magnus. Brain Res. *204:* 184–188 (1981).

Giesler, G.J.; Liebeskind, J.C.: Inhibition of visceral pain by electrical stimulation of the periaqueductal gray matter. Pain *2:* 43–48 (1976).

Glazer, E.J.; Basbaum, A.I.: Axons which take up (^{3}H)serotonin are presynaptic to enkephalin immunoreactive neurons in cat dorsal horn. Brain Res. *298:* 386–391 (1984).

Goodman, S.J.; Holocombe, V.: Selective and prolonged analgesia in monkey resulting from brain stimulation; in Bonica, Albe-Fessard, Advances in pain research and therapy, vol. 1, pp. 495–502 (Raven Press, New York 1976).

Gorlitz, B.D.; Frey, H.H.: Central monoamines and antinociceptive drug action. Eur. J. Pharmacol. *20:* 171–180 (1972).

Griersmith, B.T.; Duggan, A.W.: Prolonged depression of spinal transmission of nociceptive information by 5HT administered in the substantia gelatinosa: antagonism by methysergide. Brain Res. *187:* 231–256 (1980).

Grofova, I.; Ottersen, O.P.; Rinvik, E.: Mesencephalic and diencephalic afferents to the superior colliculus and periaqueductal gray substance demonstrated by retrograde axonal transport of horseradish peroxidase in the rat. Brain Res. *146:* 205–220 (1978).

Guilbaud, G.; Besson, J.M.; Oliveras, J.L.; Liebeskind, J.C.: Suppression by LSD of the inhibitory effect exerted by dorsal raphe nucleus on certain spinal cord interneurons in the cat. Brain Res. *61:* 417–422 (1973).

Guilbaud, G.; Oliveras, J.L.; Giesler, G.J. Jr.; Besson, J.M.: Effects induced by stimulation of the centralis inferior nucleus of the raphe on dorsal horn interneurons in cat's spinal cord. Brain. Res. *126:* 355–360 (1977).

Gybels, J.: Electrical stimulation of the central gray for pain relief in humans. A critical review; Bonica, Albe-Fessard, Liebeskind, Advances in pain research and therapy, vol. 3, pp. 499–509 (Raven Press, New York, 1979).

Hammond, D.L.; Tyce, G.M.; Yaksh, T.L.: Efflux of 5-hydroxytryptamine and noradrenaline into spinal cord superfusates during stimulation of the rat medulla. J. Physiol. Lond. *359:* 151–162 (1985).

Hammond, D.L.; Yaksh, T.L.: Antagonism of stimulation produced antinociception by intrathecal administration of methysergide or phentolamine. Brain Res. *298:* 329–377 (1984).

Hayes, R.L.; Newlon, P.G.; Rosencrans, J.A.; Mayer, D.J.: Reduction of stimulation produced analgesia by Lysergic acid diethylamide, a depressor of serotoninergic neural activity. Brain Res. *122:* 367–372 (1977).

Headley, P.M.; Duggan, A.W.; Griersmith, B.T.: Selective reduction by noradrenaline and 5-hydroxytryptamine of nociceptive responses of cat dorsal horn neurones. Brain Res. *145:* 185–189 (1978).

Hentall, I.D.; Fields, H.L.: Segmental and descending influences on intraspinal thresholds of single C-fibers. J. Neurophysiol. *42:* 1527–1537 (1979).

Hoffert, M.J.; Miletic, V.; Ruda, M.A.; Dubner, R.: Immunocytochemical identification of serotonin axonal contacts on characterized neurons in laminae I and II of the cat dorsal horn. Brain Res. *267:* 361–364 (1983).

Hokfelt, T.; Ljungdahl, A.; Steinbusch, H.; Verhofstad, A.N.; Nilsson, G.; Brodin, E.; Pernow, B.; Goldstein, M.: Immunohistochemical evidence of substance P like immunoreactivity in some 5-hydroxytryptamine containing neurons in the rat central nervous system. Neuroscience *3:* 517–538 (1978).

Hosobuchi, Y.: Tryptophan reversal of tolerance to analgesia induced by central gray stimulation. Lancet *ii:* 47 (1978).

Hosobuchi, Y.; Adams, J.E.; Linchitz, R.: Pain relief by electrical stimulation of the central gray matter in humans and its reversal by naloxone. Science *197:* 183–186 (1977).

Howe, J.R.; Yaksh, T.L.: Changes in sensitivity to intrathecal norepinephrine and serotonin after 6-hydroxydopamine (6-OHDA), 5-6-dihydroxytryptamine (5,6-DHT) or repeated monoamine administration. J. Pharmac. exp. Ther. *220:* 311–321 (1982).

Hu, J.W.; Dostrovsky, J.O.; Sessle, B.J.: Primary afferent depolarization of tooth pulp afferents is not affected by naloxone. Nature Lond. *276:* 283–284 (1978).

Johannessen, J.N.; Watkins, L.R.; Carlton, S.M.; Mayer, D.J.: Failure of spinal cord serotonin depletion to alter analgesia elicited from the periaqueductal gray. Brain Res. *237:* 373–386 (1982).

Jones, M.W.; Hodge, C.J. Jr.; Apkarian, A.V.; Stevens, R.T.: A dorsolateral spinothalamic pathway in cat. Brain Res. *335:* 188–193 (1985).

Jordan, L.M.; Kenshalo, D.R., Jr.; Martin, R.F.; Haber, L.H.; Willis, W.D.: Depression of primate spinothalamic tract neurons by iontophoretic application of 5-hydroxytryptamine. Pain *5:* 135–142 (1978).

Jordan, L.M.; Kenshalo, D.R., Jr.; Martin, R.F.; Haber, L.H.; Willis, W.D.: Two populations of spinothalamic tract neurons with opposite responses to 5-hydroxytryptamine. Brain Res. *164:* 342–346 (1979).

Kiser, R.S.; Lebowitz, R.M.; German, D.C.: Anatomic and pharmacologic differences between two types of aversive midbrain stimulation. Brain Res. *155:* 331–342 (1978).

Kneisley, L.W.; Biber, M.P.; Lavail, J.H.: A study of the origin of brain stem projections to monkey spinal cord using the retrograde transport method. Expl. Neurol. *60:* 116–139 (1978).

Kuypers, H.J.G.M.; Huisman, A.M.: The anatomy of the descending brain pathways; in Sjolund, Bjorklund, Brain stem control of spinal mechanisms, pp. 29–54 (Elsevier/Biomedical Press, Amsterdam 1982).

Lamotte, C.C.; Lanerolle, N.C. de: Ultrastructure of chemically defined neuron-systems in the dorsal horn of monkey. III. Serotonin immunoreactivity. Brain Res. *274:* 65–77 (1983).

Le Bars, D.; Calvino, B.; Villanueva, L.; Cadden, S.: Physiological approaches to counter-irritation phenomena; in Tricklebank, Curzon, Stress induced analgesia, pp. 67–101 (Wiley, New York, 1984).

Le Bars, D.; Chitour, D.; Kraus, E.; Clot, A.M.; Dickenson, A.H.; Besson, J.M.: The effect of systemic morphine upon diffuse noxious inhibitory controls (DNIC) in the rat: evidence for a lifting of certain descending inhibitory controls of dorsal horn convergent neurones. Brain Res. *215:* 257–274 (1981).

Le Bars, D.; Dickenson, A.H.; Besson, J.M.: Diffuse noxious inhibitory controls (DNIC). I. Effects on dorsal horn convergent neurones in the rat. Pain *6:* 283–304 (1979a).

Le Bars, D.; Dickenson, A.H.; Besson, J.M.: Diffuse noxious inhibitory controls (DNIC). II. Lack of effect on non convergent neurones, supraspinal involvement and theoretical implications. Pain *6:* 305–327 (1979b).

Le Bars, D.; Dickenson, A.H.; Besson, J.M.: Microinjection of morphine within the nucleus raphe magnus and dorsal horn neurones activities related to nociception in the rat. Brain Res. *189:* 467–481 (1980).

Le Bars, D.; Dickenson, A.H.; Besson, J.M.: The triggering of bulbo-spinal serotonergic inhibitory controls by noxious peripheral inputs; Sjölund, Björklund Brain stem control of spinal mechanisms, pp. 381–410 (Elsevier/Biomedical Press, Amsterdam 1982).

Le Bars, D.; Dickenson, A.H.; Besson, J.M.: Opiate analgesia and descending control systems; in Bonica, Lindblom, Iggo, Advances in pain research and therapy, vol. 5, pp. 341–372 (Raven Press, New York, 1983).

Leichnitz, G.R.; Watkins, L.R.; Griffin, G.; Murfin, R.; Mayer, D.J.: The projection from nucleus raphe magnus and other brain stem nuclei to the spinal cord in the rat. A study using the HRP blue-reaction. Neurosci. Lett. *8:* 119–124 (1978).

Lewis, J.W.; Terman, G.W.; Shavit, Y.; Nelson, L.R.; Liebeskind, J.C.: Neural, neurochemical and hormonal bases of stress-induced analgesia; In Kruger, Liebeskind, Advances in pain research and therapy, pp. 227–288 (Raven Press, New York, vol. 6, 1984).

Lewis, V.A.; Gebhart, G.F.: Evaluation of the periaqueductal central gray (PAG) as a morphine specific locus of action and examination of morphine-induced and stimulation-produced analgesia at coincident PAG loci. Brain Res. *124:* 283–303 (1977).

Liebeskind, J.C. Giesler, G.J., Jr.; Urca, G.: Evidence pertaining to an endogenous mechanism of pain inhibition in the central nervous system; in Zotterman, Sensory functions of the skin in primates with special reference to man, pp. 561–573 (Pergamon Press, Oxford 1976).

Liebeskind, J.C.; Guilbaud, G.; Besson, J.M.; Oliveras, J.L.: Analgesia from electrical stimulation of the periaqueductal gray matter in the cat: behavioral observations and inhibitory effects on spinal cord interneurons. Brain Res. *50:* 441–446 (1973).

Light, A.R.: The spinal terminations of single physiologically characterized axons originating in the pontomedullary raphe of the cat. J. comp. Neurol. *234:* 536–548 (1985).

Light, A.R.; Kavookjian, A.M.; Petrusz, P.: The ultrastructure and synaptic connections of serotonin-immunoreactive terminals in spinal laminae I and II. Somatosensory Res. *1:* 35–50 (1983).

Light, A.R.; Metz, C.B.: The morphology of spinal cord efferent and afferent neurons contributing to the ventral roots of the cat. J. comp. Neurol. *179:* 501–516 (1978).

Long, J.B.; Kalivas, P.W.; Youngblood, W.W.; Prange, A.J.; Kiser, J.S.: Possible involvement of serotoninergic neurotransmission in neurotensin but not morphine analgesia. Brain Res. *310:* 35–41 (1984).

Lovick, T.A.: The role of 5-HT, GABA and opioid peptides in presynaptic inhibition of tooth pulp input from the medial brain stem. Brain Res. *289:* 135–142 (1983).

Lovick, T.A.: Ventrolateral medullary lesions block the antinociceptive and cardiovascular responses elicited by stimulating the dorsal periaqueductal gray matter in rats. Pain *21:* 241–252 (1985).

Lovick, T.A.; West, D.C.; Wolstencroft, J.H.: Responses of raphe-spinal and other bulbar raphe neurones to stimulation of the periaqueductal grey in the cat. Neurosci. Lett. *8:* 45–49 (1978).

Lovick, T.A.; West, D.C.; Wolstencroft, J.H.: A presynaptic action of the raphe on tooth pulp fibre terminals: is this mediated by an opioid peptide? In Van Ree, Terenius, Characteristics and functions of opioids, pp. 175–177 (Elsevier/North Holland, Amsterdam 1978).

Lundberg, A.: Inhibitory control from the brain stem of transmission from primary afferents to motoneurons, primary afferent terminals and ascending pathways; in

Sjolund, Bjorklund, Brain stem control of spinal mechanisms, pp. 179–224 (Elsevier/Biomedical Press, Amsterdam 1982).

Mantyh, P.W.: Connections of midbrain periaqueductal gray in the monkey. II. Descending efferent projections. J. Neurophysiol. *49:* 582–594 (1983).

Martin, R.F.; Haber, L.H.; Willis, W.D.: Primary afferent depolarization of identified cutaneous fibers following stimulation in medial brain stem. J. Neurophysiol. *42:* 779–790 (1979).

Martin, R.F.; Jordan, L.M.; Willis, W.D.: Differential projections of cat medullary raphe neurons demonstrated by retrograde labelling following spinal cord lesions. J. comp. Neurol. *182:* 77–88 (1978).

Mason, P.; Strassman, A.; Maciewicz, R.: Pontomedullary raphe neurons: monosynaptic excitation from midbrain sites that suppress the jaw opening reflex. Brain Res. *329:* 384–389 (1985).

Maxwell, D.J.; Leranth, C.; Verhofstad, A.A.J.: Fine structure of serotonin-containing axons in the marginal zone of the rat spinal cord. Brain Res. *226:* 253–259 (1983).

Mayer, D.J.: Endogenous analgesia systems: neural and behavioral mechanisms; in Bonica, Liebeskind, Albe-Fessard, Advances in pain research and therapy, vol. 3, pp. 385–410 (Raven Press, New York, 1979).

Mayer, D.J.; Hayes, R.L.: Stimulation produced analgesia. Development of tolerance and cross-tolerance to morphine. Science *188:* 941–943 (1975).

Mayer, D.J.; Liebeskind, J.C.: Pain reduction by focal electrical stimulation of the brain. An anatomical and behavioral analysis. Brain Res. *68:* 73–93 (1976).

Mayer, D.J.; Price, D.D.: Central nervous system mechanisms of analgesia. Pain *2:* 379–404 (1976).

Mayer, D.J.; Watkins, L.R.: Multiple endogenous opiate and non opiate analgesia systems; in Kruger, Liebeskind, Advances in pain research and therapy, vol. 6, pp. 253–276 (Raven Press, New York 1984).

Mayer, D.J.; Wolfle, T.L.; Akik, H.; Carder, B.; Liebeskind, J.C.: Analgesia from electrical stimulation in the brain stem of the rat. Science *174:* 1351–1354 (1971).

McCreery, D.B., Bloedel, J.R.; Hames, E.G.: Effects of stimulation in raphe nuclei and in reticular formation on response of spinothalamic neurons to mechanical stimuli. J. Neurophysiol. *42:* 166–182 (1979).

McMahon, S.B.; Wall, P.D.: A system of rat spinal cord lamina 1 cells projecting through the contralateral dorsolateral funiculus. J. comp. Neurol. *214:* 217–223 (1983).

McMahon, S.B.; Wall, P.D.: Electrophysiological mapping of brainstem projections of spinal cord lamina 1 cells in the rat. Brain Res. *333:* 19–26 (1985).

Messing, R.B.; Lytle, L.D.: Serotonin containing neurons: their possible role in pain and analgesia. Pain *4:* 1–21 (1977).

Miletic, V.; Hoffert, M.J.; Ruda, M.A.; Dubner, R.; Shigenaga, Y.: Serotoninergic axonal contacts on identified cat spinal dorsal horn neurons and their correlation with nucleus raphe magnus stimulation. J. comp. Neurol. *228:* 129–141 (1984).

Mohrland, J.S.; McMagnus, D.Q.; Gebhart, G.F.: Lesions in nucleus reticularis gigantocellularis: effects on the antinociception produced by microinjection of morphine and focal electrical stimulation in the periaqueductal gray matter. Brain Res. *231:* 143–152 (1982).

Morrow, T.J.; Casey, K.L.: Analgesia produced by mesencephalic stimulation: effect on

bulboreticular neurons; in Bonica, Albe-Fessard, Advances in pain research and therapy, vol. 1, pp. 503–510 (Raven Press, New York 1976).
Morrow, T.J.; Casey, K.L.: Suppression of bulboreticular unit responses to noxious stimuli by analgesic mesencephalic stimulation. Somatosensory Res. *1:* 151–168 (1983).
Morton, C.R.; Duggan, A.W.; Zhao, Z.Q.: The effects of lesions of medullary midline and lateral reticular areas on inhibition in the dorsal horn produced by periaqueductal grey stimulation in the cat. Brain Res. *301:* 121–130 (1984).
Nishikawa, N.; Bennett, G.J.; Ruda, M.A.; Lu, G.W.; Dubner, R.: Immunocytochemical evidence for a serotoninergic innervation of dorsal column postsynaptic neurons in cat and monkey. Light and electron-microscopic observations. Neuroscience *10:* 1333–1340 (1983).
Olds, M.E.; Frey, J.H.: Effects of hypothalamic lesions on escape behavior produced by midbrain electric stimulation. Am. J. Physiol. *221:* 8–18 (1971).
Oleson, T.D.; Kirkpatrick, D.B.; Goodman, S.J.: Elevation of pain threshold to tooth shock by brain stimulation in primates. Brain Res. *194:* 79–95 (1980).
Oleson, T.D.; Twombly, D.A.; Liebeskind, J.C.: Effects of pain attenuating brain stimulation and morphine on electrical activity in the raphe nuclei of the awake rat. Pain *4:* 211–230 (1978).
Olgart, L.: Local mechanisms in dental pain, in Beers, Bassett, Mechanisms of pain and analgesic compounds, pp. 285–294 (Raven Press, New York 1979).
Oliveras, J.L.; Besson, J.M.; Guilbaud, G.; Liebeskind, J.C.: Behavioral and electrophysiological evidence of pain inhibition from midbrain stimulation in the cat. Expl. Brain Res. *20:* 32–44 (1974).
Oliveras, J.L.; Bourgoin, S.; Hery, F.; Besson, J.M.; Hamon, M.: The topographical distribution of serotoninergic terminals in the spinal cord of the cat: biochemical mapping by the combined use of microdissection and microassay procedures. Brain Res. *138:* 393–406 (1977).
Oliveras, J.L.; Guilbaud, G.; Besson, J.M.: The use of tooth pulp stimulation to evaluate the analgesia induced by stimulation of raphe nuclei in the cat; in Anderson, Matthews, Pain in the trigeminal region, pp. 319–330 (Elsevier/North-Holland, Amsterdam 1977).
Oliveras, J.L.; Guilbaud, G.; Besson, J.M.: A map of serotoninergic structures involved in stimulation producing analgesia in unrestrained freely moving cats. Brain Res. *164:* 317–322 (1979).
Oliveras, J.L.; Hosobuchi, Y.; Bruxelle, J.; Passot, C.; Besson, J.M.: Analgesic effects induced by electrical stimulation of the nucleus raphe magnus in the rat: interaction with morphine analgesia. Abstr. 7th Int. Congr. Pharmacology, Paris, 1978, vol. 1, p. 280.
Oliveras, J.L.; Hosobuchi, Y.; Guilbaud, G.; Besson, J.M.: Analgesic electrical stimulation of the feline nucleus raphe magnus: development of tolerance and its reversal by 5-HTP. Brain Res. *146:* 404–409 (1978).
Oliveras, J.L.; Woda, A.; Guildbaud, G.; Besson, J.M.: Inhibition of the jaw opening reflex by electrical stimulation of the periaqueductal gray matter in the awake unrestrained cat. Brain Res. *72:* 328–331 (1974).
Prieto, G.J.; Cannon, J.T.; Liebeskind, J.C.: N. Raphe magnus lesions disrupt stimulation-produced analgesia from ventral but not dorsal midbrain areas in the rat. Brain Res. *261:* 53–57 (1983).

Proudfit, H.K.; Anderson, E.G.: New-long latency bulbospinal evoked potentials blocked by serotonin antagonists. Brain Res. *65:* 542–554 (1974).

Proudfit, H.K.; Anderson, E.G.: Morphine analgesia: blockade by raphe magnus lesions. Brain Res. *98:* 612–618 (1975).

Proudfit, H.K.; Hammond, D.L.: Alterations in nociceptive threshold and morphine-induced analgesia produced by intrathecally administered amine antagonists. Brain Res. *218:* 393–399 (1981).

Proudfit, H.K.; Larson, A.A.; Anderson, E.G.: The role of GABA and serotonin in the mediation of raphe evoked spinal cord dorsal root potentials. Brain Res. *195:* 149–165 (1980).

Reynolds, D.V.: Surgery in the rat during electrical analgesia induced by focal brain stimulation. Science *164:* 444–445 (1969).

Rhodes, D.L.; Liebeskind, J.C.: Analgesia from rostral brain stem stimulation in the rat. Brain Res. *143:* 521–532 (1978).

Richardson, D.E.; Akil, H.: Pain reduction by electrical brain stimulation in man: Part II: chronic self administration in the periaqueductal gray matter. J. Neurosurg. *47:* 184–194 (1977).

Rivot, J.P.; Chaouch, A.; Besson, J.M.: Nucleus raphe magnus modulation of response of rat dorsal horn neurons to unmyelinated fiber inputs: partial involvement of serotoninergic pathways. J. Neurophysiol. *44:* 1039–1057 (1980).

Rivot, J.P.; Chiang, C.Y.; Besson, J.M.: Increase in serotonin metabolism within the dorsal horn of the spinal cord during nucleus raphe magnus stimulation, as revealed by in vivo electrochemical detection. Brain Res. *238:* 117–126 (1982).

Roberts, M.H.T.: 5-Hydroxytryptamine and antinociception. Neuropharmacology *23:* 1529–1536 (1984).

Ruda, M.A.; Allen, B.; Gobel, S.: Ultrastructural analysis of medial brain stem afferents to the superficial dorsal horn. Brain Res. *205:* 175–180 (1981).

Ruda, M.A.; Coffield, J.; Steinbusch, H.W.M.: Immunocytochemical analysis of serotoninergic axons in laminae I and II of the lumbar spinal cord of the cat. J. Neurosci. *2:* 1660–1671 (1982).

Ruda, M.A.; Gobel, S.: Ultrastructural characterization of axonal endings in the substantia gelatinosa which take up (^{3}H)serotonin. Brain Res. *184:* 57–83 (1980).

Sagen, J.; Winker, M.A.; Proudfit, H.K.: Hypoalgesia induced by the local injection of phentolamine in the nucleus raphe magnus. Blockade by depletion of spinal cord monoamines. Pain *16:* 253–263 (1983).

Sakai, K.; Touret, M.; Salvert, D.; Leger, L.; Jouvet, M.: Afferent projections to the cat locus coeruleus as visualized by the horseradish peroxidase technique. Brain Res. *119:* 21–41 (1977).

Sandkuhler, J.; Gebhart, G.F.: Relative contributions of the nucleus raphe magnus and adjacent medullary reticular formation to the inhibition by stimulation in the periaqueductal gray of a spinal nociceptive reflexes in the phenobarbital anesthetized rat. Brain Res. *305:* 77–87 (1984).

Satoh, M.; Akaike, A.; Nakazawa, T.; Takagi, H.: Evidence for involvement of separate mechanisms in the production of analgesia by electrical stimulation of the nucleus reticularis paragigantocellularis and nucleus raphe magnus. Brain Res. *194:* 525–529 (1980).

Savard, P.; Merand, Y.; Dupont, A.: The effects of *p*-chlorophenylalanine (pCPA) treatment on the substance P content measured in discrete brain nuclei of normal and neonatally-induced hypothyroid rats. Brain Res. *292:* 349–355 (1984).

Schmauss, C.; Hammond, D.L.; Ochi, J.W.; Yaksh, T.L.: Pharmacological antagonism of the antinociceptive effects of serotonin in the rat spinal cord. Eur. J. Pharmacol. *90:* 349–357 (1983).

Schmitt, P.; Abouahmed, H.; Karli, P.: Effects aversifs et appétitifs induits par stimulation mésencéphalique et hypothalamique. Brain Res. *130:* 521–530 (1977).

Shah, Y.; Dostrovsky, J.O.: Electrophysiological evidence for a projection of the periaqueductal gray matter to nucleus raphe magnus in cat and rat. Brain Res. *193:* 534–538 (1980).

Shah, Y.; Dostrovsky, J.O.: Postsynaptic inhibition of cat medullary dorsal horn neurons by stimulation of nucleus raphe magnus and other brain stem sites. Expl. Neurol. *77:* 419–435 (1982).

Skagerberg, G.; Bjorklund, A: Topographic principles in the spinal projections of serotoninergic and non-serotoninergic brain stem neurons in the rat. Neuroscience *15:* 445–480 (1985).

Tenen, S.S.: Antagonism of the analgesic effect of morphine and other drugs by *p*-chlorophenylalanine, a serotonin depletor. Psychopharmacologia *12:* 278–285 (1968).

Tohyama, M.; Sakai, K.; Touret, M.; Salvert, D.; Jouvet, M.: Spinal projections from the lower brain stem in the cat as demonstrated by the horseradish peroxidase technique. II. Projection from the dorsolateral pontine tegmentum and raphe nuclei. Brain Res. *176:* 215–231 (1979).

Vanegas, H.; Barbaro, N.M.; Fields, H.L.: Midbrain stimulation inhibits tail-flick only at currents sufficient to excite rostral medullary neurons. Brain Res. *321:* 127–133 (1984).

Vanegas, H.; Barbaro, N.M.; Fields, H.L.: Tail-flick related activity in medullospinal neurons. Brain Res. *321:* 135–141 (1984).

Villanueva, L.; Chitour, D.; Le Bars, D.: Involvement of the dorsolateral funiculus (DLF) in the descending part of the loop subserving diffuse noxious inhibitory controls (DNIC). Neurosci. Lett., suppl. *22:* p. 468 (1985).

Villanueva, L.; Peschanski, M.; Calvino, B.; Le Bars, D.: Ascending pathways in the spinal cord involved in triggering of diffuse noxious inhibitory controls (DNIC) in the rat. J. Neurophysiol. *55:* 34–55 (1986).

Vogt, M.: The effect of lowering the 5-hydroxytryptamine content of the rat spinal cord on analgesia produced by morphine. J. Physiol., Lond. *236.* 483–498 (1974).

Walberg, F.: Paths descending from the brain stem; in Sjolund, Bjorklund, Brain Stem control of spinal mechanisms, pp. 1–27 (Elsevier/Biomedical Press, Amsterdam 1982).

Wang, J.K.: Antinociceptive effect of intrathecally administered serotonin. Anesthesiology *47:* 269–271 (1977).

Watkins, L.R.; Griffin, G.; Leichnetz, G.R.; Mayer, D.J.: The somatotopic organization of the nucleus raphe magnus and surrounding brain stem structures as revealed by HRP slow-release gels. Brain Res. *181:* 1–15 (1980).

Watkins, L.R.; Griffin, G.; Leichnetz, G.R.; Mayer, D.J: Identification and somatotopic organization of nuclei projecting via the dorsolateral funiculus in rats. A retrograde tracing study using HRP slow-release gels. Brain Res. *223:* 237–255 (1981).

Watkins, L.R.; Johannessen, J.N.; Kinscheck, I.B.; Mayer, D.J.: The neurochemical basis of footshock analgesia: the role of spinal cord serotonin and norepinephrine. Brain Res. *290:* 107–117 (1984).

Wessendorf, M.W.; Proudfit, H.K.; Anderson, E.G.: The identification of serotoninergic neurons in the nucleus raphe magnus by conduction velocity. Brain Res. *214:* 168–173 (1981).

West, D.C.; Wolstencroft, J.H.: Electrophysiological identification of raphe-spinal neurones in the cat. J. Physiol. Lond. *265:* 29–30P (1977).

Willcockson, W.S.; Chung, J.M.; Hori, Y.; Lee, K.H.; Willis, W.D.: Effects of iontophoretically released peptides on primate spinothalamic tract cells. J. Neurosci. *4:* 741–750 (1984).

Willcockson, W.S.; Gerhart, K.D.; Cargill, C.L.; Willis, W.D.: Effects of biogenic amines on raphe-spinal tract cells. J. Pharmac. exp. Ther. *225:* 637–645 (1983).

Willis, W.D.: Control of nociceptive transmission in the spinal cord; in Progress of sensory physiology, vol. 3, p. 159 (Springer, Berlin 1982).

Willis, W.D.: The raphe-spinal system; in Barnes, Brain stem control of spinal cord function, pp. 141–214 (Academic Press, New York 1984).

Willis, W.D.; Gerhart, K.D.; Willcockson, W.S.; Yezierski, R.P.; Wilcox, T.K.; Cargill, C.L.: Primate raphe and reticulo-spinal neurons: effects of stimulation in periaqueductal gray or VPLc thalamic nucleus. J. Neurophysiol. *51:* 467–480 (1984).

Willis, W.D.; Haber, L.H., Martin R.F.: Inhibition of spinothalamic tract cells and interneurons by brain stem stimulation in the monkey. J. Neurophysiol. *40:* 968–981 (1977).

Yaksh, T.L.; Wilson, P.R.: Spinal serotonin terminal systems mediates antinociception. J. Pharmac. exp. Ther. *208:* 446–453 (1979).

Yaksh, T.L.; Yeung, J.C.; Rudy, T.A.: An inability to antagonize with naloxone the elevated nociceptive thresholds resulting from electrical stimulation of the mesencephalic gray. Life Sci. *18:* 1193–1198 (1976).

Yezierski, R.P.; Gerhart, K.D.; Schrock, B.J.; Willis, W.D.: A further examination of effects of cortical stimulation on primate spino-thalamic tract cells. J. Neurophysiol. *49:* 424–441 (1983).

Yezierski, R.P.; Wilcox, T.K.; Willis, W.D.: The effects of serotonin antagonists on the inhibition of primate spinothalamic tract cells produced by stimulation in nucleus raphe magnus or periaqueductal gray. J. Pharmac. exp. Ther. *220:* 266–277 (1982).

Zemlan, F.P.; Leonard, C.M.; Kow, L.; Pfaff, D.W.: Ascending tracts of the lateral columns of the rat spinal cord. A study using the silver impregnation and horseradish peroxidase techniques. Expl. Neurol. *62:* 298–334 (1978).

Zorman, E.; Belcher, G.; Adams, J.E.; Fields, H.L.: Lumbar intrathecal naloxone blocks analgesia produced by microstimulation of the ventromedial medulla in the rat. Brain Res. *236:* 77–84 (1982).

Zorman, G.; Hentall, I.D.; Adams, J.E.; Fields, H.L.: Naloxone-reversible analgesia produced by microstimulation in the rat medulla. Brain Res. *219:* 137–148 (1981).

J.M. Besson, MD, Unité des Recherches de Neurophysiologie Pharmacologique, Inserm U-161, 2, rue d'Alésia, F-75014 Paris (France)

Pain Headache, vol. 9, pp. 101–128 (Karger, Basel 1987)

The Descending Noradrenergic System and Analgesia

Yasushi Kuraishi, Masamichi Satoh, Hiroshi Takagi[1]
Department of Pharmacology, Faculty of Pharmaceutical Sciences, Kyoto University, Kyoto, Japan

Introduction

It is well known that the descending inhibitory systems play an important role in pain modulation and analgesia. These systems participate in morphine analgesia and stimulation-produced analgesia, and exert a tonic inhibition from the brain stem to spinal cord on nociceptive processes [Fields and Basbaum, 1978; Mayer and Price, 1976; Takagi, 1982; Willis, 1982]. In the past decade, much information has been gathered about their chemical characteristics, i.e. which neurotransmitter(s) mediate the descending inhibition. Noradrenaline (NA) and serotonin are the principal candidates for involvement in these systems. The noradrenergic and serotonergic terminals in the spinal cord are well known to have their cell bodies in the brain stem, but not in the spinal cord [Dahlstrom and Fuxe, 1964]. This review will focus on the descending noradrenergic system.

There are many reports, describing seemingly contradictory results regarding the involvement of NA in morphine analgesia [see introduction of Dennis et al., 1980; review by Willis, 1982]. These confusing results may, at least in part, be due to the focus of many experiments on brain NA, wherein adrenergic manipulations were applied systemically. It appears, however, that NA plays opposite roles in the brain and spinal cord. Analgesia is facilitated by NA in the spinal cord [Kuraishi et al., 1979a, b, 1983a; Reddy and Yaksh, 1980; Xie et al., 1983; Yaksh, 1979],

[1] We thank Dr. H. Akil (University of Michigan) for valuable help in manuscript preparation and G. Baldrighi (University of Michigan) for figure preparation.

but is inhibited by NA in the brain [Sewell and Spencer, 1975; Xie et al., 1983]. Therefore, this chapter concerning the descending noradrenergic system will review reports in which noradrenergic functions in the spinal cord were specifically manipulated.

Noradrenaline Inhibits Nociceptive Transmission in the Spinal Cord

The most direct evidence that the descending noradrenergic system innervating the spinal cord participates in analgesia was obtained by experiments which examined, electrophysiologically and behaviorally, whether exogenously administered NA would inhibit nociceptive transmission in the spinal cord.

Iontophoretic Studies

Headley et al. [1978] reported that NA iontophoretically applied either to the substantia gelatinosa or near the cell body of the dorsal horn neurons (laminae IV and V) in the spinal cat produced a selective reduction of nociceptive responses (to thermal noxious stimulus). Similarly, Belcher et al. [1978] demonstrated that NA ejected near the cell bodies inhibited the nociceptive responses (to thermal noxious stimulus) of the nociceptive dorsal horn neurons in the chloralose-anesthetized cat. Satoh et al. [1979b] showed that NA ejected near the cell bodies suppressed the nociceptive responses (to intraarterial injections of bradykinin, a pain-producing substance) of the lamina V type (wide-dynamic range) neurons in the spinal rabbit. Recently, Fleetwood-Walker et al. [1985] demonstrated a selective inhibition of nociceptive responses (to thermal stimuli) of the feline spinal dorsal horn neurons by NA and clonidine, but not by isoproterenol (a beta-adrenergic agonist), applied microiontophoretically near the cell bodies, and the inhibitory effect of NA was antagonized by the selective alpha$_2$-antagonists yohimbine and idazoxan. These electrophysiological results suggest an antinociceptive role of NA in the spinal dorsal horn, possibly mediated by alpha$_2$-adrenoceptors. However, an involvement of alpha$_1$-adrenoceptors cannot be excluded.

Behavioral Studies

Participation of spinal NA in antinociception was more directly verified behaviorally using freely-moving animals. Kuraishi et al. [1979b]

demonstrated that NA injected intrathecally (in doses of 0.5–10 nmol/animal) in freely-moving rats suppressed the nociceptive response to mechanical noxious stimulation, and that such suppression was antagonized by phenoxybenzamine, an alpha-adrenergic blocker. NA in doses producing antinociception did not inhibit startle response to a non-noxious stimulus induced by a strong blast of air to the hip area, indicating the selectivity of the antinociceptive action of NA. It should be noted that antinociceptive effects of other putative pain-inhibitory transmitters in the dorsal horn, serotonin and Met-enkephalin, were 1/180 and less than 1/350, respectively, as potent as that of NA. Also other putative inhibitory transmitters, gamma-aminobutyric acid and glycine, did not inhibit nociceptive response in an intrathecal dose of 300 nmol/animal. Antinociceptive action of intrathecal NA on the nociceptive responses to thermal and chemical stimulation were also shown in rats, mice and cats [Hylden and Wilcox, 1983; Kuraishi et al., 1985b; Reddy and Yaksh, 1980], although the potencies of NA varied depending on the analgesic methods used (see later).

Alpha-adrenergic agonists other than NA, including methoxamine (an adrenergic alpha$_1$-selective agonist) and clonidine (an alpha$_2$-selective agonist), also produce antinociception after an intrathecal injection [Howe et al., 1983; Hylden and Wilcox, 1983; Kuraishi et al., 1979b; Reddy et al., 1980; Takagi et al., 1980; Yaksh and Reddy, 1981], whereas intrathecal injection of a beta-adrenergic agonist, isoproterenol, has only a weak effect on nociceptive behaviors [Reddy and Yaksh, 1980; Takagi et al., 1980]. Antinociception of intrathecal NA is antagonized by alpha-blockers, but not by beta-blockers [Howe et al., 1983; Hylden and Wilcox, 1983; Kuraishi et al., 1979b; Reddy and Yaksh, 1980]. Howe et al. [1983] further suggested that activation of either alpha$_1$- or alpha$_2$-adrenoceptors in the spinal dorsal horn can produce antinociception. Early experiments (in 1950s and before) conducted about the effects of intrathecal injection of adrenergic agents have been reviewed by Rothballer [1959].

Since adrenergic agonists have been administered intrathecally for many years as a vasoconstrictor to prolong spinal anesthetic action of local anesthetics [Axelsson and Widman, 1981; Bonica et al., 1951; Converse et al., 1954; Egbert and Deas, 1960; Park et al., 1975], the question arises whether the spinal antinociception of NA alone results from vasoconstriction and ischemia in the spinal cord. The following findings, however, tend to exclude such a possibility: (1) NA has much weaker constricting action on the blood vessels in the central nervous system than on the peripheral blood vessels [Bohr et al., 1961]. When an antinociceptive dose of NA (5 nmol) or adrenaline (30 nmol) was directly applied to the pia matter of the dorsal surface of the rat spinal cord after laminectomy, we did not observe through a microscope any changes of the blood vessels on the dorsal surface of the spinal cord [unpubl. observations]. Moreover, Crawford et al. [1977] reported that intra-arterial infusion of NA was without overall effect on bloodflow of the dog spinal cord having intact blood-brain barriers. (2) Antinociceptive effect of intrathecal NA is more potent than that of intrathecal serotonin, although the latter is more potent in contracting cerebral vessels. Antinociceptive potency of intrathecal NA is 40–750 times that of intrathecal serotonin in the thermal and mechanical algesic tests in rats [Kuraishi et al., 1985b]. On the other hand, with respect to the effects of these monoamines on the isolated cerebral arteries of the dog, the maximum tension developed in response to NA was about a half that developed in response to serotonin, and the ED_{50} value of NA was 20 times as large as that of serotonin [Toda and Fujita, 1973]. Moreover, the threshold concentrations of NA and adrenaline to contract the cerebral resistance vessels of the dog were more than 10 times that of serotonin [Borh et al., 1961]. (3) Reddy et al. [1980] demonstrated that intrathecal injections of vasodilators, papaverine and bradykinin, did not inhibit the antinociceptive action of intrathecal NA in the thermal algesic tests in rats. (4) Intrathecal NA produces a selective antinociception. NA in antinociceptive doses did not inhibit non-nociceptive response [Kuraishi et al., 1979b] and did not elicit behavioral changes or motor disturbance [Kuraishi et al., 1979b; Reddy and Yaksh, 1980].

Thus, the spinal NA antinociception is probably due to the specific inhibitory action on the synaptic transmission of nociceptive information rather than the vasoconstricting action in the spinal cord.

Clinical Studies

To determine whether 'antinociception' of intrathecal NA would be 'true analgesia', the effect of NA should be examined in man, but to our knowledge there have been few such studies. Uchida and Murayama [6th Asian and Australasian Congress of Anaesthesiologists, Auckland, 1982] reported that intrathecal injection of a combination of NA (0.05–0.3 mg, 1:10,000) plus dibucaine (1.5 mg, 0.5 ml) remarkably suppressed postoperative pain with no apparent blockade of motor activity. Neither of these agents alone produced analgesia. Some time ago, however, adrenaline instead of NA was reported to be analgesic after intrathecal injection in man. Priddle and Adros [1950] used intrathecal adrenaline to produce saddle-block anesthesia in obstetrics, and reported that adrenaline (0.2–1 mg, 1:1,000) produced analgesic effects at levels between hypoalgesia and complete anesthesia with little systemic action. It is reasonable to consider now that such effects of adrenaline action in man are also mediated by alpha-adrenoceptors in the spinal cord. Recently, Tamson and Gorth [1984] reported an analgesic effect of epidural clinidine in human patients.

Mechanisms of Spinal Antinociception of Noradrenaline

Does NA inhibit nociceptive transmission pre- or postsynaptically in the dorsal horn?

Postsynaptic Inhibition

NA iontophoretically applied near the cell body has been shown to inhibit excitatory amino acid-induced activity of the nociceptive neurons in the dorsal horn [Belcher et al., 1978], and suppress both nociceptive response (induced by intraarterial injection of bradykinin) and non-nociceptive response (induced by tactile stimulation) of the lamina V type (wide-dynamic range) neurons in the dorsal horn [Satoh et al., 1979b]. Moreover, using techniques of intracellular recording and *in vitro* superfusion of slices of the rat spinal cord, North and Yoshimura [1984] demonstrated the hyperpolarizing action of NA on the neurons of the substantia gelatinosa. Although these *in vitro* experiments cannot identify nociceptive neurons, the unmyelinated C-fibers terminate mostly in the substantia gelatinosa [Kumazawa and Perl, 1977]. Together, these findings suggest a postsynaptic inhibitory action of NA on the nociceptive dorsal horn neurons.

Presynaptic Inhibition

Does NA exert presynaptic inhibition on the nociceptive primary afferents? This question could be addressed by examining the effect of NA on the release of a 'pain transmitter' from the primary afferents. The *in situ* release of substance P, a putative pain transmitter, from the rabbit dorsal horn is reproducibly increased by noxious but not non-noxious mechanical skin stimuli [Kuraishi et al., 1985a]. NA administered locally to the dorsal horn completely suppresses the noxious stimuli-induced release of substance P, and the suppressive effect of NA was more effectively antagonized by yohimbine than by prazosin [Kuraishi et al., 1985c]. Thus, NA inhibits presynaptic terminals of substance P-containing nociceptive afferents, predominantly via alpha$_2$-adrenoceptors. Regarding electrophysiological results on presynaptic action of NA, Jeftinija et al. [1981] reported that NA, applied by iontophoresis or pressure microinjection, produced increases in threshold for antidromic activation of the sural C-fibers in spinal cat, although such action was not clearly antagonized by alpha-blockers, yohimbine or phentolamine.

These findings, taken together, suggest that NA suppresses nociceptive transmission by presynaptically inhibiting the substance P-containing primary afferents (in the case of mechanical antinociception) and by postsynaptically inhibiting the nociceptive dorsal horn neurons. In addition, as described in the previous section, the NA inhibition of the nociceptive transmission in the spinal dorsal horn is probably mediated by alpha$_1$-(postsynaptic) and/or alpha$_2$-(predominantly presynaptic) adrenoceptors [Fleetwood-Walker et al., 1985; Howe et al., 1983; Hylden and Wilcox, 1983; Kuraishi et al., 1979b; Reddy and Yaksh, 1980].

Morphine Analgesia and Stimulation-Produced Antinociception

Although there are multiple sites of action of morphine analgesia in the CNS, the mechanisms of pain inhibition by systemically administered morphine have been linked to the suppressive action of this narcotic on the nociceptive transmission in the dorsal horn. This suppression can occur by two separate mechanisms: indirect inhibition through the descending inhibitory systems arising from the brain stem and direct inhibition in the spinal dorsal horn [Fields and Basbaum, 1978; Takagi, 1980]. The indirect inhibition is often considered to play a more essential role than the direct inhibition in the suppression of pain transmission by systemic administration of morphine in 'analgesic doses' [Takagi, 1982]. Participation of the descending noradrenergic system in morphine analgesia has been investigated as a development and expansion of the study on the indirect inhibitory action of this narcotic.

Since the pioneering study by Tsou and Jang [1964], studies employing intracerebral and intracerebroventricular injection technique have demonstrated multiple analgesic sites of morphine, for example, the periaqueductal gray matter (PAG), the nucleus reticularis gigantocellularis (NRGC), the nucleus reticularis paragigantocellularis (NRPG), the nucleus raphe magnus, and the spinal dorsal horn [for review see Fields and Basbaum, 1978; Mayer and Price, 1976; Takagi, 1980; Yaksh and Rudy, 1978]. On the other hand, work on brain stimulation-produced antinociception, starting from the work of Reynolds [1969], have also demonstrated the multiple active sites, the distribution of which is similar to those of the morphine analgesia [Mayer and Price, 1976; Takagi, 1982; Willis, 1982]. Among the sites involved in stimulation-produced and morphine anesthesia, the NRGC/NRPG and the PAG have been shown to produce antinociception mediated by the descending noradrenergic system. In this section, therefore, we will describe the participation of this descending noradrenergic system in antinociception by systemic administration of morphine and by manipulations of the NRGC/NRPG and the PAG.

Systemic Administration of Morphine

If morphine analgesia is produced through the descending noradrenergic system, systemically given morphine should activate this system. Shiomi and Takagi [1974] measured the concentration of normetanephrine, a metabolite of NA by catechol-0-methyltransferase, as an index of the turnover of spinal NA (or an index of the release of NA from the spinal noradrenergic terminals). They found that subcutaneous injections of morphine (2.5–10 mg/kg) increased the concentration of normetanephrine in the spinal cord, especially in its dorsal half, in a dose-dependent fashion. The concentration of NA itself in the spinal cord was unaffected. The location of the morphine action was suggested to be in the lower brain stem, since the normetanephrine increasing effect of morphine was abolished after the spinal transection at a level of C1 but not after the brain stem transection at an intercollicular level. From these results, they first proposed the hypothesis that the descending noradrenergic system participates in the production of morphine analgesia [Shiomi and Takagi, 1974]. An important role of the descending noradrenergic system in the production of morphine-induced analgesia was subsequently shown by Kuraishi et al. [1983a]. They demonstrated that the mechanical antinociceptive effect of subcutaneously administered morphine (4 mg/kg) was significantly suppressed after the reduction of NA in the caudal part of the spinal cord with intrathecal injection of 6-hydroxydopamine. Morphine analgesia was not affected by depletion of NA in the rostral part of the spinal cord or in the medulla oblongata. Moreover, systemically administered morphine abolished the noxious mechanical skin stimulation-induced release of substance P from the rabbit dorsal horn, which was recovered by the direct administration of prazosin, an alpha-blocker, to the dorsal horn [Kuraishi et al., 1983b].

In contrast to the clear participation of the descending noradrenergic system in the mechanical antinociception of systemic morphine, when the thermal stimuli were used for noxious stimulation, the antinociceptive action of systemically administered morphine seems not to be modified by adrenergic manipulations of the spinal cord [Kuraishi et al., 1983a; Proudfit and Hammond, 1981]. The cause of these discrepancies in morphine antinociception will be discussed later in the section of 'type of noxious stimuli and noradrenaline antinociception'.

Nucleus Reticularis Gigantocellularis and Paragigantocellularis

In rats, the nucleus reticularis gigantocellularis (NRGC) and the nucleus reticularis paragigantocellularis (NRPG) [Fifkova and Marsala, 1967] are located in the ventromedial part of the rostral medulla oblongata; the latter, which is located just ventral to the former, nearly corresponds to the nucleus reticularis gigantocellularis pars alpha of the rabbit [Messen and Olszewski, 1949], the magnocellularis tegmental field [Berman, 1968] or the nucleus reticularis magnocellularis [Basbaum et al., 1978] of the cat, and the nucleus reticularis gigantocellularis pars ventralis of the opossum [Goode et al., 1980]. In rats, the NRPG, drawn by Paxinos and Watson [1982], is located more lateral to that drawn by Fifkova and Marsala [1967], and probably corresponds to the nucleus paragigantocellularis lateralis described by Andrezik et al. [1981]. To avoid making the discussion complicated, we will use the term NRPG [Fifkova and Marsala, 1964] regardless of the species.

It was first demonstrated by Takagi and co-workers [Akaike et al., 1978; Takagi et al., 1976, 1977] that microinjection of morphine into the NRGC and NRPG produced behavioral antinociception in rats. Microinjection of morphine into the NRGC/NRPG of the rabbit depressed nociceptive responses (induced by injection of bradykinin into the femoral artery) of the lamina V type cells in the spinal dorsal horn [Takagi et al., 1976], suggesting the participation of the descending inhibitory systems in NRGC/NRPG antinociception.

When a mechanical stimulus was used for noxious stimulation (tail-pinch test), the antinociceptive doses of morphine injected into the NRGC/NRPG were 15–380 ng/rat [Takagi et al., 1977]. Injection of similar doses (76–380 ng/rat) of morphine into the same medullary regions produced an increase of normetanephrine in the spinal cord [Kuraishi et al., 1978], the same phenomenon caused by systemic administration of morphine [Shiomi and Takagi, 1974]. The fast onset and peak time of this action correspond well to those of the antinociception after intra-NRGC/NRPG morphine [Kuraishi et al., 1978]. The normetanephrine in the spinal cord was not increased after microinjection of morphine (380 ng/rat) into the other brain stem regions, e.g. the PAG or the nucleus raphe magnus [Kuraishi et al., 1978]; microinjection of morphine in the same dose (380 ng/rat) into these regions did not produce mechanical antinociception [Takagi et al., 1977], and higher doses were required to produce a definite antinociception after morphine injection into these regions [Satoh et al., 1983a]. Antinociception [Takagi et al., 1978] and normetanephrine increase in the spinal cord [Kuraishi et al., 1978] were also observed after microinjection of Met-enkephalin in the NRGC/NRPG. These findings lead to the suggestion that morphine injected into the NRGC/NRPG facilitates the activity of the descending noradrenergic system. Such activation of the noradrenergic system is considered to inhibit the transmission of noxious information in the dorsal horn, because the antinociceptive effect of morphine (380 ng/rat) injected into the NRGC/NRPG was inhibited either by alpha-adrenoceptor blockade with intrathecal injection of phenoxybenzamine [Kuraishi et al., 1979a] or by reduction of NA in the spinal cord by intrathecal injection of 6-hydroxydopamine [Kuraishi et al., 1983a]. The antinociceptive effect of intra-NRGC/NRPG morphine was also inhibited by systemic administration of phenoxybenzamine but not propranolol and methysergide [Kuraishi et al., 1979a].

When compared to morphine microinjection, the site supporting electrical stimulation produced antinociception is restricted to the more ventral part of the NRGC/NRPG; in fact, electrical stimulation of the NRGC produced escape behaviors rather than antinociception in animals [Casey, 1971; Keene and Casey, 1970; review by Casey and Morrow, 1983]. Akaike et al. [1978] demonstrated that electrical stimulation of the NRPG but not the NRGC produced an antinociceptive effect in rats. NRPG stimulation-produced antinociception was inhibited by phenoxybenzamine but not by methysergide [Satoh et al., 1980]. Antinociception produced by electrical stimulation of the NRGC (probably including the NRPG; see fig. 2 of Yaksh et al. [1981]) was inhibited by intrathecal injection of phentolamine but not methysergide [Yaksh et al., 1981]. Moreover, microinjection of *L*-glutamate, considered more selective than electrical stimulation for activating the cell bodies but not fibérs into the NRPG produced antinociception that was reduced by intrathecal injection of phenoxybenzamine and phentolamine, but not by propranolol and methysergide [Satoh et al., 1983b]. These results strongly suggest that the antinociception produced by activation of the NRPG cells is mediated by the activation of alpha-adrenoceptors, but not beta-adrenoceptors and serotonin receptors, in the spinal cord.

Electrical stimulation of the NRGC/NRPG suppressed the nociceptive responses of lamina V type neurons in the dorsal horn. These effects were reduced following the depletion of spinal monoamines with tetrabenazine, and the effects of tetrabenazine were reversed by administration of *L*-DOPA, a precursor of NA, but not 5-hydroxytryptophan, a serotonin precursor [Takagi et al., 1975]. Moreover, electrical stimulation of NRGC increased the release of NA but not serotonin in spinal cord superfusate [Yaksh et al., 1981]. Thus, the stimulation of cells in the NRGC/NRPG (especially NRPG) activates the descending noradrenergic system, which in turn increases the release of NA and produces the alpha-adrenoceptor mediated inhibition of transmission of noxious information in the dorsal horn.

With regard to the NRGC/NRPG, both morphine and SPA are similar in noradrenergic but not serotonergic participation, suggesting that these effects are mediated by a common system. Such a view is supported further by the following electrophysiological data. Iontophoretic application of morphine and Leu-enkephalin produced naloxone-reversible activation of the excitatory nociceptive neurons in the NRPG [Satoh et al., 1979a], and morphine potentiated the nociceptive responses of the NRPG neurons [Azami et al., 1981]. Moreover, iontophoretic application of morphine produced naloxone-reversible activation on most NRPG neurons [Mohrland and Gebhart, 1981]. Therefore, we believe that the activation of NRPG (and NRGC) neurons by morphine, electrical stimulation and *L*-glutamate results in the facilitation of the descending noradrenergic system.

Periaqueductal Gray Matter

The PAG has received much attention as a pain modulatory locus [Hosobuchi et al., 1977; Mayer and Price, 1976; Oliveras et al., 1974; Reynolds, 1969; Richardson and Akil, 1977; Satoh et al., 1983a; Tsou and Jang, 1964; Yaksh and Rudy, 1978]. Microinjection of morphine (4 to 16 μg/rat) into the PAG or electrical stimulation of the same region inhibits nociceptive responses of the interneurons in the dorsal horn of rats [Bennett and Mayer, 1979], and cats [Oliveras et al., 1974; Carstens et al., 1979]. These data suggest participation of the descending inhibitory systems in PAG antinociception. Thermal antinociception produced by microinjection of morphine (5 μg/rat) into the PAG was inhibited by intrathecal injection of phentolamine [Yaksh, 1975], and mechanical antinociception was suppressed after the reduction of spinal noradrenaline by intrathecal injection of 6-hydroxydopamine [Kuraishi et al., 1983a].

There is neuronal projection from the PAG to the NRPG [Abols and Basbaum, 1981], and manipulations of the PAG, such as morphine microinjection [Mohrland and Gebhart, 1981] and electrical stimulation [Mohrland and Gebhart, 1980], excite cells in the NRPG of rats. These findings suggest the existence of a PAG-NRGC-spinal noradrenergic system which may participate, at least in part, in the antinociception elicited from the PAG.

Is the Descending Noradrenergic System an Intrinsic Pain-Regulatory System?

As summarized above, the descending noradrenergic system clearly plays an important role in the production of antinociception by systemically or intracerebrally administered morphine and by electrical brain stimulation. Does the descending noradrenergic system function as an intrinsic pain-regulatory system? If it does, peripheral noxious stimuli should change the activity of the descending noradrenergic system and in turn alter nociceptive responses.

Acute Noxious Stimulation

Takagi et al. [1979] found that exposure of rats to noxious stimuli (placed on the hot-plate at 60 °C for 3 min, or intermittent electrical stimulation of the tail root for 15 min) increased the concentration of normetanephrine in the rat spinal cord without changes in the concentration of NA. The increase in normetanephrine was observed only in the dorsal half of the spinal cord, and remained unchanged after intercollicular transection. After spinal transection, however, it was abolished within the cord caudal to the transection site. These observations are very similar to those seen after systemically given morphine [Shiomi and Takagi, 1974]. In this context, Tyce and Yaksh [1981] reported that bilateral stimulation of the sciatic nerve, at an intensity capable of firing A-delta/C fibers, resulted in an increase in the NA concentration in the spinal cord superfusate in cats, and this effect was abolished by cold-block spinalization. Taken together, these findings suggest that somatic noxious stimulation activates the descending noradrenergic system via the lower brain stem. It should be noted that neither the normetanephrine increase in the spinal cord elicited by the 3-min thermal stimulation nor the NA increase in the spinal superfusate elicited by the intermittent 20-min electrical stimulation of the sciatic nerve were blocked by naloxone [Tyce and Yaksh, 1981; Takagi et al., 1979]. These results suggest that the activation of the descending noradrenergic system by acute noxious stimulation is not mediated by the opioid peptide system.

Sustained Noxious Stimulation

Although the facilitation of the activity of the descending noradrenergic system by acute noxious stimulation is not mediated by an opioid peptide system, it is possible that sustained noxious stimulation would elicit the release of opioid peptides (e.g. Met-enkephalin) in brain stem regions, which in turn further increases the activity of the descending noradrenergic system.

Using subcutaneous injection of formalin to produce a sustained noxious stimulus [Dubuisson and Dennis, 1977], we tested the above hypothesis [Sugimoto, Kuraishi, Satoh and Takagi, submitted]. Using the time spent licking the hind-paw in response to a formalin injection as an index of nociceptive response, the nociceptive responses persisted for about 60 min (fig. 1). Flexion of the affected limb, however, persisted for about 2 hours or more. These observations suggest that, in spite of existence of the noxious stimulus, the nociceptive behavioral responses decrease and disappear, conceivably due to the activation of the pain-inhibitory system. Since the time course of the licking responses was composed of two phases, it was divided into three periods (fig. 1); during period I we observed the first phase of nociceptive responses, which

Drugs	Route	Licking response, % of control I	II	III
Phentolamine	i.t.	+ 34	+ 68	+ 169
Propranolol	i.t.	N	N	N
Naloxone	s.c.	N	N	+ 88
Naloxone	i.c.			+ 105
Naloxone	i.t.			N

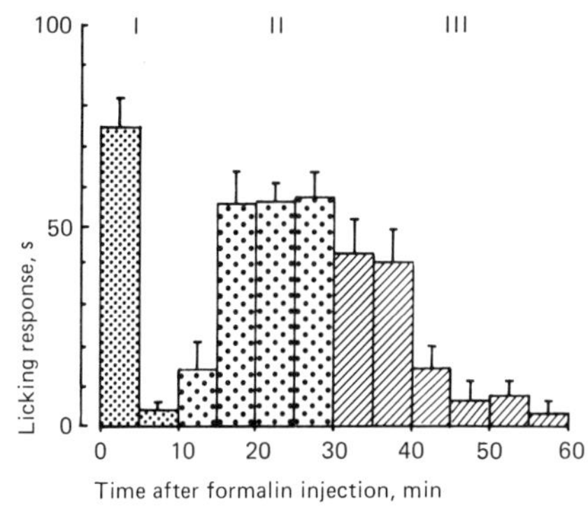

Fig. 1. Modulation of formalin-induced pain by naloxone and adrenergic agents. For noxious stimulation, 0.05 ml of dilute formalin (5% formaldehyde) subcutaneously injected to the plantar region of the hind-paw of the rat, and time course plotted. Naloxone administered subcutaneously (0.1 mg/h/animal for 3 days) into the fourth ventricle (i.c.) (30 nmol/animal, 5 min before period III) or into the spinal subarachnoid space (i.t.) (30 and 300 nmol/animal, 5 min before period III). Phentolamine and propranolol (30 nmol/animal, i.t.) injected 5 min before the formalin injection. Roman numerals = Period of the time course; Arabic numerals = significant changes (%) of the duration of licking response from the vehicle control; N = no significant alteration.

increased in period II, and decreased in period III. Intrathecal injection of phentolamine facilitated the nociceptive responses in all three periods, with the most remarkable facilitation occurring during period III, suggesting that the descending noradrenergic system was activated by sustained noxious stimulation (formalin injection). On the other hand, systemic naloxone produced a facilitation of the nociceptive responses only in period III, which was also observed following the intracerebroventricular, but not intrathecal, injection of this narcotic antagonist. In this context, the same formalin stimulation was shown to produce a gradual increase in release of Met-enkephalin from the

NRGC/NRPG over the 3-hour period after the injection [Kuraishi et al., 1984b]. Furthermore, morphine and opioid peptides activate the descending noradrenergic system via the NRGC/NRPG [Kuraishi et al., 1978, 1979a]. From these findings, we believe that sustained noxious stimulation evokes the release of opioid peptide(s), especially Met-enkephalin, from the NRGC/NRPG (and maybe PAG), which subsequently activates the descending noradrenergic system to suppress excessive input of noxious information.

Tonic Descending Inhibition

Although it is well known that the lower brain stem exerts a tonic inhibitory influence on the dorsal horn neurons [Fields and Basbaum, 1978; Pompeiano, 1973], the precise origin and neurotransmitter(s) of this tonic inhibitory system remain to be elucidated. Sagan and Proudfit [1984] reported that nociceptive threshold was lowered by intrathecal injection of alpha noradrenergic blockers with the following order of potency: yohimbine>phentolamine>prazosin. Similarly, we [Kuraishi et al., 1985c] have shown that direct administration of yohimbine, but not prazosin, to the dorsal horn on acute spinalization remarkably increased the noxious stimulation-induced release of substance P from the dorsal horn. Finally, Hall et al. [1982] showed that tonic descending inhibition on the nociceptive dorsal horn neurons was reduced by bilateral lesions of the nucleus reticularis lateralis, which roughly corresponds to the A1 NA cell group. Thus, the descending noradrenergic system may be involved in the tonic descending inhibition, at least in part, via alpha$_2$-adrenoceptors.

Cell Origin of the Descending Noradrenergic System

Although there exist direct projections from the NRPG [Basbaum et al., 1978; Goode et al., 1980] and the PAG [Kuyper and Maisky, 1975] to the spinal dorsal horn, there are no NA-containing cells in these brain stem regions [Dahlstrom and Fuxe, 1964; Hokfelt et al., 1984; Palkovits and Jacobowitz, 1974]. It is unlikely, therefore, that these direct descending systems are noradrenergic.

Although Dahlstrom and Fuxe [1964] suggested that caudal medullary catecholamine cell groups (A1–A3, especially the A1 cell group) are the main source of the descending catecholamine systems innervating the spinal cord, subsequent studies claimed that principal sources innervating the spinal cord are the pontine NA cell groups (especially A5 and A6) [Blessing et al., 1981; Loewy et al., 1979; Nygren and Olson, 1977; Satoh et al., 1977]. However, the dense distribution of NA includes the ventral horn, the intermediolateral column, and the dorsal horn [Carlsson et al., 1964; Fleetwood-Walker and Coote, 1981]. Therefore, in order to examine the descending noradrenergic systems regulating nociceptive transmission, it is necessary to know the cell origin innervating mainly the dorsal horn, esecially the laminae I, II and V, regions well known to participate in transmission and regulation of noxious information. There have been few anatomical studies of this problem [Nygren and Olson, 1977; review by Moore and Card, 1984]. In this section, we will review A1 and A6 NA cell groups, the participation of which in pain and analgesia has been pharmacologically examined.

A1 Noradrenaline Cell Group

A1 NA cells are located in the nucleus reticularis lateralis and outside the dorsal or lateral boundary of this nucleus [Blessing et al., 1980; Hokfelt et al., 1984; Moore and Card, 1984; Palkovits and Jacobowitz, 1974]. In the rostrocaudal direction, they appear at the level of the area postrema and continue caudally into the spinal cord [Hokfelt et al., 1984]. From indirect evidence, Nygren and Olson [1977] speculated that the A1 cells project to laminae I–III of the dorsal horn and the sympathetic lateral column, but not to the ventral horn. If this is the case, the findings that morphine [Shiomi and Takagi, 1974] and noxious stimulation [Takagi et al., 1979] increased the concentration of normetanephrine in the dorsal half but not in the ventral half of the dorsal horn would implicate the A1 cells as a source of the descending noradrenergic system participating in pain inhibition. Previously, Pert et al. [1980] have reported that the antinociceptive effect of morphine was reduced by administration of 6-hydroxydopamine into the nucleus reticularis lateralis. Since the rostral part of the A1 cell group projects mainly to the intermediolateral column but not to the dorsal horn [Blessing et al., 1981; McKellar and Loewy, 1982], the outer laminae of the dorsal horn might receive the NA fibers from the caudal part of the A1 cell group. In fact, electrical stimulation of the caudal but not rostral part of the nucleus reticularis lateralis inhibited selectively the excitation of the wide-dynamic range neurons (in laminae IV and V) elicited by electrical stimulation of the C fibers [Morton et al., 1983]. In contrast, bilateral electrical lesions of the caudal but not rostral part of the nucleus reticularis lateralis reduced the tonic descending inhibition of the same dorsal horn neurons [Hall et al., 1982]. Although further studies into this problem are necessary, the above findings raise the possibility that the A1 NA cell group is an origin of the descending noradrenergic system participating in antinociception.

A6 Noradrenaline Cell Group

A6 NA cells are located in the locus coeruleus [Hokfelt et al., 1984; Moore and Card, 1984; Palkovits and Jacobowitz, 1974] and are claimed to be the main source of the spinal noradrenaline fibers and terminals [Commissioning et al., 1978; Nygren and Olson, 1977; Satoh et al., 1977]. According to Nygren and Olson [1977], the NA fibers originating the A6 cell group descend through the anterior funiculus and the ventral part of the lateral funiculus, and mainly terminate in the deeper layers (laminae IV and V) and in the ventral horn (laminae VII–IX).

Electrical stimulation of the locus coeruleus produces antinociception in rats [Segal and Sandberg, 1977], and inhibits nociceptive responses of neurons in laminae IV and V [Hodge et al., 1981]. In contrast, an electrical lesion of the locus coeruleus reduces morphine antinociception [Sasa et al., 1977]. These findings suggest that the activation of neurons in the locus coeruleus is accompanied by antinociception. However, the inhibitory effect of electrical stimulation of the locus coeruleus on the spinal dorsal horn neurons is not related to a catecholaminergic coeruleo-spinal system in the cat [Hodge et al., 1983]. Moreover, the stimulation-produced antinociception of the locus coeruleus was not suppressed by the blockade of the catecholaminergic synapses with alpha-methyl-*p*-tyrosine, whereas it was reduced by the blockade of the serotonergic synapses with methysergide and *p*-chlorophenylalanine [Sandberg and Segal, 1978]. Thus, it is unlikely that the stimulation-produced antinociception of the locus coeruleus is primarily mediated by the noradrenergic system. Moreover, locus coeruleus neurons are depressed by iontophoretic application of morphine [Bird and Kuhar, 1977] and Met-enkephalin [Young et al., 1977], and by systemic administration of morphine [Korf et al., 1974]. Furthermore, innervation of the ventral as well as dorsal horn by the A6 cell group is inconsistent with the findings that morphine [Shiomi and Takagi, 1974] and noxious stimulation [Takagi et al., 1979] increased the concentration of normetanephrine in the dorsal half but not in the ventral half of the spinal cord. Thus, taken together, these results cast doubt on the role in antinociception of the descending noradrenergic system arising from the A6 cell group. The A6 cell group is a main source of the ascending as well as descending noradrenergic systems [Moore and Card, 1984], and the ascending noradrenergic system has been considered to antagonize antinociception produced by opiates and electrical brain stimulation [Akil and Liebeskind, 1975; Bodnar et al., 1978; Cicero et al., 1974; Khachaturian and Watson, 1982; Sewell and Spencer, 1975; Watson et al., 1977].

Thus, the precise cell origin of the descending noradrenergic systems involved in antinociception remains to be elucidated. For this purpose, systematic pharmacological and physiological experiments combined with histochemistry should be conducted.

Type of Noxious Stimuli and Noradrenaline Antinociception – Comparison with Serotonin

It is apparent that there are at least three kinds of pain-inhibiting neurotransmitters, including NA, in the dorsal horn; the others are serotonin [Belcher et al., 1978; Besson, in this issue; Headley et al., 1978; Kuraishi et al., 1979b, 1983a; Shiomi et al., 1978; Yaksh, 1979; Yaksh and Wilson, 1979] and Met-enkephalin [Satoh et al., 1979b; Yaksh et al., 1977]. Why are three separate systems required for pain regulation? Do they have different roles in the regulation of processing of noxious information in the dorsal horn?

The descending monoaminergic systems form a part of the spinal-reticulospinal inhibitory systems [Takagi et al., 1979; Tyce and Yaksh, 1981], while the spinal enkephalin system may form a part of the segmental inhibitory system [Yaksh and Elde, 1981].

Intrathecal injections of NA [Howe et al., 1983; Hylden and Wilcox, 1983; Kuraishi et al., 1979b; Reddy et al., 1980; Reddy and Yaksh, 1980; Takagi et al., 1980], serotonin [Kuraishi et al., 1979b; Takagi et al., 1980; Wang, 1977; Yaksh and Wilson, 1979] and opioids [Kuraishi et al., 1980; Takagi et al., 1980; Yaksh and Rudy, 1977; Yaksh et al., 1977] produce antinociception in the mechanical (tail-pinch) and thermal (hot-plate and tail-flick) algesic tests. However, the comparative study [Kuraishi et al., 1985b] on relative sensitivities of mechanical and thermal algesic tests in rats to NA, serotonin and opioids revealed that intrathecal NA inhibited responses to nociceptive mechanical (to the tail-pinch) more potently than nociceptive thermal stimuli (to the tail-flick and hot-plate). In contrast, intrathecal serotonin, Met-enkephalin and morphine more potently inhibited the responses to thermal rather than mechanical nociceptive stimuli.

With regard to morphine antinociception, the mechanical but not the thermal antinociception of systemically given morphine was suppressed following the selective reduction of spinal NA by intrathecal injection of 6-hydroxydopamine. On the other hand, thermal but not mechanical antinociception was suppressed following selective reduction of spinal serotonin by intrathecal injection of 5,6-dihydroxytryptamine [Kuraishi et al., 1983a]. Although the thermal antinociception of morphine injected into the PAG was inhibited by intrathecal injection of either phentolamine or methysergide [Yaksh, 1979], the mechanical antinociception of intra-PAG morphine was suppressed after the selective reduction of spinal NA by intrathecal injection of 6-hydroxydopamine, but not after the selective reduction of spinal serotonin by intrathecal injection of 5,6-dihydroxytryptamine [Kuraishi et al., 1983a].

More remarkable results were obtained from the experiments on the NRGC/NRPG, the activation of which excites the descending noradrenergic, but not the serotonergic, system [Kuraishi et al., 1978; Yaksh et al., 1981] (see the section 'nucleus reticularis gigantocellularis and paragigantocellularis'). When the mechanical algesic (tail-pinch) test was used, antinociceptive effects were produced by relatively small doses (15–380 ng/rat) of morphine injected into the NRGC/NRPG [Takagi et al., 1977]; when the injection sites were restricted to the NRPG, much smaller doses (0.15–15 ng/rat) were effective [Akaike et al., 1978]. On the other hand, when the thermal algesic (tail-flick) test was used, larger doses (1,000 and 5,000 ng/rat) of morphine injected into the NRPG were required to produce antinociception [Azami et al., 1982], and a ceiling effect was seen [Yaksh, personal commun.].

All these findings taken together suggest that the descending noradrenergic system exerts more potent inhibition on the mechanical nociceptive system than the thermal one in the spinal dorsal horn, while the descending serotonergic system may inhibit thermal nociception more potently than mechanical nociception. Such differences between the descending noradrenergic and serotonergic systems seem to be explained in part by the difference in presynaptic inhibition on the primary nociceptive afferents by NA and serotonin.

There is substantial evidence that substance P is a neurotransmitter of the nociceptive primary afferent [Henry, 1982; Jessell, 1982; Willis, 1985]. The *in situ* release of substance P from the rabbit dorsal horn was increased by noxious but not non-noxious mechanical stimulation, and not by other skin stimulation, such as noxious and non-noxious thermal stimulation, suggesting that this peptide is released from the terminals of the mechanical nociceptive afferents [Kuraishi et al., 1985a]. This evoked release of substance P was completely blocked by local administration of NA, an effect that was antagonized by alpha-blockers yohimbine and prazosin [Kuraishi et al., 1985c]. Serotonin did not significantly affect the evoked release of substance P [Kuraishi et al., 1984a]. Moreover, suppression of the evoked substance P release by systemic administration of morphine was completely blocked by direct administration of prazosin but not methysergide, to the dorsal horn [Kuraishi et al., 1983b].

Noxious thermal skin stimulation increased the release of somatostatin from the rabbit dorsal horn. This release was not affected by noxious mechanical and non-noxious thermal stimulation [Kuraishi et al. 1985a]. Preliminary experiments conducted in our laboratory showed that suppression of the evoked release of somatostatin by systemic administration of morphine was partially antagonized by direct administration of a serotonin antagonist to the dorsal horn [unpubl. observation].

From the data mentioned above, it is conceivable that the activation of the descending noradrenergic system by either morphine or electrical brain stimulation and intrathecal administration of NA itself presynaptically inhibits the terminals of the mechanical nociceptive afferents, while the activation of the descending serotonergic system presynaptically inhibits the terminals of the thermal nociceptive afferents. These presynaptic actions of NA and serotonin probably contribute to the production of mechanical and thermal antinociception, respectively.

Conclusions

The evidence clearly indicates that the descending noradrenergic system plays a role in physiological pain regulation and permits us to propose the following schema (fig. 2). The descending noradrenergic system constitutes a part of the spinal-reticulospinal negative feedback loop. Under normal conditions, the descending noradrenergic system

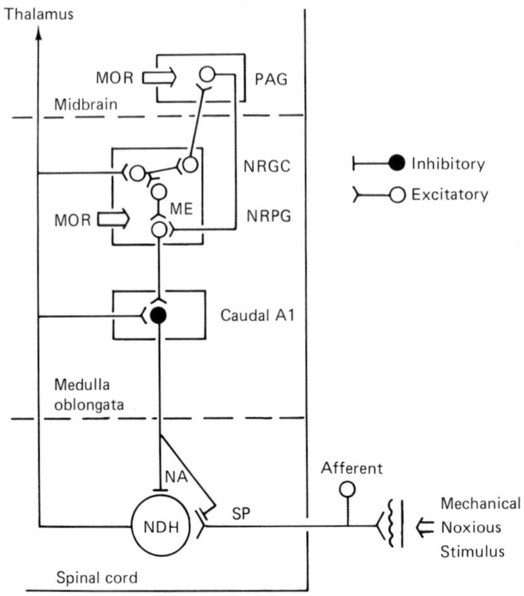

Fig. 2. Hypothesis concerning the role of the descending noradrenergic system in pain regulation and morphine analgesia. PAG = Periaqueductal gray matter; NRGC = nucleus reticularis gigantocellularis; NRPG = nucleus reticularis paragigantocellularis; A1 = A1 noradrenaline cell group; NDH = nociceptive dorsal horn neuron; MOR = morphine; ME = Met-enkephalin; NA = noradrenaline; SP = substance P.

slightly inhibits nociceptive transmission in the dorsal horn. Acute noxious stimulation activates this system, which is not mediated by the opioid peptide system. Sustained, strong noxious stimulation, however, activates the Met-enkephalin system in the lower brain stem, such as NRGC/NRPG (and perhaps PAG), and further accelerates the activity of the descending noradrenergic system to block some overflow of noxious information. The inhibition of nociceptive transmission by NA is achieved by the postsynaptic inhibition of the dorsal horn neurons and the presynaptic inhibition of the nociceptive, especially mechanical, primary afferents. This inhibition appears mediated by alpha$_1$- and/or alpha$_2$-adrenoceptors in the spinal dorsal horn. Morphine acts on the PAG and NRGC/NRPG neurons activating the descending noradrenergic system, probably playing an important role in the production of mechanical antinociception, and to lesser extent thermal antinociception.

Pain transmission, however, is not exclusively regulated by the descending noradrenergic system. As widely reviewed in this issue, many neuronal systems and many types of neurotransmitters can interact to regulate pain perception. Within the spinal cord, there may be at least four inhibitory systems involved in pain regulation; the descending noradrenergic system, the descending serotonergic system, the direct NRPG-spinal system and the spinal enkephalinergic system.

Electrophysiological studies have classified the nociceptors and their axons into three or more types [Willis, 1985]. However, we have as yet very little information as to the neurotransmitters of these nociceptive afferents and mechanisms regulating nociceptive transmission mediated by each of such transmitters. For a better understanding of pain-inhibitory as well as pain-transmitting systems, these problems should be elucidated.

References

Abols, I.A.; Basbaum, A.I.: Afferent connections of the rostral medulla of the cat. A neural substrate for midbrain-medullary interactions in the modulation of pain. J. comp. Neurol. *201:* 285–297 (1981).

Akaike, A.; Shibata, T.; Satoh, M.; Takagi, H.: Analgesia induced by microinjection of morphine into, and electrical stimulation of, the nucleus reticularis paragigantocellularis of rat medulla oblongata. Neuropharmacology *17:* 775–778 (1978).

Akil, H.; Liebeskind, J.C.: Monoaminergic mechanism of stimulation-produced analgesia. Brain Res. *94:* 279–296 (1975).

Andrezik, J.A.; Chan-Palay, V.; Palay, S.L.: The nucleus paragigantocellularis lateralis in the rat. Conformation and cytology. Anat. Embryol. *161:* 355–371 (1981).

Axelsson, K.; Widman, B.: Blood concentration of lidocaine after spinal anaesthesia using lidocaine and lidocaine with adrenaline. Acta anaesth. scand. *25:* 240–245 (1981).

Azami, J.; Llewelyn, M.B.; Roberts, M.H.T.: The contribution of nucleus reticularis paragigantocellularis and nucleus raphe magnus to the analgesia produced by systemically administered morphine, investigated with the microinjection technique. Pain *12:* 229–246 (1982).

Azami, J.; Wright, D.M.; Roberts, M.H.T.: Effects of morphine and naloxone on the responses to noxious stimulation of neurones in the nucleus reticularis paragigantocellularis. Neuropharmacology *20:* 869–876 (1981).

Basbaum, A.I.; Clanton, C.H.; Fields, H.L.: Three bulbospinal pathways from the rostral medulla of the cat. An autoradiographic study of pain modulating systems. J. comp. Neurol. *178:* 209–224 (1978).

Belcher, G.; Ryall, R.W.; Schaffner, R.: The differential effects of 5-hydroxytryptamine, noradrenaline and raphe stimulation on nociceptive and non-nociceptive dorsal horn interneurones in the cat. Brain Res. *151:* 307–321 (1978).

Bennett, G.J.; Mayer, D.J.: Inhibition of spinal cord interneurons by narcotic microinjection and focal electrical stimulation in the periaqueductal gray matter. Brain Res. *172:* 243–257 (1979).

Berman, A.L.: The brain stem of the cat. A cytoarchitectonic atlas with stereotaxic coordinates (University of Wisconsin Press, Madison 1968).

Bird, S.J.; Kuhar, M.J.: Iontophoretic application of opiates to the locus coeruleus. Brain Res. *122:* 523–533 (1977).

Blessing, W.W.; Frost, P.; Furness, J.B.: Catecholamine cell groups of the cat medulla oblongata. Brain Res. *192:* 69–75 (1980).

Blessing, W.W.; Goodchild, A.K.; Dampney, R.A.L.; Chalmers, J.P.: Cell groups in the lower brain stem of the rabbit projecting to the spinal cord, with special reference to catecholamine-containing neurons. Brain Res. *221:* 35–55 (1981).

Bodnar, R.J.; Ackermann, R.F.; Kelly, D.D.; Glusman, M.: Elevations in nociceptive thresholds following locus coeruleus lesions. Brain Res. Bull. *3:* 125–130 (1978).

Bohr, D.F.; Goulet, P.L.; Taquini, A.C.: Direct tension recording from smooth muscle of resistance vessels from various organs. Angiology *12:* 478–485 (1961).

Binica, J.J.; Backup, P.H.; Pratt, W.H.: The use of vasoconstrictors to prolong spinal anesthesia. Anesthesiology *12:* 431–441 (1951).

Carlsson, A.; Dahlstrom, A.; Fuxe, K.; Hillarp, N.-A.: Cellular localization of monoamines in the spinal cord. Acta physiol. scand. *60:* 112–119 (1964).

Carstens, E.; Yokota, T.; Zimmermann, M.: Inhibition of spinal neuronal responses to noxious skin heating by stimulation of mesencephalic periaqueductal gray in the cat. J. Neurophysiol. *42:* 558–568 (1979).

Casey, K.L.: Escape elicited by bulboreticular stimulation in the cat. Int. J. Neurosci. *2:* 29–34 (1971).

Casey, K.L.; Morrow, T.J.: Supraspinal pain mechanisms in the cat; in Kitchell, Erickson, Animal pain. Perception and alleviation, pp. 63–81 (American Physiological Society, Waverly Press, Baltimore 1983).

Cicero, T.J.; Meyer, E.R.; Smithloff, B.R.: Alpha-adrenergic blocking agents: antinociceptive activity and enhancement of morphine-induced analgesia. J. Pharmac. exp. Ther. *189:* 72–82 (1974).

Commissioning, J.W.; Hellstrom, S.O.; Neff, N.H.: A new projection from locus coeruleus to the spinal ventral columns: histochemical and biochemical evidence. Brain Res. *148:* 207–213 (1978).

Converse, J.G.; Landmesser, C.M.; Harmel, M.H.: The concentration of pentocaine hydrochloride in the cerebrospinal fluid during spinal anesthesia, and the influence of epinephrine in prolonging the sensory anesthetic effect. Anesthesiology *15:* 1–10 (1954).

Crawford, R.A.; Griffiths, I.R.; McCulloch, J.: The effect of norepinephrine on the spinal cord circulation and its possible implications in the pathogenesis of acute spinal trauma. J. Neurosurg. *47:* 567–576 (1977).

Dahlstrom, A.; Fuxe, K.: Evidence for the existence of monoamine neurons in the central nervous system. I. Demonstration of monamines in the cell bodies of brain stem neurons. Acta physiol. scand. *62:* suppl. 232, pp. 1–155 (1964).

Dennis, S.G.; Melzack, R.; Gutman, S.; Boucher, F.: Pain modulation by adrenergic agents and morphine as measured by three pain tests. Life Sci. *26:* 1247–1259 (1980).

Dubuisson, D.; Dennis, S.G.: The formalin test: a quantitative study of the analgesic effects of morphine, meperidine and brain stem stimulation in rats and cats. Pain *4:* 161–174 (1977).

Egbert, L.D.; Deas, T.C.: Effect of epinephrine upon the duration of spinal anesthesia. Anesthesiology *21:* 345–347 (1960).

Fields, H.L.; Basbaum, A.I.: Brain stem control of spinal pain-transmission neurons. A. Rev. Physiol. *40:* 217–248 (1978).

Fifkova, E.; Marsala, J.: Steretaxic atlases for the cat, rabbit and rat; in Bures, Petrans, Zacher, Electrophysiological methods in biological research, pp. 653–731 (Academic Press, New York 1967).

Fleetwood-Walker, S.M.; Coote, J.H.: Contribution of noradrenaline-, dopamine- and adrenaline-containing axons to the innervation of different regions of the spinal cord of the cat. Brain Res. *206:* 95–106 (1981).

Fleetwood-Walker, S.M.; Mitchell, R.; Hope, P.J.; Molony, V.; Iggo, A.: An $alpha_2$ receptor mediates the selective inhibition by noradrenaline of nociceptive responses of identified dorsal horn neurones. Brain Res. *334:* 243–254 (1985).

Goode, G.E.; Humbertson, A.O.; Martin, G.F.: Projections from the brain stem reticular formation to laminae I and II of the spinal cord. Studies using light and electron microscopic techniques in the North American opossum. Brain Res. *189:* 327–342 (1980).

Hall, J.G.; Duggan, A.W.; Morton, C.R.; Johnson, S.M.: The location of brain stem neurones tonically inhibiting dorsal horn neurones of the cat. Brain Res. *244:* 215–222 (1982).

Headley, P.M.; Duggan, A.W.; Griersmith, B.T.: Selective reduction by noradrenaline and 5-hydroxytryptamine of nociceptive responses of cat dorsal horn neurones. Brain Res. *145:* 185–189 (1978).

Henry, J.L.: In Porter, O'Conner, Relation of substance P to pain transmission: neurophysiological evidence, pp. 206–217 (Pitman, London 1982).

Hodge, C.J., Jr.; Apkarian, A.V.; Stevens, R.; Vogelsang, G.; Wisnicki, H.J.: Locus coeruleus modulation of dorsal horn unit responses to cutaneous stimulation. Brain Res. *204:* 415–420 (1981).

Hodge, C.J., Jr.; Apkarian, A.V.; Stevens, R.T.; Vogelsang, G.D.; Brown, O.; Franck, J.I.: Dorsolateral pontine inhibition of dorsal horn cell responses to cutaneous stimulation. Lack of dependence on catecholaminergic system in cat. J. Neurophysiol. *50:* 1220–1235 (1983).

Hokfelt, T.; Martensson, R.; Bjorklund, A.; Kleinau, S.; Goldstein, M.: Distribution of maps of tyrosine-hydroxylase immunoreactive neurons in the rat brain; in Bjorklund, Hokfelt, Handbook of chemical neuroanatomy, vol. 2. Classical transmitters in the CNS, part I, pp. 277–379 (Elsevier, Amsterdam 1984).

Hosobuchi, Y.; Adams, J.E.; Linchitz, R.: Pain relief by electrical stimulation of the central gray matter in humans and its reversal by naloxone. Science *197:* 183–186 (1977).

Howe, J.R.; Wang, J.-Y.; Yaksh, T.L.: Selective antagonism of the antinociceptive effect of intrathecally applied alpha adrenergic agonists by intrathecal prazosin and intrathecal yohimbine. J. Pharmac. exp. Ther. *224:* 552–558 (1983).

Hylden, J.L.K.; Wilcox, G.L.: Pharmacological characterization of substance P-induced nociception in mice. Modulation by opioid and noradrenergic agonists at the spinal level. J. Pharmac. exp. Ther. *226:* 398–404 (1983).

Jeftinija, S.; Semba, K.; Randic, M.: Norepinephrine reduces excitability of single cutaneous primary afferent C-fibers in the cat spinal cord. Brain Res. *219:* 456–463 (1981).

Jessell, T.M.: Substance P in nociceptive sensory neurons; in Porter, O'Connor, Substance P in the nervous system, pp. 225–240 (Pitman, London 1982).

Keene, J.J.; Casey, K.L.: Excitatory connection from lateral hypothalamic self-stimulation sites to escape sites in medullary reticular formation. Expl. Neurol. *28:* 155–166 (1970).

Khachaturian, H.; Watson, S.J.: Some perspectives on monoamine-opioid peptide interaction in rat central nervous system. Brain Res. Bull. *9:* 441–462 (1982).

Korf, J.; Bunney, B.S.; Aghajanian, G.K.: Noradrenergic neurons. Morphine inhibition of spontaneous activity. Eur. J. Pharmacol. *25:* 165–169 (1974).

Kumazawa, T.; Perl, E.R.: Primate cutaneous sensory units with unmyelinated (C) afferent fibers. J. Neurophysiol. *40:* 1325–1338 (1977).

Kuraishi, Y.; Fukui, K.; Shiomi, H.; Akaike, A.; Takagi, H.: Microinjection of opioids into the nucleus reticularis gigantocellularis of the rat. Analgesia and increase in the normetanephrine level in the spinal cord. Biochem. Pharmac. *27:* 2756–2758 (1978).

Kuraishi, Y.; Harada, Y.; Satoh, M.; Takagi, H.: Antagonism by phenoxybenzamine of the analgesic effect of morphine injected into the nucleus reticularis gigantocellularis of the rat. Neuropharmacology *18:* 107–110 (1979a).

Kuraishi, Y.; Harada, Y.; Takagi, H.: Noradrenaline regulation of pain-transmission in the spinal cord mediated by alpha-adrenoceptors. Brain Res. *174:* 333–336 (1979b).

Kuraishi, Y.; Satoh, M.; Harada, Y.; Akaike, A.; Shibata, T.; Takagi, H.: Analgesic action of intrathecal and intracerebral beta-endorphin in rats. Comparison with morphine. Eur. J. Pharmacol. *67:* 143–146 (1980).

Kuraishi, Y.; Harada, Y.; Aratani, S.; Satoh, M.; Takagi, H.: Separate involvement of the spinal noradrenergic and serotonergic systems in morphine analgesia. The differences in mechanical and thermal algesic tests. Brain Res. *273:* 245–252 (1983a).

Kuraishi, Y.; Hirota, N.; Sugimoto, M.; Satoh, M.; Takagi, H.: Effects of morphine on noxious stimuli-induced release of substance P from rabbit dorsal horn *in vivo.* Life Sci. *33:* suppl. I, pp. 693–696 (1983b).

Kuraishi, Y.; Hirota, N.; Hino, Y.; Sato, Y.; Satoh, M.; Takagi, H.: Noradrenaline inhibits noxious pinch-induced release of substance P from the rabbit dorsal horn *in situ.* Pain, suppl. 2, p. 277 (1984a).

Kuraishi, Y.; Sugimoto, M.; Hamada, T.; Kayanoki, Y.; Takagi, H.: Noxious stimuli and Met-enkephalin release from nucleus reticularis gigantocellularis. Brain Res. Bull. *12:* 123–127 (1984b).

Kuraishi, Y.; Hirota, N.; Sato, Y.; Nino, Y.; Satoh, M.; Takagi, H.: Evidence that substance P and somatostatin transmit separate information related to pain in the spinal dorsal horn. Brain Res. *325:* 294–298 (1985a).

Kuraishi, Y.; Hirota, N.; Satoh, M.; Takagi, H.: Antinociceptive effects of intrathecal opioids, noradrenaline and serotonin in rats. Mechanical and thermal algesic tests. Brain Res. *326:* 168–171 (1985b).

Kuraishi, Y.; Hirota, N.; Sato, Y.; Kaneko, S.; Satoh, M.; Takagi, H.: Noradrenergic inhibition of the release of substance P from the primary afferents in the rabbit spinal dorsal horn. Brain Res. *359:* 177–182 (1985c).

Kuypers, H.G.J.M.; Maisky, V.A.: Retrograde axonal transport of horseradish peroxidase from spinal cord to brain stem cell groups in the cat. Neurosci. Lett. *1:* 9–14 (1975).

Loewy, A.D.; McKellar, S.; Saper, C.B.: Direct projections from the A5 catecholamine cell group to the intermediolateral cell column. Brain Res. *174:* 309–314 (1979).

Mayer, D.J.; Price, D.D.: Central nervous system mechanisms of analgesia. Pain *2:* 379–404 (1976).

McKellar, S.; Loewy, A.D.: Efferent projections of the A1 catecholamine cell group in the rat. An autoradiographic study. Brain Res. *241:* 11–29 (1982).

Messen, H.; Olszewski, J.: Cytoarchitectonic atlas of the rhombencephalon of the rabbit (Karger, Basel 1949).

Mohrland, J.S.; Gebhart, G.F.: Effect of focal electrical stimulation and morphine microinjection in the periaqueductal gray of the rat mesencephalon on neuronal activity in the medullary reticular formation. Brain Res. *201:* 23–37 (1980).

Mohrland, J.S.; Gebhart, G.F.: Effect of morphine administered in the periaqueductal gray and at the recording locus on nociceptive neurons in the medullary reticular formation. Brain Res. *225:* 401–412 (1981).

Moore, R.Y.; Card, J.P.: Noradrenaline-containing neuron systems; in Bjorklund, Hokfelt, Handbook of chemical neuroanatomy, vol. 2. Classical transmitters in the CNS, part I, pp. 123–156 (Elsevier, Amsterdam 1984).

Morton, C.R.; Johnson, S.M.; Duggan, A.W.: Lateral reticular regions and the descending control of dorsal horn neurones of the cat. Selective inhibition by electrical stimulation. Brain Res. *275:* 13–21 (1983).

North, R.A.; Yoshimura, M.: The actions of noradrenaline on neurones of the rat substantia gelatinosa *in vitro.* J. Physiol., Lond. *349:* 43–55 (1984).

Nygren, L.-G.; Olson, L.: A new major projection from locus coeruleus: the main source of noradrenergic nerve terminals in the ventral and dorsal columns of the spinal cord. Brain Res. *132:* 85–93 (1977).

Oliveras, J.L.; Besson, J.M.; Guilbaud, G.; Liebeskind, J.C.: Behavioral and electrophysiological evidence of pain inhibition from midbrain stimulation in the cat. Expl Brain Res. *20:* 32–44 (1974).

Palkovits, M.; Jacobowitz, D.M.: Topographic atlas of catecholamine and acetylcholinesterase-containing neurons in the rat brain. II. Hindbrain (mesencephalon, rhombencephalon). J. comp. Neurol. *157:* 29–42 (1974).

Park, W.Y.; Balingit, P.E.; Macnamara, T.E.: Effects of patient age, pH of cerebrospinal fluid, and vasopressor on onset and duration of spinal anesthesia. Anesth. Analg. *54:* 455–458 (1975).

Paxinos, G.; Watson, C.: The rat brain in stereotaxic coordinates (Academic Press, New York 1982).

Pert, A.; Massari, V.J.; Tizabi, Y.; O'Donohue, T.L.; Jacobowitz, D.: Effects of 6-hydroxydopamine and 5,7-dihydroxytryptamine brain stem lesions on morphine analgesia in the rat; in Way, Endogenous and exogenous opiate agonists and antagonists, pp. 151–154 (Pergamon Press, New York 1980).

Pompeiano, O.: Reticular formation; in Iggo, Handbook of sensory physiology, vol. II, Somatosensory system, pp. 381–488 (Springer, Berlin 1973).

Priddle, H.D.; Andros, G.J.: Primary spinal anesthetic effects of epinephrine. Curr. Res. Anesth. Analg. *29:* 156–162 (1950).

Proudfit, H.K.; Hammond, D.L.: Alterations in nociceptive threshold and morphine-induced analgesia produced by intrathecally administered amine antagonists. Brain Res. *218:* 393–399 (1981).

Reddy, S.V.R.; Maderdrut, J.L.; Yaksh, T.L.: Spinal cord pharmacology of adrenergic agonist-mediated antinociception. J. Pharmac. exp. Ther. *213:* 525–533 (1980).

Reddy, S.V.R.; Yaksh, T.L.: Spinal noradrenergic terminal system mediates antinociception. Brain Res. *189:* 391–401 (1980).

Reynolds, D.V.: Surgery in the rat during electrical analgesia induced by focal brain stimulation. Science *164:* 444–445 (1969).

Richardson, D.E.; Akil, H.: Pain reduction by electrical brain stimulation in man. Acute administration in periaqueductal and periventricular sites. J. Neurosurg. *47:* 184–194 (1977).

Rothballer, A.B.: The effects of catecholamines on the central nervous system. Pharmac. Rev. *11:* 494–547 (1959).

Sagan, J.; Proudfit, H.K.: Effect of intrathecally administered noradrenergic antagonists on nociception in the rat. Brain Res. *310:* 295–301 (1984).

Sandberg, D.E.; Segal, M.: Pharmacological analysis of analgesia and self-stimulation elicited by electrical stimulation of catecholamine nuclei in the rat brain. Brain Res. *152:* 529–542 (1978).

Sasa, M.; Munekiyo, K.; Osumi, Y.; Takaori, S.: Attenuation of morphine analgesia in rats with lesions of the locus coeruleus and dorsal raphe nucleus. Eur. J. Pharmacol. *42:* 53–62 (1977).

Satoh, K.; Tohyama, M.; Yamamoto, K.; Sakumoto, T.; Shimizu, N.: Noradrenaline innervation of the spinal cord studied by the horseradish peroxidase method combined with monoamine oxidase staining. Expl Brain Res. *30:* 175–186 (1977).

Satoh, M.; Akaike, A.; Takagi, H.: Excitation by morphine and enkephalin of single neurons of nucleus reticularis paragigantocellularis in the rat. A probable mechanism of analgesic action of opioids. Brain Res. *169:* 406–410 (1979a).

Satoh, M.; Kawajiri, S.; Ukai, Y.; Yamamoto, M.: Selective and non-selective inhibition by enkephalins and noradrenaline of nociceptive response of lamina V type neurons in the spinal dorsal horn of the rabbit. Brain Res. *177:* 384–387 (1979b).

Satoh, M.; Akaike, A.; Nakazawa, T.; Takagi, H.: Evidence for involvement of separate mechanisms in the production of analgesia by electrical stimulation of the nucleus reticularis paragigantocellularis and nucleus raphe magnus in the rat. Brain Res. *194:* 525–529 (1980).

Satoh, M.; Kubota, A.; Iwama, T.; Wada, T.; Yasui, M.; Fujibayashi, K.; Takagi, H.: Comparison of analgesic potencies of mu, delta and kappa agonists locally applied to various CNS regions relevant to analgesia in rats. Life Sci. *33:* suppl. I, pp. 689–692 (1983a).

Satoh, M.; Oku, R.; Akaike, A.: Analgesia produced by microinjection of *L*-glutamate into the rostral ventromedial bulbar nuclei of the rat and its inhibition by intrathecal alpha-adrenergic blocking agents. Brain Res. *261:* 361–364 (1983b).

Segal, M.; Sandberg, D.: Analgesia produced by electrical stimulation of catecholamine nuclei in the rat brain. Brain Res. *123:* 369–372 (1977).

Sewell, R.D.E.; Spencer, P.S.J.: Anti-nociceptive activity of narcotic agonists and partial agonists in mice given biogenic amines by intracerebroventricular injection. Psychopharmacologia *42:* 67–71 (1975).

Shiomi, H.; Takagi, H.: Morphine analgesia and bulbospinal noradrenergic system. Increase in the concentration of normetanephrine in the spinal cord of the rat caused by analgesics. Br. J. Pharmacol. *52:* 519–526 (1974).
Shiomi, H.; Murakami, H.; Takagi, H.: Morphine analgesia and the bulbospinal serotonergic system. Increase in concentration of 5-hydroxyindoleacetic acid in the rat spinal cord with analgesics. Eur. J. Pharmacol. *52:* 335–344 (1978).
Takagi, H.: The nucleus reticularis paragigantocellularis as a site of analgesic action of morphine and enkephalin. Trends Pharmacol. Sci. *1:*182–184 (1980).
Takagi, H.; Critical review of pain relieving procedures including acupuncture; in Yoshida, Hagihara, Ebashi, Advances in pharmacology and therapeutics II, vol. 1, pp. 79–92 (Pergamon Press, New York 1982).
Takagi, H.; Doi, T.; Kawasaki, K.: Effects of morphine, *L*-DOPA and tetrabenazine on the lamina V cells of spinal dorsal horn. Life Sci. *17:* 67–72 (1975).
Takagi, H.; Doi, T.; Akaike, A.: Microinjection of morphine into the medial part of the bulbar reticular formation in rabbit and rat: inhibitory effects of lamina V cells of spinal dorsal horn and behavioral analgesia; in Kosterlitz, Opiates and endogenous opioid peptides, pp. 191–198 (Elsevier/North-Holland/Biomedical Press, Amsterdam 1976).
Takagi, H.; Satoh, M.; Akaike, A.; Shibata, T.; Kuraishi, Y.: The nucleus reticularis gigantocellularis of the medulla oblongata is a highly sensitive site in the production of morphine analgesia in the rat. Eur. J. Pharmacol. *45:* 91–92 (1977).
Takagi, H.; Satoh, M.; Akaike, A.; Shibata, T.; Yajima, H.; Ogawa, H.: Analgesia by enkephalins injected into the nucleus reticularis gigantocellularis of rat medulla oblongata. Eur. J. Pharmacol. *49:* 113–116 (1978).
Takagi, H.; Shiomi, H.; Kuraishi, Y.; Fukui, K.; Ueda, H.: Pain and the bulbospinal noradrenergic system: pain-induced increase in normetanephrine content in the spinal cord and its modification by morphine. Eur. J. Pharmacol. *54:* 99–107 (1979).
Takagi, H.; Kuraishi, Y.; Harada, Y.: Potent antinociceptive effects of intrathecal opioids and catecholamines in rats. Integrative control functions of the brain, vol. 3, pp. 97–99 (Kodansha Scientific, Tokyo 1980).
Tamsen, A.; Gorth, T.: Epidural clonidine produces analgesia. Lancet *ii:* 231–232 (1984).
Toda, N.; Fujita, Y.: Responsiveness of isolated cerebral and peripheral arteries to serotonin, norepinephrine, and transmural electrical stimulation. Circulation Res. *33:* 98–104 (1973).
Tsou, K.; Jang, C.S.: Studies on the site of analgesic action of morphine by intracerebral micro-injection. Scientia sin. *13:* 1099–1109 (1964).
Tyce, G.M.; Yaksh, T.L.: Monoamine release from cat spinal cord by somatic stimuli. An intrinsic modulatory system. J. Physiol., Lond. *314:* 513–529 (1981).
Wang, J.K.: Antinociceptive effect of intrathecally administered serotonin. Anesthesiology *47:* 269–271 (1977).
Watson, S.J.; Akil, H.; Barchas, J.D.: A possible role for the dorsal periventricular catecholamine bundle in stimulation produced analgesia, a behavioral and histochemical study. Brain Res. *130:* 335–342 (1977).
Willis, W.D.: Control of nociceptive transmission in the spinal cord; in Autrum, Ottoson, Perl, Schmidt, Progress in sensory physiology, vol. 3 (Springer, Berlin 1982).
Willis, W.D.: The pain system. The neural basis of nociceptive transmission in the mammalian nervous system (Karger, Basel 1985).

Xie, C.-W.; Tang, J.; Han, J.S.: Central norepinephrine in acupuncture analgesia: differential effects in the rat brain and spinal cord. Acta physiol. sin. *35:* 186–190 (1983).

Yaksh, T.L.: Direct evidence that spinal serotonin and noradrenaline terminals mediate the spinal antinociceptive effects of morphine in the periaqueductal gray. Brain Res. *160:* 180–185 (1979).

Yaksh, T.L.; Elde, R.D.: Factors governing release of methionine enkephalin-like immunoreactivity from mesencephalon and spinal cord of the cat *in vivo.* J. Neurophysiol. *46:* 1056–1075 (1981).

Yaksh, T.L.; Reddy, S.V.R.: Studies on the analgetic effects of intrathecal opiates, alpha-adrenergic agonists and baclofen: Their pharmacology in the primate. Anesthesiology *54:* 451–467 (1981).

Yaksh, T.L.; Rudy, T.A.: Studies on the direct spinal action of narcotics in the production of analgesia in the rat. J. Pharmac. exp. Ther. *202:* 411–428 (1977).

Yaksh, T.L.; Rudy, T.A.: Narcotic analgesics. CNS sites and mechanisms of action as revealed by intracerebral injection techniques. Pain *4:* 299–359 (1978).

Yaksh, T.L.; Wilson, P.R.: Spinal serotonin terminal system mediates antinociception. J. Pharmac. exp. Ther. *208:* 446–453 (1979).

Yaksh, T.L.; Huang, S.P.; Rudy, T.A.: The direct and specific opiate-like effect of Met^5-enkephalin and analogues on the spinal cord. Neuroscience *2:* 593–596 (1977).

Yaksh, T.L.; Hammond, D.L.; Tyce, G.M.: Functional aspects of bulbospinal monoaminergic projections in modulating processing of somatosensory information. Fed. Proc. *40:* 2786–2794 (1981).

Young, W.S.; Bird, S.J.; Kuhar, M.J.: Iontophoresis of methionine-enkephalin in the locus coeruleus area. Brain Res. *129:* 366–370 (1977).

Yasushi Kuraishi, PhD, Department of Pharmacology, Faculty of Pharmaceutical Sciences, Kyoto University, Kyoto 606 (Japan)

Opioid Systems and Analgesia

Pain Headache, vol. 9, p. 129–159 (Karger, Basel 1987)

Opioids and Pain Regulation

J. Lewis, A. Mansour, H. Khachaturian, S.J. Watson, H. Akil

Mental Health Research Institute, University of Michigan,
Ann Arbor, Mich., USA

Opioid peptides were originally isolated and characterized based on their morphine-like properties [Hughes et al., 1975], with the first demonstration of a behavioral effect of these peptides being reported by Belluzi and Stein [1976] following the central administration of enkephalin. Since that time, there have been 10 years of pharmacological and behavioral research, culminating in nearly 20 000 published reports which characterize the functional effects of opioid peptides. While it is widely agreed that opioid peptides can influence responsiveness to pain, one's confidence level rapidly decreases when statements designed to characterize their mechanisms of action are attempted. That is, despite considerable advances in our understanding of the biogenesis of the various opioid peptides, their anatomical distribution, and the characterization of their multiple receptors, a great many fundamental questions remain regarding the mechanisms by which these peptides serve as analgesics. For example: Where are their loci of action? Is there an 'analgesia receptor'? Which peptide(s) is (are) the endogenous pain-killers? What are the stimuli capable of activating these endogenous analgesia systems? Since there exist numerous reviews, even entire volumes, devoted to the topic of opioid peptides, a comprehensive review will not be attempted here. Rather, we hope to provide the reader with some background on the recent developments regarding the biogenesis and anatomical distribution of the various opioid peptides and receptors as well as discuss what we feel are the important questions regarding the function of these systems that remain to be answered.

Biogenesis and Anatomy of Three Families of Opioid Peptides

The fact that Hughes et al. [1975] had identified two rather than one endogenous opioid peptide appeared surprising at the time. Yet, by the end of the decade, the field was struggling with the existence of over ten peptides and their fragments containing either methionine- or leucine-enkephalin (e.g. Tyr-Gly-Gly-Phe-Met or -Leu) at their amino termini, and with various carboxy-terminal extensions. These included opioids with potent opioid-like pharmacological properties (e.g. β-endorphin; BE) [Bloom et al., 1978; Wei et al., 1977], as well as others with unique pharmacological properties (e.g. dynorphin A [Goldstein et al., 1979], alpha-neo-endorphin [Hayushi et al., 1980]). Some fragments were identified with both a methionine and a leucine-enkephalin, or with two methionine enkephalins, one at each terminus [Lewis, R.V. et al., 1980a, b]. While it was suspected that several of these peptides represented biosynthetic intermediates [Lewis, R.V. et al., 1980a], it was not until 1981 that several critical questions concerning the biosynthetic arrangements and relationships between these various enkephalin-containing peptides became clear. Up to that point, the only well-described precursor system was Pro-opiomelanocortin, which gives rise to BE and ACTH [Mains et al., 1977; Roberts and Herbert, 1977]. Fortunately, the 1980s have brought order to the question of opioid biogenesis, and we now have some basic facts to build upon, although we are still far from understanding the subtleties of processing and post-translational events in endogenous opioid systems.

All opioids now appear to be derived from one of three precursors: Pro-opiomelanocortin (POMC) which begets BE, ACTH, and 3 copies of the melanocyte stimulating hormone core; Pro-enkephalin which contains 1 copy of leu- and 6 copies of met-enkephalin; and Pro-dynorphin which contains dynorphin A, dynorphin B, and alpha/beta-neo-endorphin, all of which contain leu-enkephalin. Additionally, since the messenger RNA (mRNA) structures of all three precursors have also been identified [Comb et al., 1982; Nakanishi et al., 1979; Noda et al., 1982; Kakidani et al:, 1982], we now know that these precursors share a number of characteristics, suggesting a common phylogenetic origin followed by gene duplication. For example, they are similar in length, and in all three precursors the opioids are found in the carboxy-terminal domain. The amino-terminal domains of these precursors tend to be long stretches, with no well established functional significance. Further, as mentioned before, the opioid pentapeptide core (i.e. enkephalin) is located at the amino terminus of the eventual product and is often preceded by double basic residues (often lysine-arginine), thus permitting their enzymatic liberation from the precursor. Finally, the mRNAs for all three precursors have been localized primarily in endocrine and neuronal tissue [Dores et al., 1984].

Of interest to the pain researcher is whether these endogenous opioid precursors do indeed give rise to products that modulate pain. Several lines of evidence, including their anatomical localization, suggest that the answer is tentatively yes.

The brain POMC system is perhaps the most straightforward. Its anatomy is relatively simple, and its most active opioid product, the full BE 1–31 (BE 1–31), is among the most potent endogenous analgesics to be described [Bloom et al., 1978; Walker et al., 1977b; Wei et al., 1977]. The biology of BE processing in brain, however, is not completely clear. We know from studies in the pituitary that, depending on the tissue (anterior vs. intermediate lobe), BE can be modified by carboxy-terminal cleavage to produce BE 1–27 (approximately 10 times less potent than BE 1–31 in opioid receptor binding assays), or BE can be acetylated yielding N-acetyl BE 1–31 or N-acetyl BE 1–27 (both of which are totally devoid of opiate-like biological activity) [Akil et al., 1981; Deakin et al., 1980]. Currently, there is some controversy as to whether or not similar mechanisms are operative in the brain. Our work suggests that each brain region contains a unique mixture of forms of BE, usually comprising a large proportion of carboxy-terminal cleaved BE (i.e. BE 1–27 or shorter forms) and a small amount of N-acetylated BE. Thus, depending on the brain region of interest, BE-containing terminals or perikarya may store and release material ranging from extremely opioid-potent (e.g. BE 1–31), moderately opioid active (BE 1–27), to totally opioid inactive (N-acetyl BE). Therefore, to complete our understanding of the function of BE systems, it will be critical to delineate the natural course of processing and cleavage of BE, and the conditions under which one or the other of the forms may be released.

In contrast to POMC, the processing of Pro-enkephalin is much less clear. In fact, very little is known about the fate of Pro-enkephalin in the brain, except for the fact that authentic met- and leu-enkephalin are produced and stored in substantial amounts [Hughes et al., 1975]. As for the role of the enkephalins in pain inhibition, pharmacological experiments have, unfortunately, succeeded in demonstrating only small to modest analgesic effects of the enkephalins [Belluzzi et al., 1976; Urca et al., 1977]. These results, however, are likely due to the rapid cleaving and inactivation of the enkephalins in neural tissue, since the stabilized enkephalin analogues can yield potent and long-lasting analgesia [Pert et al., 1976; Walker et al., 1977b]. Moreover, the possibility that enkephalins are involved in pain regulation remains open for several reasons; (a) In addition to met- and leu-enkephalin, Pro-enkephalin gives rise to several slightly extended forms of enkephalin including met-enkephalin-arg^6-phe^7, met-enkephalin-arg^6-gly^7-leu^8 [Rossier et al., 1980], as well as larger forms such as BAM-12P, BAM-22P [Mizuno et al., 1980], and peptides F and E [Kilpatrick et al., 1981]. It may be that these peptides are more resistant to degradation by enkephalinase [Malfroy et al., 1978; Sullivan et al., 1978, 1980] and hence possess more prolonged biological activity. (b) Although receptor binding assays would suggest that enkephalin typically intereacts with delta opiate receptors, some of the Pro-enkephalin products manifest substantially more affinity for the morphine preferring mu receptor [Quirion et al., 1983], often proposed as the mediator of analgesia [Pasternak et al., 1980a, b; Ward et al., 1982]. (c) It is conceivable that the larger peptides, possibly biosynthetic intermediates, containing two enkephalin copies, may sometimes be released as products. This notion is a completely unexplored question in the brain. Thus, it is quite likely that Pro-enkephalin containing neurons may make and liberate appropriate products which, by interacting with the appropriate opioid receptor in the protected environment of the synapse, could modulate pain responsiveness under physiological conditions.

The issue of biogenesis vis-à-vis production of endogenous analgesics is still different in the case of Pro-dynorphin. The analgesic properties of alpha-neo-endorphin and dynorphin B have not, to our knowledge, been well described and the situation with dynorphin A is particularly puzzling. In our experience, a modified form of dynorphin A, dynorphin 1–13, does not produce analgesia when microinjected into the lateral ventricle or the midbrain central gray, although other behavioral effects can be observed [Walker et al., 1980; see, however, Petrie et al., 1982]. A further complication is introduced by the possibility that dynorphin A appears to have a nonopioid core which itself produces a unique set of effects, often in a direction opposite to those elicited by opioids [Walker et al., 1982a, b; see also Walker this volume]. The lack of analgesic effects of dynorphin A have most often been attributed to its selectivity as a kappa receptor ligand [Chavkin et al., 1982; Corbett et al., 1982; Young et al., 1983] or to its propensity for in vivo degradation [Leslie and Goldstein, 1982]. The potentially countervailing nonopioid effects, however, should not be ignored. It should also be noted that the processing and selective release of dynorphin A has not been fully elucidated in the brain. There is evidence that dynorphin A 1–8 is predominant in certain brain regions [Weber et al., 1982], and we have evidence for significant regional and species differences in the extent of conversion of dynorphin A 1–17 to dynorphin A 1–8 [Dores et al., 1984]. Moreover, in contrast to the kappa selectivity exhibited by dynorphin A 1–17, dynorphin A 1–8 has a substantial affinity for the mu opioid receptor [Quirion and Weiss, 1983; Corbett et al., 1982]. Thus, differential processing of dynorphin A could have significant consequences for its analgesic activity in vivo.

Anatomy of the Three Opioid Families

The distribution of these three precursors has been extensively mapped, and they have been shown to have unique distributions in the central nervous system [for recent review, see Khachaturian et al., 1985a]. The POMC system has two nuclear cell groups; a major system arising from the arcuate nucleus of the hypothalamus [Bloom et al., 1978; Watson et al., 1978], and a second system, more recently described, originating in the medullary nucleus tractus solitarius (NTS) [Schwarzberg and Nakane, 1983]. Emanating from the arcuate cell group is a long projection pathway that courses through a number of limbic structures, the medial periventricular thalamus, the midbrain central gray, and terminates in the region of the locus coeruleus [Khachaturian et al., 1985b]. Additionally, it is possible that many of these same brain areas are also innervated by POMC-containing fibers originating in the NTS, since this nucleus has extensive projections to all the above-mentioned areas [Ricardo and Koh, 1978]. Recently, we have observed BE containing fibers and terminals in the spinal cord and have determined that the source of these fibers is supraspinal [Tsou et al., 1986]. Issues critical to our understanding of the role of BE in pain regulation are the relative contribution of the arcuate nucleus vs. NTS to the POMC input to each of these projectional fields, and the possibility of differential processing of BE into forms of varying opioid potency depending on the cell of origin. Regardless of the source of the projections, the BE systems encompass brain areas classically thought to be critical for pain control, and correspond well to sites which have been independently established to yield analgesia with electrical stimulation [Watson et al., 1977].

Pro-dynorphin and Pro-enkephalin neurons are found in widespread systems with multiple cell groups involved in each [Elde et al., 1976; Khachaturian et al., 1982, 1983a, b; Watson et al., 1982]. In the present context, we shall simply point to a number of observations relevant to pain inhibitory systems. First, both Pro-enkephalin and Pro-dynorphin pathways are situated in several areas consistent with the ability to modulate pain at several levels of the neuraxis. Beginning with primary afferents in the dorsal horn of the spinal cord (particularly rich in enkephalin, but also containing dynorphin), these systems course through multiple brain stem structures, particularly the midbrain central gray, through the medial thalamus, and hypothalamus. Moreover, the widespread distribution of enkephalin-containing perikarya in cortex [Khachaturian, et al., 1983a; McGinty et al., 1984] coupled with the region-specific patterns of multiple opioid receptors [Lewis, M.E. et al., 1981, 1983c] is suggestive of a role of these peptides in the modulation of higher functions, possibly including pain perception.

Opioid peptides are not only present in areas involved in pain modulation, they are also found in areas classically associated with stress responses in general. For example, BE is present in the corticotrophs of the anterior pituitary, Pro-enkephalin is in the adrenergic chromaffin cells of the adrenal medulla, and dynorphin is co-stored with arginine-vasopressin in the magnocellular neurons of the hypothalamus [Watson et al., 1981]. (Note that vasopression is not only involved in the control of water retention through the posterior pituitary pathway, but is a well known secretogogue of ACTH from the anterior lobe corticotrophs [Vale et al., 1983].) In addition, all three opioid precursors are found in NTS, an area classically associated with regulation of vagal and other autonomic functions. Thus, the anatomy of these opioids suggests that not only do they play a role in the inhibition of pain, but are importantly involved in many of the adaptive autonomic and endocrinological responses to stress. The picture that emerges from our current view of the biology of endogenous opioid systems, in combination with the pharmacology of the opiate alkaloids, is that of a system responsible for coping at a most basic survival level.

Multiple Opioid Receptor Subtypes

Several lines of evidence have suggested the existence of multiple opioid receptor subtypes in the central and peripheral nervous system [Chang and Cuatracasas, 1981; Robson et al., 1983; Zukin and Zukin, 1984]. These subtypes have been referred to as mu, delta and kappa. More recently, some investigators have suggested the existence of additional subtypes (e.g. Sigma) as well as subclasses within the subtypes mentioned above (e.g. mu_1, mu_2, $kappa_1$, $kappa_2$). For the purposes of this chapter we will only discuss the distribution of the more widely accepted subtypes, namely, the mu, delta and kappa sites, within the classical pain pathways. In addition, as species-differences are a major consideration in describing the distribution of opioid receptors [Maurer, 1982], the following review concentrates on the rat. Reviews of the opioid receptor distributions in the human [Pfeiffer et al., 1982; Maurer et al., 1983], monkey [Wamsley et al., 1982; Wise and Herkenham, 1982; Lewis et al., 1984] and the mouse [Moskowitz and Goodman, 1984] are presented elsewhere.

The distribution of the opioid receptor subtypes can be differentiated at several levels of the neuroaxis and particularly in areas involved in the processing of nociceptive information. At the level of the spinal cord, mu binding is predominant in the superficial layers of the dorsal horn in the substantia gelantinosa [Atweh and Kuhar, 1977a; Lamotte et al., 1976] with no marked differences in the density of sites at the various levels of the spinal cord [Gouarderes et al., 1985]. Delta and kappa binding, while also concentrated in the superficial layers of the dorsal horn, are differentially localized within the various segments of the cord [Quirion et al., 1984]. Whereas delta binding is highest in the cervical segments, with relatively low densities of binding in the lumbosacral region, kappa binding shows the reverse pattern, being particularly enriched in the lumbar and sacral portions of the cord [Czlonkowksi et al., 1983, Gouarderes et al., 1985]. As one moves rostrally into the medulla, there is dense mu binding throughout the nucleus tractus solitarius [Herkenham and Pert, 1980; Lewis et al., 1985]. While there also have been reports of kappa binding within this nucleus, particularly within the rostral tip [Lynch et al., 1985], there are few data at present demonstrating delta binding using selective radiolabelled ligands.

At the level of the midbrain, one observes a moderate labelling of mu receptors in the periaqueductal gray and a dense distribution of sites in the caudal linear and median nuclei of the raphe [Atweh and Kuhar, 1977b; Goodman et al., 1980; Herkenham and Pert, 1980]. The kappa receptor distribution pattern is distinct from the mu pattern at this anatomical level and is relatively dense throughout the PAG with a moderate distribution in the caudal linear and median nuclei of the raphe [Mansour et al., 1986]. In contrast, the periaqueductal gray and raphe nuclei appear to be devoid of delta receptors.

At the level of the diencephalon, mu binding is predominant in nearly all the thalamic nuclei, yet quite sparse in the hypothalamus [Herkenham and Pert, 1982; Quirion et al., 1983]. The thalamic nuclei with fairly low mu binding include the paraventricular and ventroposterior nuclei. Kappa binding, on the other hand, is relatively dense in the hypothalamic nuclei, including the paraventricular nucleus and dorsomedial nucleus, and only present in the midline thalamic nuclei (central medial, paracentral, and ventromedial) and habenula [Lynch et al., 1985; Mansour et al., 1987]. Delta binding is largely absent in most of the thalamus and the hypothalamus [Goodman et al., 1980].

The distribution of opioid receptor subtypes in the cortex varies with cortical division and lamina. For a more detailed review of the distribution of mu sites in the monkey and rat cortex the reader is referred to Lewis et al. [1981, 1983]. Because of its intimate connections to the limbic system, the distribution of opioid receptor subtypes will be reviewed in the cingulate cortex. In the anterior cingulate cortex there is a high number of mu receptors in layers I, V and VI. Delta binding, on the other hand, is more diffuse with a moderate density of binding in layers II, III, V and VI. Kappa sites appear to be reduced compared to the other subtypes and display a moderate labelling in the deeper layers (III, V and VI). A parasagittal dark-field autoradiogram of mu binding sites is presented in Figure 1 and highlights some of the anatomical regions thought to be involved in the mediation of opioid analgesia.

Is There an 'Analgesia Receptor'

Considerable effort has been devoted to elucidating the specific neural substrates acted upon by opiates in the production of analgesia, and recently attempts have been made to ascribe analgesic functions to particular subtypes of opioid receptors. These tasks are complex for several reasons: Systematically administered opiates may exert analgesic

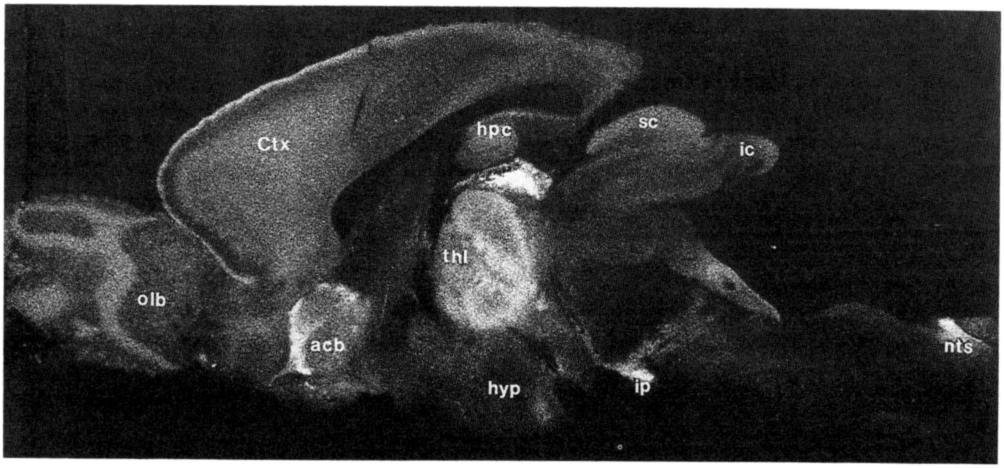

Fig. 1. Dark-field autoradiogram of mu ([^{3}H]DAGO) binding in rat brain. Notice the dense areas of binding, depicted by the light areas, in the cingulate cortex, thalamus, interpeduncular nucleus and nucleus tractus solitarius. acb = nucleus accumbens; ctx = cortex; hpc = hippocampus; hyp = hypothalamus; ic = inferior colliculus; ip = interpeduncular; nts = nucleus tractus solitarius; olb = olfactory bulb; sc = superior colliculus; thl = thalamus.

effects at multiple levels of the neuraxis [Yaksh, 1981; Yaksh and Rudy, 1978; Yeung and Rudy, 1980]; pain induced by different stimuli (e.g. thermal vs. mechanical), responds differentially to various narcotic drugs [Schmauss et al., 1983]; and even within a single neural locus, thought to be critical for the inhibition of pain (e.g. central gray and dorsal horn of the spinal cord), multiple opioid receptor subtypes can be identified. Despite the difficulty of these questions, their resolution may have important clinical value. For example, if adequate pain relief can be obtained by intrathecal administration of opiates, many untoward side effects of opiates can be minimized [Yaksh, 1981, 1984]. Also if it can be determined that two different opiate drugs exert their analgesic effects via different subsets of receptors, and if both drugs are equipotent inhibitors of pain, it may be possible to use them clinically in alternation to obviate the development of tolerance due to excessive stimulation of a single receptor group. Moreover, since all types of experimental pain do not respond equally well to all types of opiates, identification of specific opiate receptor types associated with particular types of nociception may assist in prescription of the most appropriate analgesic drug. Research directed at these questions has taken several approaches and is beginning to provide important insights. The following represents only highlights of this field of inquiry.

Using pharmacological approaches, a variety of investigators have measured the analgesic potency of a host of receptor-specific agonist drugs. At least in the case of opiates acting at a supraspinal level, the consensus of these studies seems to indicate an important involvement of mu receptors in analgesia [Pasternak et al., 1980b]. Drugs with relatively high affinity for the mu site, such as morphiceptin, morphine, and sufentanil, appear to be much more potent analgesics than delta (i.e. the enkephalins and their synthetic analogs) [Vaught et al., 1982] or kappa agonists [Piercey et al., 1982]. In support of this hypothesis are findings that the analgesic effects of mu agonists, such as morphine, are markedly reduced by the mu-specific antagonists (e.g. naloxazone [Pasternak et al., 1980b] or β-FNA [Ward et al., 1982]), but not by a delta-antagonist [Chaillet et al., 1983]. The findings are also corroborated by in vitro binding assays which reasonably accurately predict a compound's analgesic efficacy based on its potency at the mu receptor [Pert and Snyder, 1973].

It is difficult to assign an analgesic role to a single receptor subtype in the spinal cord. Multiple opioid receptors have been identified on intrinsic neurons in the dorsal horn of the spinal cord [Traynor et al., 1982] as well as on terminals of primary sensory afferents [Fields et al., 1980; LaMotte et al., 1976; Ninkovic et al., 1981]. Although intrathecal administration of all classes of opiates (mu, delta, kappa) can cause analgesia, there are qualitative differences among their actions [Yaksh, 1981]. Briefly, when responsiveness to noxious thermal stimuli is assessed, mu and delta, but not kappa, agonists cause significant analgesia [Schmauss and Yaksh, 1983; Tyers, 1980]. By contrast, when visceral pain tests are used, mu and kappa, but not delta, agonists are effective analgesics [Schmauss and Yaksh, 1983]. Regarding the spinal action of kappa agents, it should be noted that, compared to brain, the spinal cord shows a high proportion of kappa receptors [Traynor et al., 1982; Lewis and Akil, unpubl. observations] and also has high concentrations of the putative kappa ligand, dynorphin [Botticelli et al., 1981; Chavkin et al., 1982]. Also, studies comparing the antagonisms of mu and kappa drugs by naloxone on visceral tests revealed different pA2 values, suggesting that these agents inhibit pain by different receptor mechanisms

[Schmauss and Yaksh, 1983]. Finally, some workers have found that animals rendered tolerant to the analgesic effects of morphine manifest cross-tolerance to kappa, but not to delta drugs [Tung and Yaksh, 1982]. In work by others, employing the specific kappa agonist U50-488H, both U50-488H and morphine caused analgesia on thermal tests and, although tolerance developed to both compounds, there was no cross-tolerance between drugs [Von Voightlander et al., 1982]. Thus, spinal mechanisms of analgesia appear complex and probably involve all opiate receptor subtypes.

Another approach to discern the analgesia receptor uses in vitro receptor binding assays. To this end, Sadee and colleagues have recently presented a series of studies labelling and characterizing opioid receptor subtypes in vivo and compared these results with in vitro binding characteristics. They have identified 3 subtypes of opioid-binding sites [Rosenbaum et al., 1984a]. Their data suggest that site 1 resembles the mu-site, site 2 may be a combination of delta- and kappa-receptors, and site 3 represents a naloxone-binding site, previously not identified in vitro and having very low affinity for sufentanil, etorphine, or diprenorphine. The identification of the naloxone site, 'lambda-receptor', in vivo, but not in vitro, reinforces the need for in vivo experiments. Finally, Rosenbaum et al. [1984b] have used this in vivo labelling technique to examine the binding of analgesic doses of opiates. They conclude that analgesia is correlated with occupancy of site 1 receptors and that significant pain inhibition occurs when approximately 2% of these receptors are occupied [Rosenbaum et al., 1984b].

Pursuing the same line of thought, we have recently begun to investigate the effects of in vivo injections of opiates, attempting to elucidate the subclass of receptors occupied following administration of behaviorally relevant doses of drug. We have found that morphine-induced analgesia is best correlated with occupation of mu, not delta or kappa, sites. Administration of morphine (5 mg/kg, s.c.) inhibits binding of radioactive ligands to mu-sites by approximately 70% without appreciably altering binding to delta- or kappa-sites. Systemic administration of bremazocine causes both analgesia and occupation of mu-, delta-, and kappa-sites. Presumably, the analgesia is due to occupation of kappa-sites, since bremazocine appears to act as an antagonist at delta-, and mu-receptors. Finally, in preliminary experiments examining receptor occupation following administration of naloxone, we have obtained results consistent with pharmacological characterizations of the binding of this drug. That is, in low doses (1 mg/kg s.c.) naloxone appears to be a selective mu-antagonist. Doses of approximately 10 mg/kg are required to achieve significant blockade of delta-receptors, and kappa-receptors appear relatively insensitive to this drug.

Following a subcutaneous injection of morphine, however, a peculiar phenomenon occurs. Administration of morphine appears to cause an increase in the number of available brain opioid receptors. Pert and Snyder [1976], using only a single radioactive ligand, dihydromorphine, to label receptors, reported an increase in binding following morphine administration. We have recently found that administration of morphine does in fact cause an increase in whole brain opiate binding, but that this increase is specifically one of mu-receptors, not delta or kappa [Lewis, J.W., et al., 1984]. This effect is rapid and occurs within 10 min of an analgesic dose of morphine [Pert and Snyder, 1976]. The increase in mu-binding may require in vivo administration of the drug, since we have been unable to reproduce this phenomenon in rat brain homogenates incubated with morphine in vitro. The physiological significance of this phenomenon, however, remains to be elucidated. We find, as did Pert and Snyder [1976], that it is not correlated with analgesia, since proliferation of mu-receptors persists in morphine-tolerant rats challenged with morphine. One possible behavioral effect with which this increase in receptors may be compatible, however, was reported by Tulanay and Takemori [1974a, b]. They found that in the mouse, a single injection of morphine significantly enhanced the antagonistic effects of naloxone to a subsequent injection of morphine. Perhaps this is in some measure due to the increased receptors generated by the previous morphine injection.

An important caveat to bear in mind regarding interpretation of all of these attempts to ascribe particular functions to particular receptors is illustrated by findings such as the following: Bowen et al. [1981] have suggested that opioid receptor types may interconvert one to another; a series of studies by Rothman and Westfall [1981] have demonstrated allosteric intermodulation among opioid receptor types; and Holaday et al. [1985] have shown that interference with the mu- and kappa-receptors can cause an alteration in function of antagonist drugs at the delta-site. Although the specific conclusions about the nature of opiate receptor function deriving from the work of these investigators is different, they share a common theme. The opioid receptors may be dynamic, interacting entities.

Are Opioid Peptides of Peripheral Origin Involved in Analgesia?

The pituitary-adrenal and the sympatho-adrenal axes are rich in opioid peptides and have long been recognized for their important roles in the adaptation to stress. We, and many others, have shown that exposure to stress activates opioid-mediated mechanisms of pain inhibition (see below), and if pain suppression may be thought of as an adaptive response to certain stress conditions, it is not unreasonable to assume participation of these hormonal systems in stress analgesia. In fact, several pieces of evidence are available to indicate that pituitary hormones contribute importantly to the mediation of stress analgesia: the pituitary is the richest source of BE, thought to be the most potent analgesic of the opioid peptides; stress causes the release of pituitary opioids [Guillemin et al, 1979; Akil et al., 1985], and stress-induced analgesia (particularly those forms sensitive to naloxone blockade) is attenuated in hypophysectomized animals [Lewis, J.W. et al., 1981a, b; Bodnar et al., 1979; MacLennan, 1982].

Peripheral opioid peptides are also found in the chromaffin cells of the adrenal medulla. The adrenal medulla contains Pro-enkephalin and co-releases enkephalin-like peptides and catecholamines in response to sympathetic activation or trauma [Viveros et al., 1982]. Although the precise physiological function of these peptides remains to be determined, we have suggested that they are importantly involved in certain forms of stress analgesia. The analgesic response to intermittent footshock stress is antagonized by naloxone [Akil et al., 1976; Lewis, J.W. et al., 1980] and is markedly reduced by adrenal demedullation or denervation of the adrenal medulla [Lewis, J.W. et al., 1982a]. Because demedullation and ganglionectomy have as great an effect as removal of the entire adrenal gland, and because both basal and stressed adrenocortical function was unimpaired in adrenal denervated animals yet these rats failed to manifest opioid stress analgesia, this form of stress analgesia appears dependent on adrenal medullary, not cortical, function. Moreover, it appears to be the enkephalin-like peptides, not catecholamines, that are involved in mediation of this form of stress analgesia. Rats given a dose of reserpine that depletes medullary catecholamines and increases the adrenal content and stimulation-induced release of enkephalins [Viveros et al., 1980] significantly augment opioid stress analgesia [Lewis, J.W. et al., 1982a]. Thus, this form of stress analgesia appears to be due to release of enkephalin-like peptides by stress.

Biochemical evidence we have obtained in a collaborative study with Dr. O.H. Viveros supports these behavioral observations [Lewis J.W. et al., 1982b]. The amount of opiate-like material in the adrenal medulla was quantified using a receptor-binding assay. Adrenal medullary enkephalin content was significantly reduced by exposure to footshock stress that elicits an opioid-mediated analgesic response, but not by footshock stress that causes a nonopioid form of stress analgesia. Similarly, Alessi et al. [1982] have reported a decrease in adrenal content of peptide F, a Pro-enkephalin product, following exposure to a similar form of footshock stress. Medullary enkephalin-like peptide content was dramatically increased in reserpine treated rats, and this new elevated content was also reduced by exposure to intermittent footshock. Finally, rats made tolerant to the analgesic effects of footshock no longer showed depletion of adrenal enkephalin-like peptides after stress [Lewis, J.W. et al., 1982c]. These converging lines of evidence strongly implicate adrenal enkephalin-like peptides in opioid stress analgesia. Several important questions, however, such as the locus of the opiate receptor mediating this analgesia, remain to be answered before this hypothesis can be fully accepted.

A recent finding in our laboratory may begin to provide an interface for the central and peripheral systems involved in stress analgesia. We have found that electrical stimulation of the nucleus tractus solitarius (NTS), a medullary nucleus rich in opioid peptides and receptors causes opioid-mediated analgesia in rats [Lewis, J.W. et al., 1985]. That this nucleus may be important for stress analgesia is suggested by the following; this nucleus is well known to be involved in regulation of autonomic systems [Galosy et al., 1983] which are affected by stress; the NTS has ascending and descending anatomical connections with several brain loci classically thought to be involved in pain inhibition, including stress analgesia. Finally, the NTS receives neural input from the vagus, and Maixner and Randich [1983] have shown that vagotomy attenuates stress analgesia. Thus, it may be that exposure to stress first elicits the release of adrenal enkephalins (see above). These peptides cause peripheral hemodynamic changes that are detected and relayed to the CNS via the vagus. When this information arrives at the NTS, neural systems involved in homeostasis and pain inhibition are activated.

How Are Opioid Pain Inhibitory Systems Activated?

These systems have now been shown to be activated by a variety of stimuli, the most obvious being administration of opiate drugs. Additionally, analgesia can be elicited by direct electrical stimulation of these neural systems and recently, it appears that they are activated by acupuncture as well. The reader is referred to the reviews by Cannon and Liebeskind and by Tsou et al. [this volume] for details on these phenomena. Our most recent work has focussed on activation of opioid analgesia systems by exposure to stress. Although this is obviously not a useful therapeutic modality, it has proven quite useful as a tool for 'driving' these systems in order to study their anatomy, physiology, and biochemistry. A summary of this line of investigation is provided below.

Given the existence of multiple, endogenous analgesia systems, some opioid- and others nonopioid-mediated, when are they employed? Anecdotal evidence would suggest that exposure to stressful or to noxious stimulation may elicit pain inhibition as an adaptive coping response. In 1976, this line of reasoning was extended to the laboratory by two research groups, Akil et al. and Hayes et al. Both groups reported that an analgesic response could be observed in rats following exposure to stress but their conclusions regarding the neurochemical mechanisms underlying this phenomenon were quite different. Akil et al. [1976a] found that rats subjected to 30 min of inescapable, intermittent footshock displayed a pronounced analgesic response measured by the tail-flick test. This stress analgesia was attenuated by pretreatment with naloxone, and showed tolerance with repeated stress exposure, suggesting involvement of opioid peptides [Akil et al., 1976a, Madden et al., 1977]. In a subsequent study, it was demonstrated that brain opioid levels are elevated following footshock stress, and that as analgesic tolerance develops, the effect of stress on brain opioids dissipates [Madden et al., 1977]. These findings indicate the existence of central nervous systems, opioid-mediated, pain-inhibitory mechanisms and suggest an environmental stimulus, stress, can cause their activation. The findings of Hayes et al. [1976, 1978] also indicated that several stressors elicited analgesia, but they found stress-induced analgesia to be naloxone-resistant. Since these original reports, several other investigators have demonstrated stress-induced analgesia following a host of stressors [Bodnar et al., 1978a; Christie et al., 1981; Amir and Amit, 1978; Chance, 1979; Watkins and Mayer, 1982; Miczek et al., 1982; Chesher and Chan, 1977].

At the present time, the biochemical evidence describing the role for opioid peptides in stress-induced analgesia is fragmentary. In general, exposure to stressors that cause opioid-mediated analgesia has been associated with alterations in one or another, central or peripheral, index of opioid activity. For example, footshock stress has been shown to decrease hypothalamic leu-enkephalin [Rossier et al., 1978], decrease pituitary opioids [Baizman et al., 1978; Guillemin et al., 1977; Millan et al., 1980], decrease hypothalamic BE [Rossier et al., 1977], decrease brain enkephalin [McGivern et al., 1983], and decrease spinal cord dynorphin [Millan et al., 1981]. Unfortunately, these findings of alterations in total peptide content are difficult to interpret. For example, if the content of any opioid decreases, is this due to activation of the system and subsequent release, or to a decrease in peptide biosynthesis secondary to a lack of activation? Clearly, more dynamic studies of the opioids in brain are needed before the ability of stress to activate these systems can be evaluated.

Our studies with stress and the POMC systems have revealed both the ability of stressors to activate POMC cells in pituitary and brain, and the adaptability of these cells to such demands. We have employed the intermittent footshock paradigm which had been shown to bring about opioid-mediated stress analgesia [Akil et al., 1976a; Lewis, J.W. et al., 1980]. In addition, we have examined the effects of the stressor administered either acutely (one 20- or 30-min session prior to sacrifice) or phasically (one daily session for 2 weeks, and either allowing the animal to rest 24 h prior to sacrifice [chronic stress/rest] or administering a last stress session and then sacrificing the rat [chronic stress/acute stress]). These procedures have allowed us to examine the short- and long-term adaptive events which take place in pituitary and brain after such treatment. In a series of experiments on the pituitary anterior and intermediate lobe POMC cells, we have examined stress-induced changes in content, release into plasma, rate of translation of POMC in short-term primary cultures, and rates of conversion of POMC into its products such as B-LPH and BE. These studies have shown that while both the anterior lobe and the intermediate lobe respond to footshock stress, they do so differently. Perhaps of greatest relevance to the topic of pain control is that the anterior lobe releases opioid active BE 1-31 in response to acute stress. Interestingly, the BE content of the gland only decreases by a small amount in spite of a massive release measured in plasma. This is likely due to an increased rate of biosynthesis of POMC and of post-translational processing of POMC into its products [Shiomi et al., 1982].

We believe that this rapid adaptation, in the form of enhanced enzymatic activity, may account for the difficulty in detecting substantial changes in content in brain, and in fact, we have on occasion seen an increase in brain BE levels of rats exposed to footshock [Lewis, J.W. et al., 1981a]. Thus, depending on the relative magnitude of release versus enhancement in biosynthesis, and depending on the rate of transport from cell bodies to terminal areas, stimulation of a POMC system could conceivably result in a decrease, and increase, or no change in POMC content.

Following repeated exposure to stress, the intermediate lobe becomes induced and releases primarily a nonopioid form of BE (N-cetyl BE), while the anterior lobe exhibits a different strategy for meeting continued demand. Rather than the 'emergency procedure' of increasing its post-translational biosynthetic rate, there is an increase in specific messenger RNA (mRNA) for POMC, an increase in total peptide content, and specifically an increase in the most releasable peptide forms, BE 1-31 [Akil et al., in preparation]. Thus, the system now has the ready potential to release more, and challenges, such as further stress, can be met without a subsequent increase in biosynthesis. Oddly enough, because of this change in dynamics, it is easier to detect a decrease in peptide content in the adapted, chronically stressed anterior pituitary than in those subjected only to acute stress. In midbrain, a similar pattern of results emerges. While acute stress causes little reliable change in BE level, chronic stress followed by one day of rest leads to an increase in BE content (due to increased mRNA?) and in these chronically stressed rats, acute stress causes a decrease in content from 150% of normal back to normal levels. Taken together with the findings in anterior pituitary, these data suggest that the midbrain does, in fact, respond to footshock, but that detection of a response to acute stress may be obscured by factors such as immediate enzymatic increase. Thus, we have to await the shift to an increase in POMC mRNA and a return of normal post-translational enzymatic activity in order to document that a change took place. Interestingly, the form of BE that is preferentially released after chronic exposure to footshock is BE 1-27, the peptide form which is one tenth the potency of BE 1-31. Recent data from Nicolas and Li [1985] suggest that this form of BE may actually be functioning as an opioid antagonist, and it is therefore reasonable to speculate that part of the basis for the development of tolerance to stress-induced analgesia is due to the shift of the system from the release of opioid-potent BE to opioid-impotent BE 1-27.

Thus, as we increase our understanding of the dynamics of the opioids, we hope to determine whether they indeed respond to even more subtle stimuli, be they stressful, painful, emotionally arousing, or otherwise. It should also be noted that even a full understanding of the presynaptic peptidergic events is not sufficient to account for the animal's behavior, even at the simplest level. For example, while we can show that both pituitary and brain adapt to repeated stress by increasing their BE stores and releasing the opioids reliably, we know that, behaviorally, these animals are no longer analgesic and appear tolerant to the stress. Since this tolerance cannot be attributed to the inability of the system to produce and release its products, the most parsimonious explanation is a change in opioid receptor sensitivity or efficacy. Yet, we have been totally unsuccessful, however, in demonstrating any changes in number or affinity of brain mu, delta, or kappa sites in various brain or spinal regions of rats exposed to chronic stress.

Attempts have been made to assess increased occupation of opiate receptors by endogenous ligands following exposure to stressors that elicit analgesia: either warm [Christie et al., 1981], or cold [Pert and Bowie, 1979] water forced swims or conditioned fear [Chance et al., 1978] caused a decrease in opiate binding relative to controls, suggesting that these stressors, in fact, elicited the release of opioid peptides.

Recently, we have begun to characterize receptor occupancy following exposure to the form of inescapable footshock stress that elicits opioid-mediated analgesia. In vitro binding assays indicate that exposure to this stressor causes occupation of all three types of receptors, although the greatest occupation is of the mu site. These findings are consistent with observations that this stress-induced analgesia is partially antagonized by very low doses of naloxone, but is rarely completely blocked even following administration of high doses of antagonist. The relationship between any of these changes and stress-induced analgesia is, however, strictly correlational.

Conclusions

Multiple endogenous opioids and their multiple receptors are strategically located to modulate pain. Ample physiological, pharmacological, and biochemical evidence points to their critical involvement in pain regulation. That this pain regulation is tightly coupled to the basic organism's response to stressful stimuli is supported both by opioid anatomy and by the ready activation of opioid mechanisms by both noxious and stressful events. The selective activation of one or more of these opioid mechanisms by various noxious events remains to be elucidated. The interface between opioid and non-opioid mechanisms of pain control is only now being explored. Furthermore, the possible role of higher order opioid mechanisms, e.g. limbic and cortical, – in modifying the pain experiences has not been elucidated. Yet, the complexity and richness of these endogenous opioids would give credence to the possibility that they would modulate pain responsiveness at every level, from the initial sensory input, to the integration with other sensory and autonomic information, the affective response, the consequent shift in attentional mechanism, the cognitive interpretation and even to the subsequent memory of the experience.

References

Akil, H.; Hughes, J.; Richardson, D.E.; Barchas, J.D.: Enkephalin-like material elevated in ventricular cerebrospinal fluid of pain patients after analgetic focal stimulation. Science *201:* 463–465 (1978a).

Akil, H.; Madden, J.; Patrick, R.L.; Barchas, J.D.: Stress-induced increase in endogenous opiate peptides: Concurrent analgesia and its partial reversal by naloxone; in Kosterlitz, Opiates and endogenous opioid peptides, pp. 63–70 (Elsevier/North Holland, Amsterdam 1976a).

Akil, H.; Mayer, D.J.; Liebeskind, J.C.: Comparison chez le rat entre l'analgesie induite par stimulation de la substance gris peri-aqueductale et l'analgesie morphinique. C. hebd. Séanc. Acad. Sci, Paris *274:* 3603–3605 (1972).

Akil, H.; Mayer, D.J.; Liebeskind, J.C.: Antagonism of stimulation-produced analgesia by naloxone, a narcotic antagonist. Science *191:* 961–962 (1976b).

Akil, H.; Shiomi, H.; Matthews, J.: Induction of the intermediate pituitary by stress. Synthesis and release of a non-opioid form of beta-endorphin. Science *227:* 424–426 (1985).

Akil, H.; Richardson, D.E.; Barchas, J.D.; Li, C.H.: Appearance of beta-endorphin-like immunoreactivity in human ventricular cerebrospinal fluid upon analgesic electrical stimulation. Proc. nat. Acad. Sci. USA *75:* 5170–5172 (1978b).

Akil, H.; Richardson, D.E.; Barchas, J.D.: Pain control by focal brain stimulation in man: Relation to enkephalins and endorphins. 11th Miles Int. Symp.: Mechanisms of Pain and Analgesic Compounds, 1979, pp. 239–247.

Akil, H.; Watson, S.J.; Young, E.; Lewis, M.E.; Khachaturian, H.; Walker, J.M.: Endogenous opioids: biology and function. A. Rev. Neurosci. *7:* 223–255 (1984).

Akil, H.; Young, E.; Watson, S.J.; Coy, D.: Opiate binding properties of naturally occurring N- and C-terminus modified beta-endorphin. Peptides *2:* 289–292 (1981).

Alessi, N.; Taylor, L.; Akil, H.: Peptide F (Pro-enkephalin fragment) Radioimmunoassay and stress-induced changes in adrenal. Life Sci. *31:* 1875–1878 (1982).

Amir, S.; Amit, Z.: Endogenous opioid ligands may mediate stress-induced changes in the affective properties of pain related behavior in rats. Life Sci. *23:* 1143–1152 (1978).

Amir, S.; Amit, Z.: The pituitary gland mediates acute and chronic pain responsiveness in stressed and non-stressed rats. Life Sci. *24:* 439–448 (1979).

Atweh, S.F.; Kuhar, M.J.: Autoradiographic localization of opiate receptors in rat brain. I. Spinal cord and lower medulla. Brain Res. *124:* 53–67 (1977a).

Atweh, S.F.; Kuhar, M.J.: Autoradiographic localization of opiate receptors in rat brain. III. The telencephalon. Brain Res. *134:* 393–405 (1977b).

Baizman, E.R.; Cox, B.M.; Osman, O.H.; Goldstein, A.: Experimental alterations of endorphin levels in rat pituitary. Neuroendocrinology *28:* 402–424 (1979).

Bardo, M.T.; Bhatnagar, R.K.; Gebhart, G.F.: Chronic naltrexone increases opiate binding in brain and produces supersensitivity to morphine in the locus coeruleus of the rat. Brain Res. *289:* 223–234 (1983).

Basbaum, A.I.; Fields H.L.: Endogenous pain control systems. Brainstem spinal pathways and endorphin circuitry. A. Rev. Neurosci. *7:* 309–338 (1984).

Beitz, A.J.: The nuclei of origin of brain stem enkephalin and substance P projections to the rodent nucleus raphe magnus. Neuroscience *7:* 2753–2768 (1982).

Belcher, G.; Smock, T.; Fields, H.L.: Effects of intrathecal ACTH on opiate analgesia in the rat. Brain Res. *247:* 373–377 (1982).

Belluzzi, J.D.; Grant, N.; Garsky, V.; Sarantakis, D.; Wise, C.D.; Stein, L.: Analgesia induced in vivo by central administration of enkephalin in rat. Nature, Lond. *260:* 625 (1976).

Bloom, F.; Rossier, J.; Battenberg, E.; Bayon, A.; French, E.; Henrickson, S.J.; Siggins, G.R.; Segal, D.; Browne, R.; Ling, N.; Guillemin, R.: Cellular localization, electrophysical and behavioral effects. Adv. biochem. Psychopharmacol. The Endorphins *18:* 89–109 (1978).

Bodnar, R.J.; Glusman, M.; Brutus, M.; Spiaggia, A.; Kelly, D.D.: Analgesia induced by cold-water stress. Attenuation following hypophysectomy. Physiol. Behav. *23:* 53–62 (1979).

Bodnar, R.J.; Kelly, D.D.; Spiaggia, A.; Ehrenberg, C.; Glusman, M.: Dose-dependent reductions by naloxone of analgesia induced by cold-water stress. Pharmacol. Biochem. Behav. *8:* 667–672 (1978).

Bonnet, K.A.; Grotin, J.; Gioamnini, T.; Cortes, M.; Simon, E.J.: Opiate receptor heterogeneity in human brain regions. Brain Res. *221:* 437–440 (1981).

Botticelli, L.J.; Cox, B.M.; Goldstein, A.: Immunoreactive dynorphin in mammalian spinal cord and dorsal root ganglion. Proc. Natn. Acad. Sci. USA *78:* 7783–7786 (1981).

Bowen, W.D.; Gentleman, S.; Herkenham, M.; Pert, C.B.: Interconverting mu and delta forms of the opiate receptor in rat striatal patches. Proc. natn. Acad. Sci. USA *78:* 4818–4822 (1981).

Cannon, J.T.; Prieto, G.J.; Lee, A.; Liebeskind, J.C.: Evidence for opioid and non-opioid forms of stimulation-produced analgesia in the rat. Brain. Res. *243:* 315–321 (1982).

Chaillet, P.; Couland, A.; Fournie-Zaluski, M.C.; Gacel, G.; Roques, B.P.; Costentin, J.:

Pain control by endogenous enkephalins is mediated by mu opioid receptors. Life Sci. *33:* suppl. 1, pp. 685–688 (1983).
Chance, W.T.: Autoanalgesia: opiate and non-opiate mechanisms. Neurosci. Biobehav. Rev. *4:* 55–67 (1980).
Chance, W.T.; Rosecrans, J.A.: Lack of effect of naloxone on auto-analgesia. Pharmacol. Biochem. Behav. *11:* 643–646 (1979b).
Chance, W.T.; White, A.C.; Krynock, G.M.; Rosecrans, J.A.: Conditional fear-induced decreases in the binding of (^{3}H)-N-leu-enkephalin to rat brain. Brain Res. *141:* 371–374 (1978).
Chang, K.J.; Cautrecasas, P.: Heterogeneity and properties of opiate receptors. Fed. Proc. *40:* 2729–2734 (1981).
Chang, K.J.; Cooper, B.R.; Hazum, E.; Cuatrecasas, P.: Multiple opiate receptors: Different regional distribution in the brain and differential binding of opiates and opioid peptides. Molec. Pharmacol. *16:* 91–104 (1979).
Chang, K.J.; Hazum, E.; Cuatrecasas, P.: Novel opiate binding sites for benzomorphan drugs. Proc. natn. Acad. Sci. USA *78:* 4141–4145 (1981).
Chapman, C.R.; Colpitts, Y.M.; Benedetti, C.; Kitaeff, R.; Gehrig, J.D.: Evoked potential assessment of acupunctural analgesia. Attempted reversal with naloxone. Pain *9:* 183–197 (1980).
Chavkin, C.; James, I.; Goldstein, A.: Dynorphin is a specific endogenous ligand of the kappa opioid receptor. Science *215:* 413–415 (1982).
Chesher, G.B.; Chan, B.: Footshock-induced analgesia in mice. Its reversal by naloxone and cross-tolerance with morphine. Life Sci. *21:* 1569–1574 (1977).
Christie, M.J.; Chesher, G.B.; Bird, K.D.: The correlation between swim-stress induced antinociception and [^{3}H]leu-enkephalin binding to brain homogenates in mice. Pharmacol. Biochem. Behav. *15:* 853–857 (1981).
Comb, J.; Herbert, E.; Crea, R.: Partial characterization of the mRNA that codes for enkephalins in bovine adrenal medulla and human pheochromocytoma. Proc. natn. Acad. Sci. USA *79:* 360–364 (1982).
Corbett, A.D.; Peterson, S.J.; McKnight, A.T.; Magnan, J.; Kosterlitz, H.: Dynorphin (1-8) and dynorphin (1-9) are ligands for the kappa subtype of opiate receptor. Nature, Lond. *299:* 79–81 (1982).
Czlonkowski, A.; Costa, T.; Przewlocki, R.; Pasi, A.; Herz, A.: Opiate receptor binding sites in human spinal cord. Brain Res. *267:* 392–396 (1983).
Deakin, J.F.; Dostrovsky, J.O.; Smyth, D.G.: Influence of N-terminal acetylation and C-terminal proteolysis on the analgesic activity of beta-endorphin. Biochem. J. *189:* 501–506 (1980).
Dores, R.M.; Akil, H.: Region and species specific species differential post-translational processing of pro-dynorphin. Int. Narcotics Conf., Cambridge 1984.
Dores, R.M.; Akil, A.; Watson, S.J.: Strategies for studying opioid peptide regulation at the gene, message, and protein levels. Peptides *5:* suppl. 1, pp. 9–17 (1984).
Drugan, R.C.; Grau, J.W.; Maier, S.F.; Madden, J.; Barchas, J.D.: Cross-tolerance between morphine and the long-term analgesic reaction to inescapable shock. Pharmacol. Biochem. Behav. *14:* 677–682 (1981).
Duka, T.; Schubert, P., Wuster, M.; Stoiber, R.; Herz, A.: A selective distribution pattern of different opiate receptors in certain areas of rat brain as revealed by in vitro autoradiography. Neurosci. Lett. *21:* 119–124 (1981a).

Duka, T.; Wuster, M.; Schubert, P.; Stoiber, R.; Herz, A.: Selective localization of different types of opiate receptors in hippocampus as revealed by in vitro autoradiography. Brain Res. *205:* 181–186 (1981b).

Elde, R.; Hokfelt, T.; Johansson, O.; Terenius, L.: Immunohistochemical studies using antibodies to leucine-enkephalin. Initial observations on the nervous system of the rat. Neuroscience *1:* 349–351 (1976).

Fields, H.L.; Emson, P.C.; Leigh, B.K.; Gilbert, R.F.T.; Iverson, L.L.: Multiple opiate receptor sites on primary afferent fibres. Nature, Lond. *284:* 351–353 (1980).

Galosy, R.A.; Clarke, L.K.: Vasco, M.R.; Crawford, I.L.: Neurophysiology and neuropharmacology of cardiovascular regulation and stress. Neurosci. Biobehav. Rev. *5:* 137–175 (1981).

Gebhart, G.F.: Opiate and opioid peptide effects on brain stem neurons. Relevance to nociception and antinociception mechanisms. Pain *12:* 93–140 (1982).

Gee, C.; Roberts, J.; Thompson, R.; Watson, S.J.: Identification of POMC neurons in the rat hypothalamus by (in situ) hybridization. Nature, Lond. *306:* 374–376 (1983).

Goldstein, A.; Tachibana, S.; Lowney, L.I.; Hunkapiller, M.; Hood, L.: Dynorphin (1-13), an extraordinarily potent opioid peptide. Proc. Natn. Acad. Sci. USA *76:* 6666–6670 (1979).

Goodman, R.R.; Snyder, S.: Opiate receptors localized by autoradiography to deep layers of cerebral cortex. Relation to sedative effects. Proc. natn. Acad. Sci. USA *79:* 5703–5707 (1982).

Goodman, R.R.; Snyder, S.H.; Kuhar, M.J.; Young, W.S., III: Differentiation of delta and mu opiate receptor localizations by light microscopic autoradiography. Proc. natn. Acad. Sci. USA *77:* 6239–6243 (1980).

Gouarderes, C.; Cros, J.; Quirion R.: Autoradiographic localization of mu, delta and kappa opioid receptor binding sites in rat and guinea pig spinal cord. Neuropeptides *6:* 331–342 (1985).

Grau, J.W.; Hyson, R.L.; Maier, S.F.; Madden, J., IV; Barchas, J.D.: Long-term stress-induced analgesia and activation of the opiate system. Science *213:* 1409–1410 (1981).

Guillemin, R.; Vargo, T.; Rossier, J.; Minick, S.; Ling, N.; Rivier, C.; Vale, W.; Bloom, F.: B-Endorphin and adrenocorticotropin are secreted concomitantly by the pituitary gland. Science *197:* 1367–1369 (1977).

Hayashi, Y.; Minamino, N.; Mizuno, K.; Kangawa, K.; Matsuo, H.; Chino, N.; Nakagawa, Y:, Sakakibara, S.: Morphinomimetic activities of alpha-neo-endorphin and its related peptides. Jap. J. Pharm. *30:* 215P (1980).

Hayes, R.L.; Bennett, G.J.; Newlon, P.G.; Mayer, D.J.: Analgesic effects of certain noxious and stressful manipulations in the rat. Soc. Neurosci. Abstr. *2:* 939 (1976).

Herkenham, M.; Pert, C.B.: Light microscopic localization of brain opiate receptors. A general autoradiographic method which preserves tissue quality. J. Neurosci. *8:* 1129–1149 (1978).

Herkenham, M.; Pert, C.B.: In vitro autoradiography of opiate receptors in rat brain suggests loci of 'opiatergic' pathways. Proc. natn. Acad. Sci. USA *77:* 5532–5536 (1980).

Herkenham, M.; Pert, C.B.: Light microscopic localization of brain opiate receptors. A general autoradiographic method which preserves tissue quality. J. Neurosci. *2:* 1129–1149 (1982).

Holaday, J.W.; Tortella, F.C.: Multiple opioid receptors: Possible functions of mu and delta binding sites in vivo; in Muller, Genazzini, Central and peripheral endorphins: basic and clinical aspects, pp. 237–250 (Raven Press, New York 1984).
Hosobuchi, Y.; Adams, J.E.; Linchitz, R.: Pain relief by electrical stimulation of the central gray matter in humans and its reversal by naloxone. Science *197:* 183–186 (1977).
Hosobuchi, Y.; Rossier, J.; Bloom, F.E.; Guillemin, R.: Stimulation of hyman periaqueductal gray for pain relief increases immunoreactive B-Endorphin in ventricular fluid. Science *203:* 279–281 (1979).
Hughes, J.: Isolation of an endogenous compound from the brain with pharmacological properties similar to morphine. Brain Res. *88:* 295–308 (1975).
Hughes, J.; Smith, T.W.; Kosterlitz, H.W.; Fothergill, L.A.; Morgan, B.A.; Morris, H.R.: Identification of two related pentapeptides from the brain with potent opiate agonist activity. Nature, Lond. *258:* 577–579 (1975).
Kakidani, H.; Furutani, Y.; Takehashi, H.; Noda, M.; Morimoto, Y.; Hirone, T.; Asai, M.; Inayama, S.; Nakanishi, S.; Numa, S.: Cloning and sequence analysis of cDNA for porcine beta-neo-endorphin/dynorphin precursor. Nature, Lond. *298:* 245–249 (1982).
Khachaturian, H.; Lewis, M.E.; Hollt, V.; Watson, S.J.: Telencephalic enkephalinergic systems in the rat brain. J. Neurosci. *3:* 844–855 (1983a).
Khachaturian, H.; Lewis M.E.; Schafer, M.K-H.; Watson, S.J.: Anatomy of the CNS opioid systems. Trends Neurosci. *8:* 111–119 (1985a).
Khachaturian, H.; Lewis, M.E.; Tsou, K.; Watson, S.J.: Beta-endorphin alpha-MSH, ACTH and related peptides; Hokfelt, Bjorklund, Handbook of chemical neuroanatomy, vol. IV. GABA and neuropeptides in the CNS, part 1, pp. 216–272 (Elsevier/Biomedical Press, Amsterdam 1985b).
Khachaturian, H.; Lewis, M.E.; Watson, S.J.: Enkephalin systems in diencephalon and brain stem of the rat. J. comp. Neurol. *219:* 310–320 (1983b).
Khachaturian, H.; Lewis, M.E.; Watson, S.J.: Co-localization of pro-enkephalin peptides in the same neurons in the rat brain. Brain Res. *279:* 369–373 (1983c).
Khachaturian, H.; Watson, S.J.; Lewis, M.E.; Coy, D.; Goldstein, A.: Dynorphin immunocytochemistry in the rat central nervous system. Peptides *3:* 941–954 (1982).
Kilpatrick, D.L.; Taniguchi, T.; Jones, B.N.; Stern, A.S.; Shively, J.E.; Hullihen, J.; Kumura, S.; Stein, S.; Udenfriend, S.: A highly potent 3200-dalton adrenal opioid peptide that contains both a [Met]- and [Leu]enkephalin in sequence. Proc. natn Acad. Sci. USA *78:* 3265–3268 (1981).
LaMotte, C.; Pert, C.B.; Snyder, S.H.: Opiate receptor binding in primate spinal cord. Distribution and changes after dorsal root section. Brain Res. *112:* 407–412 (1976).
Leslie, F.; Goldstein, A.: Degradation of dynorphin-(1-13) by membrane-bound rat-brain enzymes. Neuropeptides *2:* 185–196 (1982).
Lewis, J.W.; Baldrighi, G.; Watson, S.J.; Akil, H.: Electrical stimulation of the nucleus tractus solitarius (NTS) causes opioid mediated analgesia in the rat. Soc. Neurosci. Abstr. *11:* 637 (1985).
Lewis, J.W.; Cannon, J.T.; Liebeskind, J.C.: Opioid and non-opioid mechanisms of stress analgesia. Science *208:* 623–625 (1980).
Lewis, J.W.; Cannon, J.T.; Liebeskind, J.C.; Akil, H.: Alterations in brain B-endorphin immunoreactivity following acute and chronic stress. Pain *1:* suppl., p. 263 (1981a).

Lewis, J.W.; Chudler, E.H.; Cannon, J.T.; Liebeskind, J.C.: Hypophysectomy differentially affects morphine and stress analgesia. Proc. West. Pharmacol. Soc. *23:* 85–88 (1981b).

Lewis, J.W.; Lewis, M.E.; Akil, H.: Occupation of brain opiate receptors following in vivo administration of morphine. Soc. Neurosci. Abstr. *10:* 519 (1984).

Lewis, J.W.; Lewis, M.E.; Loomus, D.J.; Akil, H.: Acute systemic administration of morphine selectively increases mu opioid receptor binding in the rat brain. Neuropeptides *5:* 117–120 (1984).

Lewis, J.W.; Sherman, J.E.; Liebeskind, J.C.: Opioid and non-opioid stress analgesia: Assessment of tolerance and cross-tolerance with morphine. J. Neurosci. *1:* 358–363 (1981c).

Lewis, J.W.; Stapleton, J.M.; Castiglioni, A.J.; Liebeskind, J.C.: Stimulation-produced analgesia and intrinsic mechanisms of pain suppression; in Fink, Whalley, Neuropeptides – basic and clinical aspects. Proc. 11th Pfizer Int. Symp., pp. 41–49 (Churchill Livingstone, Edinburgh 1982a).

Lewis, J.W.; Terman, G.W.; Watkins, L.R.; Mayer, D.J.; Liebeskind, J.C.: Opioid and non-opioid mechanisms of footshock-induced analgesia. Role of the spinal dorsolateral funiculus. Brain Res. *267:* 139–144 (1983).

Lewis, J.W.; Tordoff, M.G.; Liebeskind, J.C.; Viveros, O.H.: Evidence for adrenal medullary opioid involvement in stress analgesia. Soc. Neurosci. Abstr. *8:* 788 (1982b).

Lewis, J.W.; Tordoff, M.G.; Sherman, J.E.; Liebeskind, J.C.: Adrenal medullary enkephalin-like peptides may mediate opioid stress analgesia. Science *217:* 557–559 (1982c).

Lewis, M.E.; Khachaturian, H.; Watson, S.J.: Comparative distribution of opiate receptors and three opioid peptide neuronal systems in rhesus monkey central nervous system. Life Sci. *33:* suppl. 1, pp. 239–242 (1983b).

Lewis, M.E.; Lewis, M.S.; Dores, R.M.; Lewis, J.W.; Khachaturian, H.; Watson, S.J.; Akil, H.: Characterization of multiple opioid receptors and peptides in rat and guinea pig substantia nigra. Biophys. J. *47:* 54a (1985).

Lewis, M.E,; Mishkin, M.; Bragin, E.; Brown, R.M.; Pert, C.B.; Pert, A.: Opiate receptor gradients in monkey cerebral cortex. Correspondence with sensory processing hierarchies. Science *211:* 1166–1169 (1981).

Lewis, M.E.; Pert, A.; Pert, C.B.; Herkenham, M: Opiate receptor localization in rat cerebral cortex. J. comp. neurol. *216:* 339–358 (1983).

Lewis, R.V.; Stern, A.S.; Kimura, S.; Rossier, J.; Stein, S.; Udenfriend, S.: An about 50,000-dalton protein in adrenal medulla. A common precursor of (Met)- and (Leu)-enkephalin. Science *208:* 1459–1460 (1980a).

Lewis, R.V.; Stern, A.S.; Kimura, S.; Stein, S.; Udenfriend, S.: Enkephalin biosynthetic pathway. Proteins of 8000 and 14,000 daltons in bovine adrenal medulla. Proc. natn. Acad. Sci. USA *77:* 5018–5020 (1980b).

Li, C.H., Chung, D.: Isolation and structure of an untriakontapeptide with opiate activity from camel pituitary glands. Proc. natn. Acad. Sci. USA *73:* 1145–1148 (1976).

Lynch, W.C.; Watt, J.; Krall, S; Paden, C.M.: Autoradiographic localization of kappa opiate receptors in CNS taste and feeding areas. Pharmacol. Biochem. Behav. *22:* 669–705 (1985).

MacLennan, A.J.; Drugan, R.C.; Hyson, R.L.; Maier, S.F.; Madden, J.; Barchas, J.D.: Corticosterone. A critical factor in an opioid form of stress-induced analgesia. Science *215:* 1530–1532 (1982a).

MacLennan, A.J.; Drugan, R.C.; Hyson, R.L.; Maier, S.F.; Madden, J.; Barchas, J.D.: Dissociation of long-term analgesia and the shuttle box escape deficit caused by inescapable shock. J. comp. Physiol. Psychol. *96:* 904–912 (1982b).

Madden, J.; Akil, H.; Patrick, R.L.; Barchas, J.D.: Stress-induced parallel changes in central opioid levels in pain responsiveness in the rat. Nature, Lond. *265:* 358–360 (1977).

Maier, S.F.; Davies, S.; Grau, J.W.; Jackson, R.L.; Morrison, D.H.; Moye, T.; Madden, J.; Barchas, J.D.: Opiate antagonists and the long-term analgesic reaction induced by inescapable shock in rats. J. comp. Physiol. Psychol. *94:* 1172–1183 (1980).

Mains, R.E.; Eipper, B.A.; Ling, N.: Common precursor to corticotropins and endorphins. Proc. natn. Acad. Sci. USA *74:* 3014–3018 (1977).

Maixner, W.; Randich, A.: Role of the right vagal nerve trunk in antinociception. Brain Res. *298:* 374–377 (1984).

Malfroy, B.; Swerts, J.P.; Guyon, A.; Roques, B.P.; Schwartz, J.C.: High-affinity enkephalin-degrading peptidase in brain is increased after morphine. Nature, Lond. *276:* 523–526 (1978).

Mansour, A.; Lewis, M.E.; Khachaturian, H.; Akil, H.; Watson, S.J.: Pharmacological and anatomical evidence of selective mu, delta and kappa opioid receptor binding in rat brain. Brain Res. *399:* 69–79 (1986).

Mansour, A.; Lewis, M.E.; Khachaturian, H.; Akil, H.; Watson, S.J.: Autoradiographic differentiation of the mu, delta and kappa opioid receptors in the rat forebrain and midbrain. J. Neurosci. (in press, 1987).

Maurer, R.: Multiplicity of opiate receptors in different species. Neurosci. Lett. *30:* 303–307 (1982).

Maurer, R.; Cortes, R.; Probst, A.; Palacios, J.M.: Multiple opiate receptor in human brain. An autoradiographic investigation. Life Sci. *33:* suppl. I, pp. 231–234 (1983).

Mayer, D.J.; Hayes, R.L.: Stimulation-produced analgesia. Development of tolerance and cross-tolerance to morphine. Science *188:* 941–943 (1975).

Mayer, D.J.; Price, D.D.: Central nervous system mechanisms of analgesia. Pain *2:* 379–404 (1976).

Mayer, D.J.; Price, D.D.; Rafii, A.; Barber, J.: Acupuncture hypalgesia: Evidence for activation of a central control system as a mechanism of action; in Bonica, Albe-Fessard, Advances in pain research and therapy, vol. I, pp. 751–754 (Raven Press, New York, 1976).

Mayer, D.J.; Wolfle, T.L.; Akil, H.; Carder, B.; Liebeskind, J.C.: Analgesia from electrical stimulation in the brainstem of the rat. Science *174:* 1351–1354 (1971).

McGinty, J.F.; Van der Kooy, D.; Bloom, F.E.: The distribution and morphology of opioid peptide immunoreactive neurons in the cerebral cortex of rats. J. Neurosci. *4:* 1104–1117 (1984).

McGivern, R.F.; Mousa, S.; Couri, D.; Berntson, G.G.: Prolonged intermittent footshock stress decreases met and leu enkephalin levels in brain with concomitant decreases in pain threshold. Life Sci. *33:* 47–54 (1983).

Miczek, K.A.; Thompson, M.L.; Schuster, L.: Opioid-like analgesia in defeated mice. Science *215:* 1520–1522 (1982).

Millan, M.J.; Przewlocki, R.; Herz, A.: A non-B-endorphinergic adrenohypophyseal mechanism is essential for an analgetic response to stress. Pain *8:* 343–353 (1980).

Millan, M.J.; Tsang, Y.F.; Przewlocki, R.; Hollt, V.; Herz, A.: The influence of footshock stress upon pain, pituitary, and spinal cord pools of immunoreactive dynorphin in rats. Neurosci. Lett. *24:* 75–79 (1981).

Miller, R.J.: Multiple opiate receptors for multiple opioid peptides. Med. Biol. *60:* 1–6 (1982).

Mizuno, K.; Minamino, N.; Kangawa, K.; Matsuo, H.: A new familiy of endogenous 'big' [Met]enkephalins from bovine adrenal medulla. Purification and structure of docosa- (BAM22P) and eicosapeptide (BAM20P) with very potent opiate activity. Biochem. biophys. Res. Commun. *97:* 1283–1290 (1980).

Moskowitz, A.S.; Goodman, R.R.: Light microscopic autoradiographic localization of mu and delta opioid binding sites in the mouse central nervous system. J. Neurosci. *4:* 1331–1342 (1984).

Mulcahy, J.; Lee, H.S.; Basbaum, A.I.: Coexistence of immunoreactive enkephalin (I-Enk) and dynorphin (I-Dyn) in the nucleus of the solitary tract of the rat. Soc. Neurosci. Abstr. *9:* 439 (1983).

Nakanishi, S.; Inoue, A.; Kita, T.; Nukamura, M.; Chung, A.C.Y.; Cohen, S.N.; Numa, S.: Nucleotide sequence of cloned cDNA for bovine corticotropin-beta-lipotropin precursor. Nature, Lond. *278:* 423–427 (1979).

Nicolas, P.; Li, C.H.: B-Endorphin-(1-27) is a naturally occurring antagonist to etorphine-induced analgesia. Proc. natn. Acad. Sci. USA *82:* 3178–3181 (1985).

Ninkovic, M.; Hunt, S.P.; Emson, P.; Iversen, L.L.: The distribution of multiple opiate receptors in bovine brain. Brain Res. *214:* 163–167 (1981).

Ninkovic, M.; Hunt, S.P.; Gleave, J.R.W.: Localization of opiate and histamine Hl-receptors in the primate sensory ganglia and spinal cord. Brain Res. *241:* 197–206 (1982).

Noda, M.; Teranishi, Y.; Takahashi, H.; Toyosato, M.; Notake, M.; Nakanishi, S.; Numa, S.: Isolation and structural organization of the human preproenkephalin gene. Nature, Lond. *297:* 431–434 (1982).

O'Donahue, T.; Handelmann, G.; Miller, R.; Jacobowitz, D.: N-acetylation regulates the behavioral activity of alpha-melanotropin in a multi-neurotransmitter neuron. Science *215:* 1125–1127 (1982).

Pasternak, G.W.; Childers, S.R.; Snyder, S.H.: Naloxazone, a long acting opiate antagonist. Effects in intact animals and on opiate receptor binding in vitro. J. Pharmac. exp. Ther. *214:* 455–462 (1980a).

Pasternak, G.W.; Childers, S.R.; Snyder, S.H.: Opiate analgesia. Evidence for mediation by a subpopulation of opiate receptors. Science *208:* 514–516 (1980b).

Pederson, R.C.; Brownie, A.C.; Ling, N.: Proadrenocorticotropin/endorphin-derived peptides: Coordinate action on adrenal steroidogenesis. Science *104:* 4–46 (1980).

Penner, E.R.; Terman, G.W.; Liebeskind, J.C.: Cross-tolerance between opioid mediated stimulation-produced and stress-induced analgesia. Soc. Neurosci. Abstr. *8:* 619 (1982).

Pert, C.B.; Bowie, D.L.: Behavioral manipulations of rats causes alterations in opiate receptor occupancy; in Usdin, MacMillan, Endorphins in mental health research (New York, 1979).

Pert, C.B.; Pert, A.; Chang, J.K.; Fong, B.T.W.: [D-Ala^2]-Met-Enkephalinamide. A potent, long-lasting synthetic pentapeptide analgesic. Science *194:* 330–332 (1976).

Pert, C.B.; Snyder, S.H.: Opiate receptor: Demonstration in nervous tissue. Science *179:* 1011–1014 (1973).

Pert, C.B.; Snyder, S.H.: Opiate receptor binding enhancement by opiate administration in vivo. Biochem. Pharmac. *25:* 847–853 (1976).

Pert, C.B.; Taylor, D.: Type 1 and type 2 opiate receptors. A subclassification scheme based upon GTPs differential effects on binding; in Way, Endogenous and exogenous opiate agonists and antagonists, pp. 87–90, (Pergamon Press, New York, 1980).

Pert, A.; Walter, M.: Comparison between naloxone reversal of morphine and electrical stimulation induced analgesia in the rat mesencephalon. Life Sci. *19:* 1023–1032 (1976).

Petrie, E.C.; Tiffany, S.T.; Baker, T.B.; Dahl, J.L.: Dynorphin (1-13): Analgesia, hypothermia, cross-tolerance with morphine and beta-endorphin. Peptides *3:* 41–47 (1982).

Pfeiffer, A.; Pasi, A.; Mehraein, P.; Herz, A.: Opiate receptor binding sites in human brain. Brain Res. *248:* 87–96 (1982).

Piercey, M.F.; Lahti, R.A.; Schroeder, L.A.; Einspahr, F.J.;Barsuhn, C.: U-50488H, a pure kappa receptor agonist with spinal analgesic loci in the mouse. Life Sci. *31:* 1197–1200 (1982).

Pomeranz, B.; Chiu, D.: Naloxone blockade of acupuncture analgesia: Endorphin implicated. Life Sci. *19:* 1757–1762 (1976).

Quirion, R.: Pain, nociception and spinal opioid receptors. Prog. Neuro-Psychopharm. Biol. Psychiat. *8:* 571–579 (1984).

Quirion, R.; Weiss, A.S.: Peptide E and other proenkephalin-derived peptides are potent kappa opiate receptor agonists. Peptides *4:* 445–449 (1983).

Quirion, R.; Zajac, J.M.; Morgat, J.L.; Roques, B.P.: Autoradiographic distribution of mu and delta receptors in rat brain using highly selective ligands. Life Sci. *33:* suppl. I, pp. 227–230 (1983).

Reynolds, D.V.: Surgery in the rat during electrical analgesia induced by focal brain stimulation. Science *164:* 444–445 (1969).

Ricardo, J.A.; Koh, E.T.: Anatomical evidence of direct projections from the nucleus of the solitary tract to the hypothalamus, amygdala, and other forebrain structures in the rat. Brain Res. *153:* 1–26 (1978).

Richardson, D.E.; Akil, H.: Pain reduction by electrical brain stimulation in man. (part one): Acute administration in periaqueductal and periventricular sites. J. Neurosurg. *47:* 184–194 (1977a).

Richardson, D.E.; Akil, H.: Pain reduction by electrical brain stimulation in man (part two). Chronic self administration in the periventricular gray matter. J. Neurosurg. *47:* 184–194 (1977b).

Roberts, J.L.; Herbert, E.: Characterization of a common precursor to corticotropin and beta-lipotropin. Identification of beta-lipotropin peptides and their arrangement relative to corticotropin in the precursor synthesized in a cell-free system. Proc. natn. Acad. Sci. USA *74:* 5300–5304 (1977).

Robson, L.E.; Paterson, S.J.; Kosterlitz, H.W.: Opiate receptors; in S. Ivensen, L.L. Iverson and S. Snyder, Handbook of psychopharmacology, vol. 14, pp. 13–80 (Plenum Press, New York, 1983).

Rosenbaum, J.S.; Holford, N.H.G.; Richards, M.L.; Aman, R.A.; Sadee, W.: Discrimination of three types of opioid binding sites in rat brain in vivo. Molec. Pharmacol. *25:* 242–248 (1984).

Rosenbaum, J.S.; Holford, N.H.G.; Sadee, W.: Opiate receptor binding-effects relationships: Sufentanil and etorphine produce analgesia at the mu-site with low fractional receptor occupancy. Brain. Res. *291:* 317–324 (1984).

Rossier, J.; Audigier, Y.; Ling, N.; Cros, J.; Udenfriend, S.: Met-enkephalinARG6-PHE7, present in high amounts in brain of rat, cattle, and man, is an opioid agonist. Nature, Lond. *288:* 88–90 (1980).

Rossier, J.; French, E.D., Rivier, C.; Ling, N.; Guillemin, R.; Bloom, F.E.: Foot-shock induced stress increases B-endorphin levels in blood but not brain. Nature, Lond. *270:* 618–620 (1977).

Rossier, J.; Guillemin, R.; Bloom, F.E.: Foot-shock induced stress decreases leu5-enkephalin immunoreactivity in rat hypothalamus. Eur. J. Pharmacol. *48:* 465–466 (1978).

Rothman, R.B.; Westfall, T.C.: Morphine allosterically modulates the binding of [^{3}H]leucine enkephalin to a particulate fraction of rat brain. Molec. Pharmacol. *21:* 538–547 (1981).

Schmauss, C.; Yaksh, T.L.: In vivo studies on spinal opiate receptor systems mediating antinociception. II. Pharmacological profiles suggesting a differential association of *mu, delta,* and *kappa* receptors with visceral chemical and cutaneous thermal stimuli in the rat. J. Pharmac. exp. Ther. *228:* 1–12 (1983).

Schwartzberg, D.G.; Nakane, P.K.: ACTH-related peptide containing neurons within the medulla oblongata of the rat. Brain Res. *276:* 351–356 (1983).

Sherman, J.E.; Strub, H.; Lewis, J.W.: Morphine analgesia. Enhancement by shock-associated cues. Behav. Neurol. *98:* 293–309 (1984).

Shiomi, H.; Akil, H.: Pulse-chase studies of the POMC/beta-endorphin system in the pituitary of acutely and chronically stressed rats. Life Sci. *31:* 2271–2273 (1982).

Simantov, R.; Kuhar, M.J.; Uhl, G.; Snyder, S.H.: Opioid peptide enkephalin. Immunohistochemical mapping in rat central nervous system. Proc. natn. Acad. Sci. USA *714:* 2167–2171 (1977).

Simon, E.J.; Hiller, J.M.; Edelman, I.: Stereospecific binding of the potent narcotic analgesic [^{3}H]-etorphine to rat-brain homogenate. Proc. natn. Acad. Sci. USA *70:* 1947–1949 (1973).

Sjolund, B.H.; Eriksson, M.B.E.: The influence of naloxone on analgesia produced by peripheral conditioning stimulation. Brain Res. *173:* 295–301 (1979).

Sullivan, S.; Akil, H.; Barchas, J.D.: [In vitro] degradation of enkephalin. Evidence for cleavage at the gly-phe bond. Commun. Psychopharmacol. *2:* 525–531 (1978).

Sullivan, S.; Akil, H.; Blacker, D.; Barchas, J.D.: Enkephalinase. Selective inhibitors and partial characterization. Peptides *1:* 31–35 (1980).

Takagi, H.: The nucleus reticularis paragigantocellularis as a site of analgesic action of morphine and enkephalin. Trends Pharm. Sci. *1980:* 182–184.

Terenius, L.: Characteristics of the 'receptor' for narcotic analgesics in synaptic plasma membrane fraction from the rat brain. Acta pharmac. tox. *33:* 377–384 (1973).

Terenius, L.: Endorphins and modulation of pain. Adv. Neurol. *33:* 50–64 (1982).

Terman, G.W.; Lewis, J.W.; Liebeskind, J.C.: Multiple endogenous analgesia systems activated by stress in the rat; in Yokota, Dubner, Current topics in pain research and therapy. Proc. Int. Symp. Pain, pp. 81–94 (Excerpta Medica, Amsterdam 1982).

Traynor, J.R.; Kelly, P.D.; Rance, M.J.: Multiple opiate binding sites in rat spinal cord. Life Sci. *31:* 1377–1380 (1982).

Tricklebank, M.D.; Hutson, P.H.; Curzon, G.: Analgesia induced by brief footshock is inhibited by 5-hydroxytryptamine but unaffected by antagonists of 5-hydroxytryptamine or by naloxone. Neuropharmacology *21:* 51–56 (1982).

Tsou, K.; Jang, C.S.: Studies on the site of analgesic action of morphine by intracerebral microinjection. Sci. Sin. *13:* 1099–1109 (1964).

Tsou, K.; Khachaturian, H.; Akil, H.; Watson, S.J.: Immunocytochemical localization of pro-opimelanocortin-derived peptides in the adult rat spinal cord. Brain Res. *378:* 28–35 (1986).

Tulunay, F.C.; Takemori, A.E.: The increased efficacy of narcotic antagonists induced by various narcotic analgesics. J. Pharmac. exp. Ther. *190:* 395–400 (1974a).

Tulunay, F.C.; Takemori, A.E.: Further studies on the alteration of analgesic receptor-antagonist interaction induced by morphine. J. Pharmac. exp. Ther. *190:* 401–407 (1974b).

Tung, A.S.; Yaksh, T.L.: In vivo evidence for multiple opiate receptors mediating analgesia in the rat spinal cord. Brain. Res. *247:* 75–83 (1982).

Tyers, M.B.: A classification of opiate receptors that mediate antinociception in animals. Br. J. Pharmacol. *69:* 503–512 (1980).

Urca, G.; Frenk, H.; Liebeskind, J.C.; Taylor, A.N.: Morphine and enkephalin: Analgesic and epileptic properties. Science *197:* 83–86 (1977).

Vale, W.; Vaughn, J.; Smith, M.; Yamamoto, G.; Rivier, J.; Rivier, C.: Effects of synthetic ovine corticotropin releasing factor, glucocorticoids, catecholamines, neuro-hypophysical peptides and other substances on cultured corticotropic cells. Endocrinology *113:* 1121–1131 (1983).

Vaught, J.L.; Rothman, R.B.; Westfall, T.C.: *Mu* and *delta* receptors. Their role in analgesia and in the differential effects of opioid peptides on analgesia. Life Sci. *30:* 1443–1455 (1982).

Viveros, O.H.; Diliberto, E.J., gr.; Hazum, E.; Chang, K.J.: Enkephalins as possible adrenomedullary hormones: Storage, secretion, and regulation of synthesis; in Costa, Trabucchi, Neural peptides and neuronal communication, pp. 191–204 (Raven Press, New York, 1980).

Von Voigtlander, P.F.; Lahti, R.A.; Lundens, J.H.: U-50,488. A selective and structurally novel non-*Mu (Kappa)* opioid agonist. J. Pharmac. Exp. Ther. *224:* 7–12 (1982).

Walker, J.M.; Akil, H.; Watson, S.J.: Evidence from homologous actions of pro-opiocortin products. Science *210:* 1247–1249 (1981).

Walker, J.M.; Berntson, G.G.; Sandman, C.A.; Coy, D.H.; Schally, A.V.; Kastin, A.J.: Analog of enkephalin having prolonged opiate-like effects in vivo. Science *196:* 85–87 (1977a).

Walker, J.M.; Ghessari, A.; Peters, B.; Watson, S.J.; Akil, H.: Interaction among pro-opiomelanocortin products. Modulation of pain sensitivity. Peptides (under revision).

Walker, J.M.; Katz, R.J.; Akil, H.: Behavioral effects of dynorphin 1-13 in the mouse and rat. Initial observations. Peptides *1:* 341–345 (1980).

Walker, J.M.; Moises, H.; Coy, D.; Baldrighi, G.; Akil, H.: Non-opiate effects of dynorphin and Des-Tyr-dynorphin. Science *218:* 1136–1138 (1982a).

Walker, J.M.; Sandman, C.A.; Bertntson, G.G.; McGivern, R.; Coy, D.H.; Kastin, A.J.: Endorphin analogs with potent and long-lasting analgesic effects. Pharmacol. Phys. Behav. *20:* 539–542 (1977b).

Walker, J.M.; Tucker, D.E.; Coy, D.H.; Walker, B.B.; Akil, H.: Des-tyrosine-dynorphin antagonizes morphine analgesia. Eur. J. Pharmacol. *85:* 121–122 (1982b).

Walmsely, J.K.; Zarbin, M.A.; Young, W.S., III; Kuhar, M.J.: Distribution of opiate receptors in the monkey brain. An autoradiographic study. Neuroscience. *7:* 595–613 (1982).

Ward, S.J.; Portoghese, P.S.; Takemori, A.E.: Pharmacological characterization in vivo of the novel opiate, B-funaltrexamine (B-FNA). J. Pharmac. exp. Ther. *220:* 494 (1982).

Watkins, L.R.; Cobelli, D.A.; Faris, P.; Aceto, M.D.; Mayer, D.J.: Opiate vs. non-opiate footshock-induced analgesia (FSIA). The body region shocked is a critical factor. Brain Res. *242:* 299–308 (1982a).

Watkins, L.R.; Cobelli, D.A.; Newsome, H.H.; Mayer, D.J.: Footshock-induced analgesia is dependent neither on pituitary nor sympathetic activation. Brain Res. *245:* 81–96 (1982b).

Watkins, L.R.; Mayer, D.J.: The organization of endogenous opiate and non-opiate pain control systems. Science *216:* 1185–1192 (1982).

Watson, S.J.; Akil, H.; Barchas, J.D.: Possible role for the dorsal periventricular catecholamine bundle in stimulation produced analgesia: A behavioral and histochemical study. Brain Res. *130:* 335–342 (1977).

Watson, S.J.; Akil, H.; Fischli, W.; Goldstein, A.; Zimmerman, E.; Nilaver, G.; van Wimersma Greidanus, T.B.: Dynorphin and vasopressin. Common localization in magnocellular neurons. Science *216:* 85–87 (1981).

Watson, S.J.; Akil, H.; Richard, C.W., III; Barchas, J.D.: Evidence for two separate opiate peptide neuronal systems. Nature, Lond. *275:* 226–228 (1978).

Watson, S.J.; Khachaturian, H.; Akil, H.; Coy, D.H.; Goldstein, A.: Comparison of the distribution of dynorphin systems and enkephalin systems in brain. Science *218:* 1134–1136 (1982).

Weber, E.; Evans, C.J.; Barchas, J.D.: Predominance of the amino-terminal betapeptide fragment of dynorphin in rat brain regions. Nature, Lond. *299:* 77–79 (1982).

Wei, E.T.; Tseng, L.F.; Loh, H.H.; Li, C.H.: Comparison of the behavioral effects of beta-endorphin and enkephalin analogs. Life Sci. *21:* 321–327 (1977).

Wise, S.P.; Herkenham, M.: Opiate receptor distribution in the cerebral cortex of the rhesus monkey. Science *218:* 387–389 (1982).

Woolf, C.J.; Mitchell, D.; Myers, R.A.; Barrett, G.D.: Failure of naloxone to reverse peripheral transcutaneous electroanalgesia in patients suffering from acute trauma. S. Afr. med. J. *53:* 179–180 (1978).

Yaksh, T.L.: Spinal opiate analgesia. Characteristics and principles of action. Pain *11:* 293–346 (1981).

Yaksh, T.L.: Multiple spinal opiate receptor systems in analgesia; in Kruger, Liebeskind, Advances in pain research and therapy, vol. 6 (Raven Press, New York, 1984).

Yaksh, T.L.; Rudy, T.A.: Narcotic analgesics. CNS sites and mechanisms of action as revealed by intracerebral injection techniques. Pain *4:* 299–359 (1978).

Yaksh, T.L.; Yeung, J.C.; Rudy, T.A.: An inability to antagonize with naloxone the elevated nociceptive thresholds resulting from electrical stimulation of the central gray. Life Sci. *18:* 1193–1198 (1976).

Yeung, J.C.; Rudy, T.A.: Multiplicative interaction between narcotic agonists expressed at spinal and supraspinal sites of antinociceptive action as revealed by concurrent intrathecal and intracerebroventricular injections of morphine. J. Pharmac. exp. Ther. *21:* 633–642 (1980).

Young, E.; Walker, J.M.; Houghton, R.; Akil, H.: [^{3}H]-dynorphin A binds selectively to guinea pig brain. Eur. J. Pharmacol. *91:* 327–328 (1983).
Young, R.; Feldman, R.A.; Kroening, R.; Fulton, W.; Morris, J.: Electrical stimulation of the brain in the treatment of chronic pain in man; in Kruger, Liebeskind, Advances in pain research and therapy, vol. IV: neural mechanisms of pain, pp. 289–304 (Raven Press, New York 1984).
Zukin, R.S.; Zukin, S.R.: The case for multiple opiate receptors. Trends Neurosci. *May:* 160–164 (1984).

J. Lewis, MD, Mental Health Research Institute, University of Michigan,
205 Washtenaw Place, Ann Arbor, MI 48109–0720 (USA)

Pain Headache, vol. 9, pp. 160–177 (Karger, Basel 1987)

Functional Aspects of Multitransmitter Neurons

Studies of Interactions among Pro-Opiomelanocortin Products

J. Michael Walker[a], *Ali Ghessari*[a], *Betty A. Peters*[a], *Stanley J. Watson*[c], *Nabil Seidah*[b], *Michel Chretien*[b], *Huda Akil*[c]

[a] Department of Psychology, Brown University, Providence, R.I., USA; [b] Protein and Pituitary Hormone Laboratory, Clinical Research Institute of Montreal, Quebec, Canada; [c] University of Michigan, Mental Health Research Institute, Ann Arbor, Mich., USA

Multitransmitter Neurons

One of the most striking discoveries in the field of opioid peptides was that of the costorage and apparent corelease of several peptides from a single neuron. The costorage of neuromodulators has proven to be far more common than originally conceived; the literature is now replete with examples [Hokfelt et al., 1980; Chan-Palay and Palay, 1984]. As a consequence, the questions of costorage are among the first asked when dealing with newly mapped transmitter pathways. Indeed, the questions about costorage are very important because this phenomenon implies a much greater degree of complexity in the synaptic 'message' than previously imagined. Regarding the study of pain, the costorage of transmitters must be considered since all three gene families of opioid peptides secrete several opioid as well as some nonopioid peptides. Consequently, a complete understanding of endogenous systems which modulate pain sensitivity will require an understanding of the actions and interactions of the arrays of peptide products produced and presumably secreted by cells of these circuits.

By virtue of having the same precursor (pro-opiomelanocortin; POMC), the costorage of the ACTH/MSH-like peptides with β-endorphin remains one of the clearest as well as one of the earliest examples of

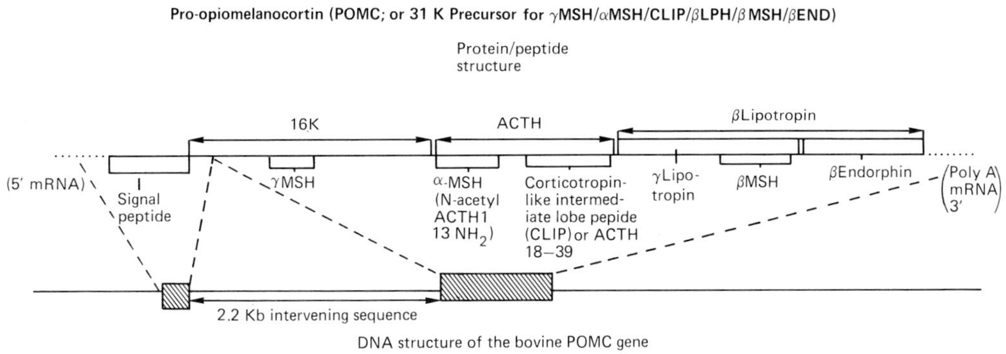

Fig. 1. Schematic diagram of POMC.

the multiple release phenomenon [Mains and Eipper, 1977]. As is now well known, the entire sequence of the precursor for β-endorphin has been identified [Nakanishi et al., 1979] (fig. 1). This precursor gives rise to numerous possible products (table I), presumably each of which may have some function in the nervous system. Studies of the functional aspects of the peptide products of the β-endorphin precursor in relation to pain sensitivity form the basis of this review.

POMC: Structural Features and Post-Translational Processing

Pro-opiomelanocortin is present in both the pituitary and the brain and is processed into several biologically active peptides. In the anterior pituitary, the processing results in the formation of ACTH and β-lipotropin (β-LPH) and some β-endorphin [Benjannet et al., 1980; Mains et al, 1977; Roberts and Herbert, 1977]. Further processing in the brain and the intermediate lobe of the pituitary results in the formation of β-endorphin and α-MSH with very little β-LPH [Gianoulakis, et al., 1979; Zakarian and Smyth, 1979]. Immunohistochemical data showed that pituitary cells package these substances within the same granules [Pelletier and Dube, 1977]. In the brain, neurons containing pro-opiomelanocortin are found in the arcuate nucleus of the hypothalamus and project to many brain areas both rostrally and caudally [McGinty and Bloom, 1980; Pelletier and Dube, 1977; Watson et al., 1978; Roberts and Herbert, 1977].

Table I. Putative products of POMC

Amino-terminal region
γ-MSH_{1-12}, γ-MSH_{1-27} (γ_3-MSH)
Lys-γ_3-MSH, POMC 1-74
Midregion
Adrenocorticotropin (ACTH), α-melanocyte-stimulating hormone (α-MSH), des-acetyl-α-MSH, di-acetyl-α-MSH, corticotropin-like intermediate lobe peptide (CLIP)
Carboxyl region
β-Lipotropin, γ-lipotropin, β-melanocyte-stimulating hormone, β-endorphin, β-$endorphin_{1-27}$, β-$endorphin_{1-26}$ N-acetyl-β-endorphin, N-acetyl-β-$endorphin_{1-27}$, N-acetyl-β-$endorphin_{1-26}$ γ-endorphin, α-endorphin

Partial list of possible cleavage products of POMC. Not all products are produced in all tissues, and recent data suggest that various experimental factors or drug treatments can affect the ratios of the various products.

cDNA sequencing provided the messenger RNA coding for POMC [Nakanishi et al., 1979]. This work confirmed and extended previous anatomical and biochemical studies, revealing the arrangement depicted in figure 1. β-Endorphin is located at the C-terminus of pro-opiomelanocortin, ACTH1-39 in the midportion, and a heretofore unknown sequence (sometimes termed 16K) at the N-terminal. A homology to the sequence, Met-Glu-His-Phe-Arg-Trp, which is known to be part of ACTH, α-MSH, and β-MSH is repeated at a third site in the new N-terminal peptide, and was therefore termed γ-MSH (the γ-MSH core is Met-Gly-His-Phe-Arg-Trp).

As with the ACTH and β-endorphin, several forms of γ-MSH occur in the pituitary. Moreover, the anterior and intermediate lobe process this region of the precursor to different points. In keeping with other pro-opiomelanocortin products, the intermediate lobe cleaves the precursor into multiple forms of γ-MSH-like immunoreactivity of varying molecular weights, while the anterior lobe primarily produces larger species [Benjannet et al., 1980; Shibasaki et al., 1980; Tanaka et al., 1980]. The presence of dibasic amino acids surrounding γ-MSH-(1-12) suggested that it might be a release product, and this sequence was used in our experiments. However, recent studies suggest that glycosylated Lys-γ3-MSH and a glycosylated POMC1-74 product might be the major form in the cell soma [Emson and Eipper, in press]. Since the possibility of further processing in the terminals cannot be discounted, the precise release products are uncertain.

POMC and Pain

Typically, studies of the role of POMC products on pain focus on β-endorphin. This, of course, is not surprising because β-endorphin is one of the most potent analgesics ever studied [Loh et al., 1976] (fig. 2). Its actions might be thought of as hormone-like in that the analgesic effects of very low doses of β-endorphin can last for several hours. This effect is reminiscent of the long-lasting analgesic effects of electrical stimulation of the periaqueductal gray (PAG) which can also last for extended periods after the cessation of the electrical current [Liebeskind, 1976]. The fact that electrical stimulation of the PAG results in increased β-endorphin levels in the CSF has led to the conclusion that simulation-induced β-endorphin release might, in part, give rise to electrical stimulation produced analgesia (SPA) [Akil et al., 1978]. This possibility is consistent with the dense plexus of β-endorphin immunoreactivity in the PAG, the strong presence of opiate receptors in the area, and the potent analgesic effects of locally administered opiates [Watson et al., 1977; Simon and Hiller, 1978; Tsou and Jang, 1964; Yaksh and Rudy, 1978].

Nonopioid POMC Products and Pain

The above observations suggest a clear role of β-endorphin in pain inhibition. But what about the other products of POMC? β-Endorphin 1-31 forms only about $\frac{1}{10}$ of the POMC sequence. In fact, one could argue that POMC is really an ACTH/MSH precursor which happens to have an opiate core at the carboxyl region. This conclusion might be justifiable because there are three MSH-like sequences in POMC. As noted above, these include the γ-MSH region, the α-MSH region, and the β-MSH region. As with β-endorphin, the levels of α-MSH are increased in the CSF of patients receiving electrical stimulation to the PAG for the relief of pain [Akil and Watson, 1980]. Do these peptides have effects upon pain sensitivity?

To examine the question of the relations of the nonopiate POMC peptides to pain, we have investigated several peptides related to ACTH and MSH for their analgesic effects following local administration in the periaqueductal gray matter of the rat [Walker et al., 1980, 1981]. These experiments generally involved surgically implanting a small diameter cannula guide above the PAG, and following a week's recovery period,

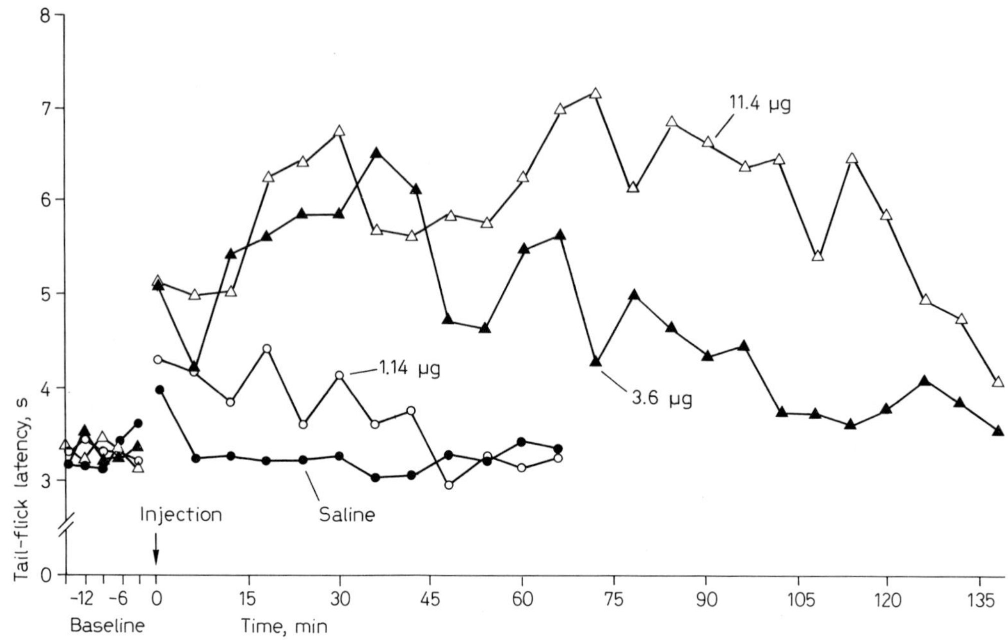

Fig. 2. Effects of β-endorphin on tail-flick latency. Various doses of β-endorphin were microinjected into the periaqueductal gray of rats.

the animals were tested using the tail-flick test before and after microinjection (in the PAG) of one of several ACTH and MSH-like peptides. Although we did not carry out other pain tests in a systematic fashion, the sensitivity of the rats to pricks or pinches was also examined and were consistent with the results of the tail-flick testing.

Early studies clearly showed the potent effects of several MSH-like peptides derived from POMC. Potent but short-lasting elevations of tail-flick latency were seen with α-MSH and des-acetyl-α-MSH. The effects of ACTH 1-24 were found to be quite a bit more potent than the shorter MSH fragments. Likewise, ORG 2766 was found to be far more potent than the natural products. ORG 2766 ($[MetO_2]^4$-$[D\text{-}Lys]^8$-$[Phe]^9$-ACTH 4-9) is a substituted analog of ACTH 4-9 which shows greatly enhanced potency in its effects on memory compared to the parent compound [de Wied et al., 1975]. As shown in figure 3, ORG 2766 produces markedly more potent analgesic effects than any of the other ACTH and MSH fragments tested. Presumably this enhancement of activity is due to

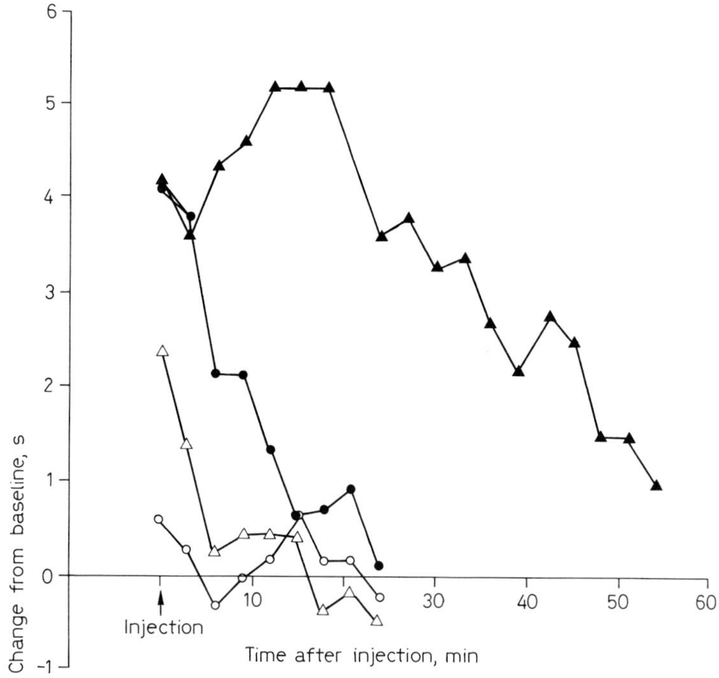

Fig. 3. Effects of α-MSH des-(ac)-α-MSH and ORG 2766 on tail-flick latency, expressed as change from baseline. The peptide (35 nmol) was injected into the periaqueductal gray via permanently implanted cannulas. ○ = saline; △ = α-MSH; ● = [des-acetyl]α-MSH; ▲ = ORG 2766.

increased resistance to enzymatic breakdown. These experiments suggested that the nonopioid members of POMC could also inhibit pain sensitivity, although not as potently as β-endorphin. The results also indicated that the same set of neurons can give rise to both opiate-mediated and nonopiate mediated analgesias.

The high potency of ORG 2766 and its apparent similarity to the endogenous substances in the production of analgesia made it a good tool for the further study of ACTH/MSH analgesia. One question of great interest was that of the interaction of the ACTH/MSH peptides with β-endorphin, since the two classes of substances do appear to be released together. Consequently, we examined the interaction between the ACTH 4-9 analog, ORG 2766, and β-endorphin in the production of analgesia. In these experiments, β-endorphin was microinjected alone or in combination with a subanalgesic dose of ORG 2766 or an analgesic

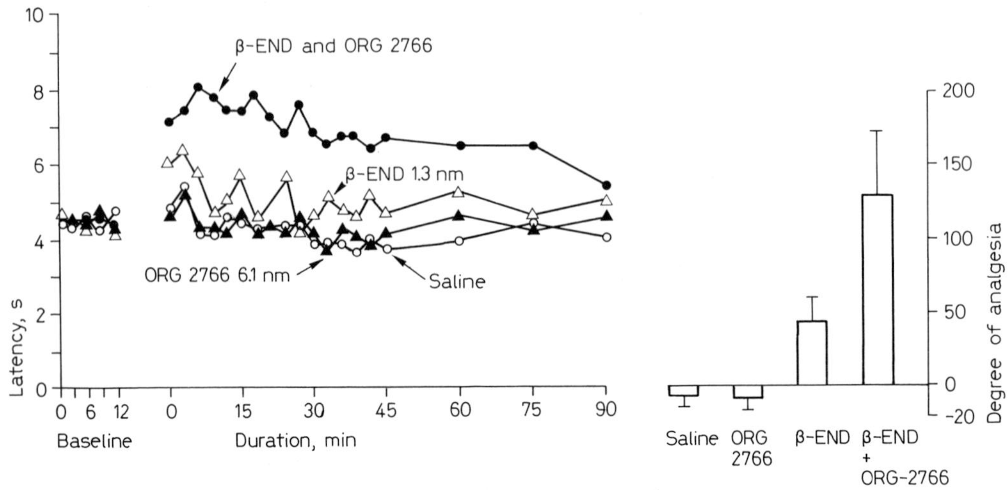

Fig. 4. Left panel: Effect of β-endorphin on tail-flick latency in the presence and absence of a sub-analgesic dose of ORG 2766, and analog of $ACTH_{4\text{-}9}$. Right panel: Degree of analgesia ± SEM was derived from the same data using methods described in the text. β-endorphin was significantly more potent in the presence of ORG 2766.

dose of ORG 2766. At both doses of ORG 2766, β-endorphin analgesia was enhanced, suggesting an additive or synergistic interaction. As a control for nonspecific effects due to the presence of another peptide, a group of rats was injected with β-endorphin as above, in combination with poly-*L*-lysine. We found that the analgesic effects of β-endorphin, which were enhanced by ORG 2766, were unaffected by poly-*L*-lysine (fig. 4–6). These findings support the possibility of physiological interactions between α-MSH/ACTH-like peptides and β-endorphin.

γ-MSH and Pain Sensitivity

$\gamma\text{-MSH}_{1\text{-}12}$ and $\gamma\text{-MSH}_{1\text{-}13}$ were tested for analgesia, but these substances failed to affect pain sensitivity in a significant way. In the absence of any clear direct effects, we hypothesized that γ-MSH might modulate the activities of other POMC products. Consequently, we carried out tests where ACTH analgesia was assessed in the presence and absence of γ-MSH. We found that both $\gamma\text{-MSH}_{1\text{-}13}$ and $\gamma\text{-MSH}_{1\text{-}12}$

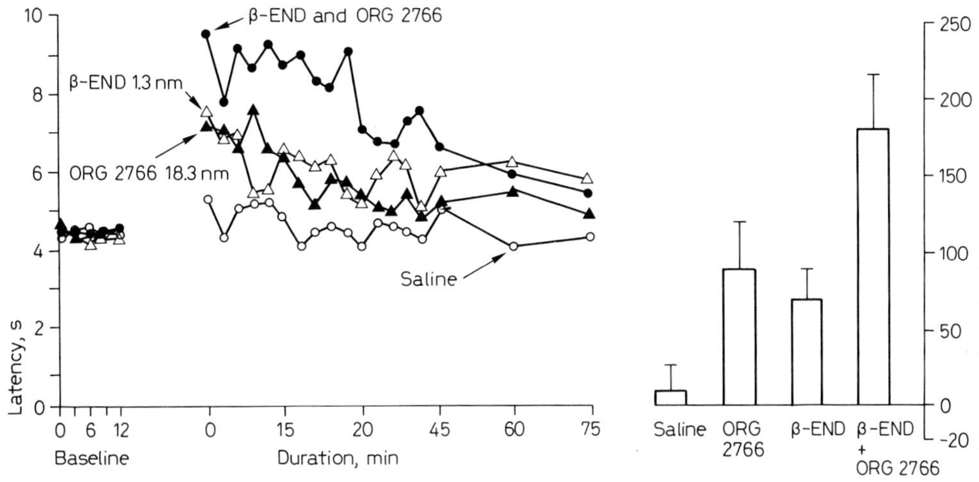

Fig. 5. Left panel: Effect of β-endorphin on tail-flick latency in the presence and absence of an analgesic dose of ORG 2766. Right panel: Degree of analgesia ± SEM was derived from the data on left using methods described in the text. β-endorphin was significantly more potent in the presence of ORG 2766.

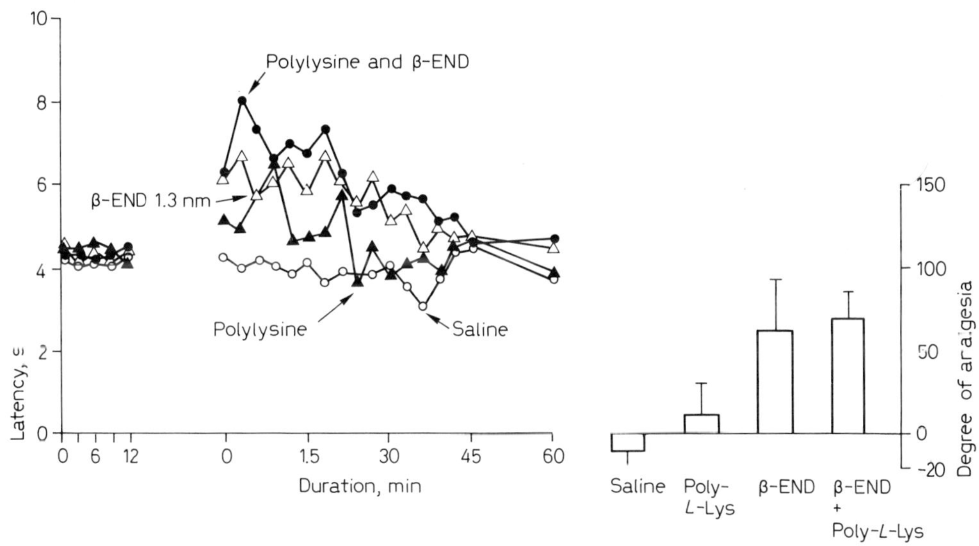

Fig. 6. Left panel: Effect of β-endorphin on tail-flick latency in the presence and absence of an unrelated protein, poly-*L*-lysine. Right panel: Degree of analgesia ± SEM was derived from data on left. Poly-*L*-lysine had no effect of its own and no significant effect on β-endorphin analgesia.

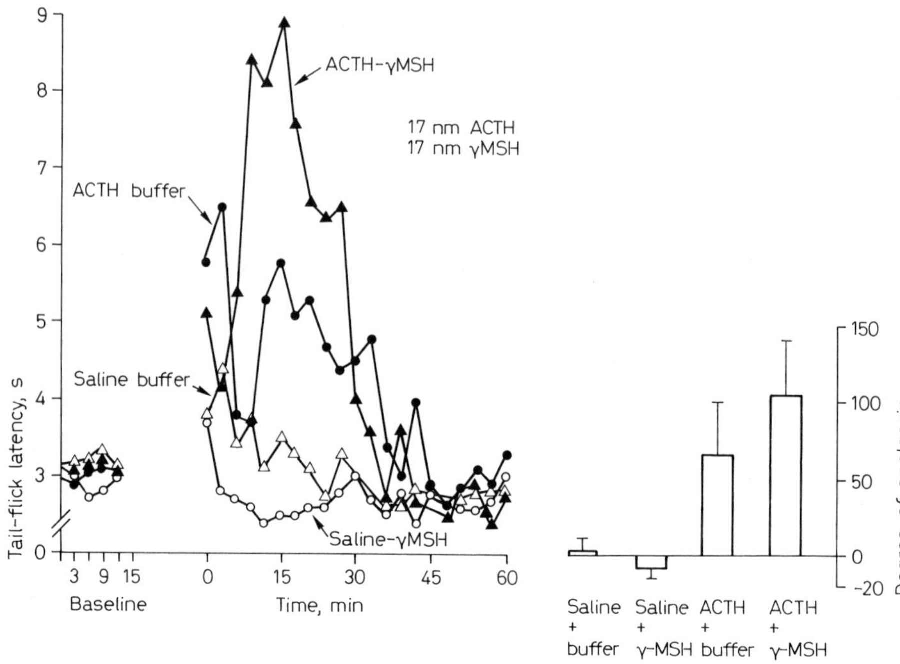

Fig. 7. Left panel: Effect of ACTH-(1-24) (17 nmol) on tail-flick latency in the presence and absence of γ-MSH-(1-12) (17 nmol). Right panel: Degree of analgesia was ± SEM derived from the data on the left using methods described in the text. Analgesia induced by ACTH-(1-24) was significantly greater in the presence of γ-MSH-(1-12).

enhanced the analgesia produced by ACTH (fig. 7, 8). It therefore seemed possible that γ-MSH might function naturally to mediate the activity of α-MSH/ACTH-like substances.

Our experiments suggest that a variety of α-MSH/ACTH-like peptides (derived from the midportion of POMC) produce analgesia and potentiate β-endorphin-induced analgesia. The γ-MSH like peptides from the amino terminal region of POMC differ structurally from α-MSH and ACTH and produce no analgesic response. However, these products appear to enhance the analgesic actions of the α-MSH/ACTH-like peptides. The interactions of γ-MSH with β-endorphin remain unknown.

In a seminal article on the structure-activity relationship of ACTH and MSH-like peptides, Schwyzer and Eberle [1977] suggested that dif-

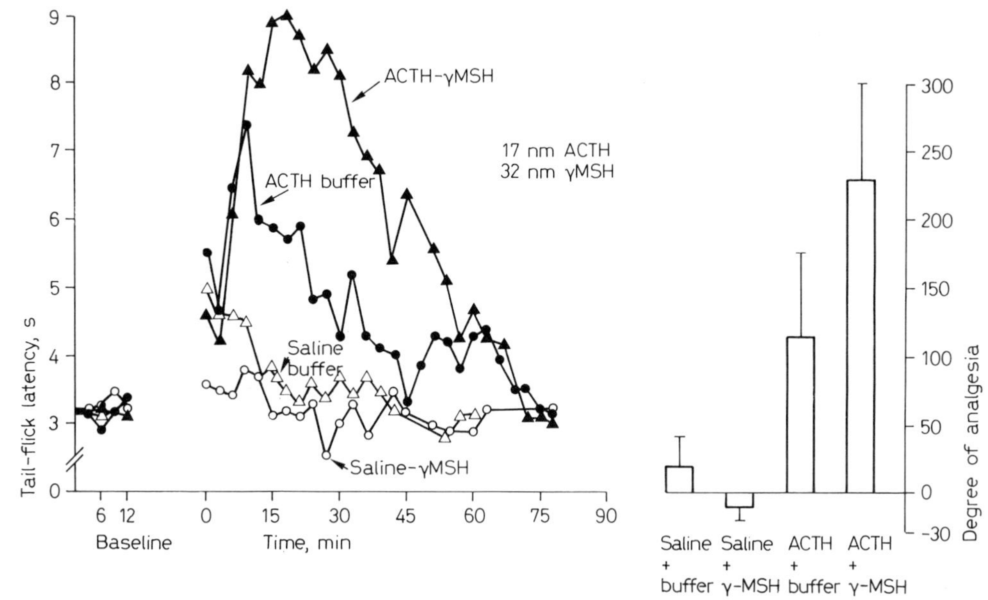

Fig. 8. Left panel: Effect of ACTH-(1-24) (17 nmol) on tail-flick latency in the presence and absence of γ-MSH-(1-12) (32 nmol). Right panel: Degree of analgesia ± SEM was derived from the data on the left using methods described in the text. ACTH was significantly more potent in the presence of γ-MSH.

ferent regions of a peptide chain encode different biological activities. These investigators suggested that a portion of the peptide might serve as an address sequence which is critical for receptor binding, whereas other parts of the sequence may be critical for biological activity or efficacy and still others may protect against enzymatic breakdown. Highly relevant to the synergism between ACTH and γ-MSH discussed above was the evidence for sequences which have little biological activity in themselves, but potentiate the action of other sequences. This potentiation is also observed in the periphery where γ-MSH potentiates the steroidogenic action of ACTH, but alone has little or no effect [Pederson et al., 1980]. If γ-MSH is proved to be a 'potentiator peptide' it might be thought of as a gain control which can govern the efficacy of the synaptic actions of other POMC products.

All of the interactions that we have seen among POMC products are additive or synergestic. We have not observed any antagonistic responses among any of these products. However, this may be reflective of their activities at the level of the PAG, but may not be true of all neuronal systems. A number of investigators have observed antagonism between ACTH-like peptides and opiates [Smock and Fields, 1981; Jacquet, 1978; Gispen et al., 1975]. These investigations have in common the use of ventricular administration of the peptides. Therefore, while the precise site of action is unknown, these experiments clearly document the presence of some antagonistic actions.

Antagonistic interactions have also been observed with fragments other than those tested in our experiments. One such candidate is β-endorphin 1-27 which was found to antagonize morphine analgesia [Nicholas and Li, 1985]. The precise mechanism for this effect is rather unclear, since the fragment along with the other fragments which are cleaved at the carboxyl terminal region and/or acetylated at the amino-terminal show greatly diminished potency at opiate receptors [Deakin et al., 1980; Akil et al., 1981]. However, these findings are potentially of great significance, because recent data indicate that with repeated exposure to stress rat POMC neurons show an increased preference toward producing this sort of molecular species. It has also been suggested that a dipeptide (glycylglutamine), formed as a consequence of these carboxyl terminal cleavages, has potent inhibitory effects after iontophoretic application in the vicinity of neurons in the reticular formation [Smyth et al., 1983]. The interaction of this peptide with β-endorphin or the ACTH-like analgesics is completely unknown at this time.

Mechanisms of Action

The potent effect of naloxone on β-endorphin-induced analgesia clearly indicates the role of opiate receptors in these actions. Binding studies indicated that β-endorphin is potent at μ- and δ-receptors but is not active at kappa receptors [Patterson et al., 1983]. It has also been suggested that β-endorphin interacts with a specialized ε-receptor that can be studied best in the rat vas deferens [Schulz et al., 1979]. In this preparation almost all opiates other than β-endorphin are completely inactive. However, it is unknown whether such receptors exist in the brain. Studies of the binding of ^{3}H-β-endorphin and ^{125}I-β-endorphin do not offer much support either, since any number of opiates which do not affect its actions in the rat vas deferens do inhibit binding in rat brain homogenates [Akil et al., 1980].

Although some nonopiate effects of des-Tyr fragments derived from β-endorphin have been reported, there is no hint that they might have a role in pain inhibition [de Wied et al., 1978; Hazum et al., 1979]. The mechanism of opioid antagonism by β-endorphin (1-27) is also unknown. The effect could be mediated by opiate receptor anatagonism or, perhaps just as likely, through physiological interactions utilizing some other mechanism. Regarding the C-terminal dipeptide, its mechanism of action is also unknown.

Mechanisms of ACTH/MSH Actions

The mechanism of action of the ACTH and MSH-like peptides has been investigated, but far more work is needed to characterize the biochemical events underlying the behavioral and physiological effects of these substances in the brain. Since ACTH 1-24 can induce some naloxone reversible effects, and since it binds with low affinity to opiate receptors [Gispen et al., 1975, 1977; Terenius et al., 1976; Akil, et al., 1980], it seemed possible that opiate receptors might mediate some of the effects of the ACTH-like peptides discussed above. We therefore attempted to reverse the analgesic effects of ORG 2766 with naloxone, and we tested for cross tolerance to morphine. Neither treatment had any effect so we concluded that some other mechanism must be responsible.

The obvious possibility that we entertained was that of an ACTH/α-MSH receptor in the brain. The likelihood of such a receptor seemed great in view of the findings above and the report by Verhoef and Witter [1977] that septal uptake of a ^{3}H-ACTH$_{1\text{-}9}$ analog was specifically reduced by other ACTH-like substances, notably the 1-24 fragment. Our own studies of the binding ^{3}H-α-MSH to bovine brain homogenates tended to support this view. In spite of the apparently small number of these receptors, we were able to observe some specific binding (about 30%) which appeared to be highest in the bovine septum. Although further work is needed, it does appear that the CNS actions of the α-MSH/ACTH like substances might be edited by an MSH receptor in the brain. Whether other MSH-like peptides interact with these possible α-MSH receptors is uncertain.

Functions of Corelease

The experiments discussed above support the concept of synergistic or additive interactions among POMC products in the CNS. These experiments focussed on presumed postsynaptic events but other possibilities should be entertained as well. A few hypotheses under consideration are discussed below:

Postsynaptic vs. Autoreceptor Function

The possibility has been raised that some transmitters may exert their effects postsynapically while others may provide feedback via autoreceptors. One such candidate is dynorphin which is costored with vasopressin in the magnocellular sypraoptic and paraventricular system of the hypothalamus. Recent results indicate that intraventricularly applied dynorphin or other kappa-opiate agonists cause diuresis by inhibiting the release of vasopressin from these cells [Leander, 1983]. Therefore, it appears that dynorphin may act via autoreceptors to inhibit the release of vasopressin. Under natural conditions, this suggests that vasopressin has postsynaptic functions and that dynorphin acts locally as part of a self-regulatory autoreceptor system.

Differential Release as a Function of Firing Rate

One interesting possibility is that neurons might differentially release an excitatory and an inhibitory transmitter as a function of the rate of discharge of the neuron. The evidence for this arises from studies of the preganglionic innervation of the urinary bladder, where acetylcholine is colocalized with enkephalin. De Groat et al. [in press], reported that the usual postganglionic discharge elicited by stimulation of a branch of the pelvic nerve is inhibited by repetitive (20–30 Hz, 3–5 s) stimulation of another branch of the same nerve. This response was sensitive to naloxone. The experiment suggests that at low firing rates the nerve has an excitatory effect and that at higher rates the release of enkephalin predominates resulting in inhibitory actions. Experiments such as this are of course very difficult to carry out in the CNS, but the possibility that the release of opioids is dependent upon the frequency of discharge has profound implications for the understanding of synaptic transmission. In addition, the long-standing question of why naloxone alone has so few effects in the normal organism seems to be answered if it is the case that opioids are released only at high rates of discharge.

Cardiovascular Effects

The work of Lassen et al. (1978), as well as others, have shown the dramatic changes in local cerebral blood flow as a function of cerebral activity. For example, subjects exhibit marked changes in blood flow in Brocas's area during the production of speech. Obviously, the brain has a method for informing the local vasculature of the functional state of the neighboring neurons. Could it be that some neuronal release products serve as neurotransmitters relaying messages to other neurons, while other products relay a message to the local vasculature? Although speculative, it does appear possible that some nonopioid effects of dynorphin and enkephalins could engage the cardiovascular system. The evidence for nonopioid effects of a variety of 'opioid peptides' is rather well documented [de Wied et al., 1978; Walker et al., 1985; Hazum et al., 1979]. One very early paper [Knoll, 1976] identified some nonopioid effects of enkephalin on arteries, while other work which has shown long-lasting impairments in motor function [Faden and Jacobs, 1984] (paralysis, circling, barrel rolling) from a number of peptides. These long-lasting pathological changes could have involved traumatic effects on the vasculature which ultimately produced tissue destruction.

Undoubtedly, one could generate many other hypotheses about the function of the multiple release of several transmitters from a single cell. If history repeats itself, the field of opioid peptides will lead the way in this field. Perhaps the end result will be more efficacious treatments for pain.

Acknowledgements

This work was supported by PHS grants IF32DA05183 and NS21745 to J. Michael Walker, DA02265 to Huda Akil and Stanley Watson, and by NSF grant BNS8004512 to Stanley Watson and Huda Akil. We are deeply grateful to Guilo Baldrighi, Betty Peters, Diane Pace and Cindy Beaulieu for their expert technical assistance. We also thank Organon, Ciba-Geigy, Dr. C.H. Li, and Dr. David H. Coy for the generous gifts of peptides.

References

Akil, H., Hewlett, W.A., Barchas, J.D.; Li, C.H.: Binding of ^{3}H-β-endorphin to rat brain membranes: Characterization of opiate properties and interaction with ACTH. Eur. J. Pharmacol., *64:* 1–8 (1980).

Akil, H.; Watson, S.J.: Neuromodulatory functions of the brain pro-opiocortin system; in Costa, Trabucchi, Neural peptides and neuronal communications, pp. 435–445 (Raven Press, New York 1980).

Akil, H.; Ueda, T.; Lin, H.L.; Lin, H.L.; Lewis, J.W.; Walker, J.M.: Shiomi, H.; Liebeskind, J.C.; Watson, S.J.: Multiple forms of β-endorphin (B–E) in pituitary and brain. Effect of stress; in Takagi, Simon, Advances in endogenous and exogenous opioid peptides, pp. 116–118 (Kodansha Press, Tokyo 1981).

Akil, H.; Young E.; Watson, S.J.; Opiate binding properties of naturally occurring N- and C-Terminus modified beta-endorphins. Peptides *2:* 289–292 (1981).

Akil, H.; Shiomi, H.; Walker, J.M.; Watson, S.J.: Characterization of multiple forms of β-endorphin in pituitary and brain: effect of stress; in Costa, Tarbucchi, Regulatory peptides: functional and pharmacological aspects (Raven Press, New York, in press).

Akil, H.; Richardson, D.E.; Barchas, J.D.; Li, C.H.: Appearance of β-END-like immunoreactivity in human ventricular cerebrospinal fluid upon analgesic electrical stimulation. Proc. natn. Acad. Sci. USA *75:* 170–172 (1978).

Benjannet, S.; Seidah, N.G.; Routhier, R.; Chretien, M.: A novel human pituitary peptide containing the γ-MSH sequence. Nature, Lond. *285*; 415–416 (1980).

Bloom, F.E.; Rossier, J.; Battenberg, E.; Bayon, A.; French, E.; Henriksen, S.J.; Siggins, G.R.; Segal, D.; Browne, R.; Ling, N.; Guillemin, R.: β-Endorphin: cellular localization, electrophysiological and behavioral effects; in Costa, Tabucchi, The endorphins. Advances in biochemical psychopharmacology, pp. 89–101 (Raven Press, New York 1978).

Botticelli, L.J.; Wurtman, R.J.: Corticotropin regulates trans-synaptically the activity of septo-hippocampal neurons. Nature, Lond. *289:* 75–76 (1981).

Chan-Palay, V.; Palay S.L.: Coexistence of neuroactive substances in neurons (Wiley, New York 1984).

D'Amour, F.E.; Smith, D.L.: A method for determining loss of pain sensation. J. Pharmac. exp. Ther. *72:* 74–79 (1941).

Deakin, J.F.; Dostrovsky, J.O.; Smyth, D.G.: Influence of N-terminal acetylation and C-terminal proteolyesis on the analgesic activity of β-END. Biochem. J. *89:* 501–506 (1980).

De Groat, W.C.; Kawatani, M.; Booth, A.M.; Whitney, T.: Enkephalinergic modulation of cholinergic transmission in parasympathetic ganglia of the urinary bladder of the cat. Abstr. Conf. Dynamics of Cholinergic Function, Ogleby 1984.

De Groat, W.C.; Kawatani, M.; Hisamitsu, T.; Lowe, I.; Morgan, C.; Roppolo, J.; Booth, A.M.; Nadelhaft, I.; Kuo, D.; Thor, K.: The role of neuropeptides in the sacral autonomic reflect pathways of the cat. J. autonom. nerv. Syst. *7:* 339-350 (1983).

De Groot, J.: The rat forebrain in sterotaxic coordinates. Verh. Hon. Ned. Akad. Wet *52:* 1 (1959).

De Wied, D.; Kovacs, G.L.; Bohus, B.; Van Ree, J.M.; Greven, H.M.: Neuroleptic activity of the neuropeptide β-LPH-$_{62-77}$ ([Destyr] γ-endorphin; DTγE). Eur. J. Pharmacol. *49:* 247 (1978).

De Wied, D.; Witter, A.; Greven, H.M.: Behaviorally active ACTH analogs. Biochem. Pharmac. *24:* 1463–1468 (1975).

Eipper, B.; Mains, R.E.: Existence of a common precursor to ACTH and endorphin in the anterior and intermediate lobes of the rat pituitary. J. supramol. Struct. *8:* 247–262 (1978).

Emerson, R.B.; Eipper, B.A.: Characterization of pro-ACTH/endorphin-derived peptides in rat hypothalamus. J. Neurosci. (in press, 1986).

Faden, A.I.; Jacobs, T.P.: Dynorphin related peptides cause motor dysfunction in the rat through a nonopiate action. B. J. Pharmacol. *81:* 271–276 (1984).

Gianoulakis, C.; Seidah, N.G.; Chretien, M.: In vitro biosynthesis and chemical characterization of ACTH and ACTH fragments by the rat pars intermedia; in Leong Way, Endogenous and exogenous opiate agonists and antagonists, pp. 289–292 (Pergamon Press), New York 1979).

Gispen, W.H.; Reith, M.E.A.; Schotman, P.; Wiegant, V.W.; Zwiers, H.; Wied, D. de: CNS and ACTH-like peptides. Neurochemical response and interaction with opiates; in Miller, Sandman, Kastin, Neuropeptide influences on the brain and behavior (Raven Press, New York, 1977).

Gispen, W.H.; Van Wimersma Greidanus, Tj. B.; Waters-Erzin, C.; Zimmerman, E.; Krivoy, W. A.; De Wied, D.: Influence of peptides on reduced response of rats to electric footshock after acute administration of morphine. Eur. J. Pharmacol. *33:* 99 (1975).

Hazum, E.; Chang, K.-J.; Cuatrecasas, P.: Specific nonopiate receptors for β-endorphin. Science *205:* 1033–1035 (1979).

Hokfelt, T.; Lundberg, J.M.; Schulteberg, M.; Johansson, O.; Ljungdahl, A.; Rehfeld, J.: Coexistence of peptides and putative transmittes in neurons; in Costa, Trabucchi, Neural Peptides and neuronal communication, pp. 1–23 (Raven Press, New York 1980).

Jacquet, Y.F.: Opiate effects after adrenocorticotropin or β-endorphin in the periaqueductal gray matter of rats. Science *201:* 1032–1034 (1978).

Kastin, A.J.; Olson, R.D.; Schally, A.V.; Coy, D.H.: CNS effects of peripherally administered brain peptides. Life Sci. *25:* 401–414 (1979).

Knoll, J.: Neuronal peptide (enkephalin) receptors in the ear artery of the rabbit. Eur. J. Pharmacol. *39:* 403–407 (1976).

Lassen, N.A.; Ingvar, D.H.; Skinhoj, E.: Brain function and blood flow. Scient. Am. *239:* 62–71 (1978).

Leander, J.D.: A kappa opioid effect. Increased urination in the rat. J. Pharmac. exp. Ther. *224:* 89–94 (1983).

Liebeskind, J.C.: Pain modulation by central nervous system stimulation; Boznia, Albe-Fessard, Advances in pain research and therapy, vol. 1, pp. 445–453 (Raven Press, New York 1976).

Ling, N.; Ying S.; Minich, S.; Guillemin, R.: Systhesis and biological activity of four γ-melanotropin hormones derived from the cryptic region of the adrenocortincotropin/β-lipotropin precursor. Life Sci. *25:* 1772–1780 (1979).

Loh, H.H.; Tseng, L.F.; Wei, E.; Li, C.H.: β-Endorphin is a potent analgesic agent. Proc. natn. Acad. Sci. *73:* 2895–2898 (1976).

Mains, R.E.; Eipper, B.A.; Ling, N.: Common precursor to corticotropins and endorphins. Proc. natn. Acad. Sci. USA *74:* 3014–3018 (1977).

Nakanishi, S.; Inoue, A.; Kita, T.; Nakamura, M.; Chang, A.C.Y.; Cohen, S.N.; Numa, S.: Nucleotide sequency of cloned cDNA for bovine corticotropin-β-lipotropin precursor. Nature, Lond. *278:* 423–427 (1979).

Nicholas, P.; Li, C.H.: Beta-endorphin (1-27) is a naturally occuring antagonist to etorphine-induced analgesia. Proc. natn. Acad. Sci. USA *82:* 317–318 (1985).

Paterson, S.J.; Robson, L.E.; Kosterlitz, H.W.: Classification of opioid receptors. Br. med. Bull. *39:* 31–36 (1983).

Pederson, R.C.; Brownie, A.C.; Ling, N.: Pro-adrenocorticotropin/endorphin-derived peptides. Coordinate action on adrenal steroidogenesis. Science *208:* 1044–1046 (1980).

Pelletier, G.: Ultrastructural localization of a fragment (16K) of the common precursor for adrenocorticotropin (ACTH) and β-lipotropin (β-LPH). Neurosci. Lett. *16:* 85–90 (1980).

Pelletier, G.; Dube, D.: Electron microscopic immunohistochemical localization of α-MSH in the rat brain. Am. J. Anat. *150:* 201–206 (1977).

Prezewlocki, R.; Shearman, G.T.; Herz, A.: Mixed opioid/nonopioid effects of dynorphin and dynorphin-related peptides after intrathecal injection in rats. Neuropeptides *3:* 233–240 (1983).

Roberts, J.D.; Herbert, E.: Characterization of a common precursor to corticotrophin and β-lipotropin. Proc. natn. Acad. Sci. *74:* 5300–5304 (1977).

Robinson, T.E.: Hippocampal rhythmic slow activity (RSA; V THETA). A critical analysis of selected studies and discussion of possible species-differences. Brain. Res. Rev. *2:* 69–101 (1980).

Schwyzer, R.; Eberle A.: On the molecular mechanisms of α-MSH receptor interactions; in Swaab, van Wimersma Greidanus, Frontiers of hormone research, melanocyte stimulating hormone: control chemistry and effects; pp. 18–25 (Karger, Basel, 1977).

Shibasaki, T.; Ling, N.; Guillemin, R.: A radioimmunoassay for γ-melanotropin and evidence that the smallest γ-melanotropin is amidated at the COOH-terminal. Biochem. biophys. res. Commun. *96:*1393–1399 (1980).

Shibasaki, T.; Ling, N.; Guillemin, R.: Pituitary immunoreactive γ-melanotropins are glycosylated oligopeptides. Nature. Lond. *285:* 416–417 (1980).

Simon, E.J.; Hiller, J.M.: The opiate receptors. A. Rev. Pharmac. Tox. *18:* 71–94 (1978).

Smock, T.; Fields, H.L.: ACTH-(1-24) blocks opiate induced analgesia in the rat. Brain Res. *212:* 202–206 (1981).

Smyth, D.G.; Parish, D.C.; Normanton, J.R.; Wolstencroft, J.H.: The C-terminal dipeptide of β-endorphin. A neuropeptide with inhibitory activity. Life Sci. *33:* suppl. 1, pp. 575–578 (1983).

Tanaka, I.; Nakai, Y.; Jinganni, H.; Fukata, J.; Nakao, K.; Oki, S.; Nakanishi, S.; Numa, S.; Imura, H.: Existence of γ-melanotropin (γ-MSH)-like immunoreactivity in bovine and human pituitary glands. Biochem. biophys. res. Commun. *94:* 211–217 (1980).

Terenius, L.: Somatostatin and ACTH are peptides with partial agonist-like selectivity for opiate receptors. Eur. J. Pharmacol. *33:* 395 (1976).

Tsou, K.; Jang, C.S. Studies of the site of analgesic, action of morphine by intracerebral microinjection. Sci. Sin. *8:* 1099–1109 (1964).

Urca, G.; Frenk, H.; Liebeskind, J.C.; Taylor, A.C.: Morphine and enkephalin: analgesic and epileptic properties. Science *197:* 83–66 (1977).

Van Ree, J.M.; Bohus, B.; Csontos, K.M.; Gispen, W.H.; Greven, H.M.; Nijkamp, F.P.; Opmeer, F.A. de Rotte, G.A.; Wimersma Greidanus, T.B. van; Witter, A.; Wied, D. de: Behavioral profile of γ-MSH; Relationship with ACTH and β-endorphin action. Life Sci. *28:* 2875–2888 (1981).

Verhoef, J.; Witter, A.; Weid, D. de: Specific uptake of a behavioral potent ^{3}H-ACTH$_{1-9}$

analog in the septal area after intraventricular injection in rats. Brain Res. *131:* 117–128 (1977).

Walker, J.M.; Berntson, G.G.; Sandman, C.A.; Coy, D.H.; Kastin, A.J.; Schally, A.V.: An analog of enkephalin having prolonged opiate-like effects in vivo. Science *196:* 85–86 (1977).

Walker, J.M.; Akil, H.; Watson, S.J.: Evidence for homologous action of pro-opiocortin products. Science *210:* 1247–1249 (1980).

Walker, J.M.; Katz, R.J.; Akil, H.: Behavioral effects of $dynorphin_{1\text{-}13}$ in the mouse and rat. Initial observations. Peptides *1:* 341–345 (1980).

Walker, J.M.; Berntson, G.G.; Sandman, C.A.; Kastin, A.J.; Akil, H.: Induction of analgesia by central administration of ORG 2766, an analog of $ACTH_{4\text{-}9}$. Eur. J. Pharmacol. *69:* 71–79 (1981).

Walker, J.M.; Moises, H.; Coy, D.H.; Baldrighi, G.; Akil, H.: Nonopiate effects of dynorphin and des-Tyr-dynorphin. Science *218:* 1136–1138 (1982).

Walker, J.M.; Tucker, D.; Walker, B.; Akil, H.: Des-tyrosine-dynorphin inhibits morphine analgesia. Eur. J. Pharmacol. *85:* 121–122 (1982).

Walker, J.M.; Moises, H.C.; Friederich, M.W.: A review of some nonopioid actions of dynorphin. Endocoids. *1985:* 309–312.

Watson, S.J.; Akil, H.; Richard, C.W.; Barchas, J.D.: Evidence for two separate opiate peptide neuronal systems. Nature, Lond. *275:* 226–228 (1978).

Watson, S.J.; Akil, H.: α-MSH in rat brain: occurrence within and outside of β-endorphin neurons. Brain Res. *182:* 217–223 (1980).

Watson, S.J.; Akil, H.; Fischli, W.; Goldstein, A.; Zimmermann, E.; Nilauer, G.; Wimersma-Greidanus, Tj. B. van: Dynorphin and vasopressin. Common localization in magnocellular neurons. Science *216:* 85–87 (1982).

Wiegant, V.M.; Gispen, W.H.; Terenius, L.; Wied, D. de: ACTH-like peptides and morphine: interaction at the level of the CNS. Neuroendocrinology *2:* 63–69 (1977).

Yaksh, T.L.; Rudy, T.A.: Narcotic analgesics. Pain *4:* 299–359 (1978).

Zakarian, S.; Smyth, D.: Distribution of active and inactive forms of endorphins in rat pituitary and brain. Proc. natn. Acad. Sci. *76:* 1972–1976 (1979).

Zimmermann, E.; Krivoy, W.A.: Antagonism between morphine and the polypeptides ACTH, $ACTH_{1\text{-}24}$, & α-MSH in the nervous system; in Zimmermann, Gispen, Marks, de Weid, Drug effects on neuroendocrine regulation. Progress in brain research, vol. 30, p. 383 (North-Holland, Amsterdam 1973).

J. Michael Walker, PhD, Department of Psychology,
Brown University, Providence, RI 02912 (USA)

Pain Headache, vol. 9, pp. 178–193 (Karger, Basel 1987)

Modulation of Morphine Analgesia by Opioid Peptides

Implications for Antinociceptive Processes

Jeffry L. Vaught[a], *Ronald W. Barrett*[b, 1]

[a] Department of Biological Research, McNeil Pharmaceutical, Spring House, Pa., USA; [b] Addiction Research Foundation, Palo Alto, Calif., USA

Introduction

It has been over 10 years since the initial report by Hughes [1975] of the isolation of an endogenous compound from the brain with pharmacological properties similar to those of morphine. Shortly thereafter this same group isolated, purified, sequenced and initiated the characterization of methionine and leucine enkephalin [Hughes et al., 1975]. Since that time, hundreds of reports have been published providing a solid foundation from which a rapidly developing pharmacological and biochemical profile is emerging characterizing this class of neuropeptides. This information has led to suggestions of opioid peptide involvement in a variety of physiological and pathophysiological conditions including nociception, mediation of satisfaction and reward [Belluzzi and Stein, 1977], schizophrenia [Bloom et al., 1976], hypovolemic or endotoxic shock [Holaday and Faden, 1978], memory process [Rigter, 1978], and general cardiovascular function [Laubie et al., 1977] to name a few [for review, see Olson et al., 1981]. One area of opioid peptide research which has received little attention is the potential of these peptides to act as 'neuromodulators', as opposed to their acting consistently as 'neurotransmitters'. Barker et al. [1978a, b] were among the first to provide evidence of a neuromodulatory-like action of leucine enkephalin.

[1] The authors wish to express their appreciation to Mrs. Pat Brennan for her careful preparation of this manuscript.

We have been particularly interested in the potential modulatory actions of the enkephalins and have shown that, despite their similarity to pure narcotic agonists in in vitro bioassay systems, marked differences exist in their ability to modify analgesia induced by morphine and related opioid alkaloids and peptides [Vaught and Takemori, 1978, 1979a, b; Vaught et al., 1982; Barrett and Vaught, 1982]. Subanalgesic doses of leucine enkephalin and certain of its analogs enhance, while methionine enkephalin and certain of its analogs attenuate the analgesic effects of opioids.

The existence of distinct enkephalin-preferring or δ-receptors and morphine-preferring or μ-receptors in the CNS has caused us to question whether the actions of the enkephalins on opioid-induced analgesia are mediated through an interaction with one or both of these subtypes of opioid receptors. Evidence from our laboratory [Vaught et al., 1982; Barrett and Vaught, 1982] and that of Lee et al. [1980] supports the hypothesis that a specific receptor mediated process is, in fact, the initial event in this opioid peptide-induced modulation of analgesia. We have speculated that this modulation is mediated via a specific interaction with the δ opioid receptor.

In this chapter we will present a compilation of in vivo data which supports our hypothesis of a δ-receptor modulation of μ-receptor mediated analgesia. Our purpose is to suggest that these data may provide an alternative way of viewing the role of opioid peptides in nociception and perhaps other physiological processes.

General Approach

The series of studies described below tested either the direct analgetic properties of the opioids or their ability to modulate morphine analgesia. These studies were carried out either in morphine naive or in morphine-tolerant mice, to test their modulatory influences under these two conditions.

To assess analgesia, the tail-flick assay, as has been previously described [Vaught and Takemori, 1979a], was used in all instances. Ten, fifteen and twenty minutes after intracerebroventricular (i.c.v.) administration of drug (in a maximum volume of 5 μl), tail-flick reaction times were determined and compared to predrug reaction times. A reaction time for a drug-treated animal greater than three standard deviations from the mean of the control reaction times for all the animals in the group was the criterion for an analgesic response.

To assess the effects of sub-analgesic doses of various opioid peptides on morphine-induced analgesia, all combinations of drugs were injected simultaneously (i.c.v.), as previously described [Vaught et al., 1982].

A minimum of three doses with 10 mice/dose was used to construct dose-response curves. ED_{50} values, their 95% confidence limits and the significance of the potency ratio between two ED_{50} values were determined by the method of Litchfield and Wilcoxon [1949].

A single dose of morphine was used to produce acute tolerance [Vaught and Takemori, 1979a]. Morphine sulfate (100 mg/kg) was administered subcutaneously (s.c.) and analgesic assays performed as described previously when reaction times had returned to predrug control values (generally by 3.5 h). Those mice whose latencies failed to return to predrug control values were discarded.

For chronic tolerance studies, mice were each implanted with one 75-mg morphine pellet in the subcutaneous space of the neck, under light anesthesia. The morphine pellets were previously wrapped in a loose-fitting nylon mesh to facilitate their subsequent removal. After 72 h, pellets were removed (again under light ether anesthesia) and mice were allowed to recover for 3 h prior to testing. The evaluation of analgesia was performed as described above.

Male Swiss-Webster (Blue Spruce Farms, Altamount, NY) or Swiss CD-1 (Charles Rivier, Kingston, NY) mice weighing between 18 and 24 g were used in all experiments. All mice were supplied with food and water ad libitum and were housed for at least 1 day prior to use, and were used only once.

All drugs and peptides (table I) were dissolved in either saline or distilled water on the same day they were used. Methionine enkephalin was initially dissolved in a minimal amount of 0.01 *N* acetic acid, neutralized with NaOH and diluted to volume with saline.

Analgesic Potencies of Opioid Peptides and Their Effects on Morphine Analgesia in Morphine Naive Mice

For ease of presentation, most of the peptides have been abbreviated as shown in table I. All of the synthetic enkephalin analogs as well as

Table I. Peptide abbreviations

LE: Leu-enkephalin	DPDPE: (D-Pen^2, D-Pen^5)-enk
ME: Met-enkephalin	DAMA: (D-Ala^2)-met-enkamide
DADLE: (D-Ala^2, D-Leu^5)-enk	DMPA: (D-Met^2, Pro^5)-enkamide
DSTLE: (D-Ser^2, Thr^6)-Leu-enk	DAPA: (D-Ala^2, Pro^5)-enkamide
DTTLE: (D-Thr^2, Thr^6)-Leu-enk	DAGO: Tyr-D-Ala-Gly-MePhe-Gly-ol
DPLPE: (D-Pen^2, L-Pen^5)-enk	

Drugs and peptides used in these experiments were morphine sulfate (Merk and Co., Rahway, N.J.), leucine enkephalin, methionine enkephalin, [D-Ala^2] methionine enkephalinamide, [D-Ala^2, D-Leu^5] enkephalin (Calbiochem, La Jolla, Calif.), morphiceptin, [D-Thr^2, Thr^6] leucine enkephalin, Tyr-D-Ala-Gly-MePhe-Gly-ol (Peninsula Laboratories, San Carlos, Calif.), [D-Ser^2, Thr^6] leucine enkephalin (a generous gift from Dr. B.P. Roques), [D-Ala^2, Pro^5] enkephalinamide and [D-Met^2, Pro^5] enkephalinamide (generous gifts from Dr. Y. Audigier). [D-Pen^2, D-Pen^5]-enkephalin (a generous gift from Drs. H. Mosberg and F. Porreca) and [D-Pen^2, L-Pen^5]-enkephalin (Cambridge Research Biochemicals Ltd, UK). Seventy-five-milligram morphine pellets were made as previously described [Gibson and Tingstadt, 1970].

Table II. Analgesic potency of morphine and various opioid peptides following their i.c.v. administration to mice

Opioid	Receptor preference	ED_{50} [1, 2], nmol. i.c.v.
Metkephamide	δ	0.03(0.02–0.04)
β-Endorphin	δ/μ	0.05(0.03–0.08)[3]
DAGO	μ	0.05(0.01–0.18)
DAPA	μ	0.05(0.02–0.11)[4]
DMPA	μ	0.08(0.04–0.15)
DADLE	δ	0.29(0.16–0.53)[4]
DTTLE	δ	0.36(0.06–2.3)
DSTLE	δ	1.02(0.56–2.65)[4]
DPLPE	δ	1.04(0.73 2.0)
DAMA	δ	3.04(1.9–5.9)
Morphine	μ	3.06(1.5–8.3)[5]
DPDPE	δ	6.09(1.8–16.0)
Morphiceptin	μ	19.02.(10.7–34.9)[4]

[1] Mouse tail-flick test.
[2] ED_{50} determined at time of peak analgesia which was 10–15 min for all compounds.
[3] Value taken from Vaught et al., [1982].
[4] Value taken from Barrett and Vaught [1982].
[5] Morphine ED_{50} is a representative value of several ED_{50} determinations. ED_{50} values were never significantly different from day to day.

Table III. Effects of subanalgesic doses of opioid peptides administered simultaneously with morphine on morphine-induced analgesia

	ED_{50} [1] nmol i.c.v.
Morphine	3.6 (1.5–8.3)[2]
+LE[3]	0.86 (0.44–1.70)[4]
+DADLE[3]	0.80 (0.49–10.8)[4, 5]
+DSTLE[3]	0.90 (0.34–2.27)[4, 5]
+DTTLE[3]	0.80 (0.04–1.42)[4]
+Metkephamide[3]	0.91 (0.2–1.5)[4]
+DPLPE[3]	0.72 (0.02–1.27)[4]
+DPDPE[3]	0.78 (0.41–4.3)[4]
+ME[3]	13.3 (6.18–25.9)[6]
+DAMA[3]	16.8 (10.1–28.1)[6]
+Morphiceptin[3]	4.2 (2.3–7.7)[5]
+DAPA[3]	3.3 (1.0–10.6)[5]
+DMPA[3]	5.3 (2.9–9.4)
+DAGO[3]	3.6 (1.4–6.6)

[1] Mouse tail-flick test.
[2] Morphine ED_{50} is a representative value of several ED_{50} determinations. On each day, a control ED_{50} to morphine was performed to which treatment groups were compared. Morphine ED_{50} values were never significantly different from day to day.
[3] The dose of peptide combined with varying doses of morphine can be found in the text.
[4] Value is significantly less than that for morphine alone, $p<0.05$.
[5] Value taken from Barrett and Vaught [1982].
[6] Value is significantly greater than that for morphine alone, $p<0.05$.

morphiceptin and β-endorphin produced a dose-dependent analgesia when administered i.c.v. to morphine naive mice table II. The peptides studied differ markedly in their reported specificity towards μ- and δ-receptors. As is shown, several of the putative δ-selective peptides, as well as the highly δ-selective peptides, DPDPE and DPLPE, are more potent than or equipotent to morphine and other μ-selective peptides. Based on these results, it does not appear that analgesic potency is reflective of receptor specificity.

In experiments designed to investigate the possible modulatory activity of these peptides, a subanalgesic dose of peptide (extrapolated from its analgesic dose-response curve) was administered (i.c.v.) with morphine. In all instances peptide alone produced no increase in tail-flick latencies. LE (68.3 nmol), DSTLE (0.073 nmol), DADLE (0.012 nmol), DTTLE (0.04 nmol), metkephamide (0.001 nmol),

Table IV. Effects of LE and ME on morphiceptin-induced analgesia

Drug	ED_{50} (95% confidence limits), nmoles/mouse
Morphiceptin	19.4 (10.7–34.9)
+68.3 nmol LE	8.1 (5.2–12.6)[1]
+66.2 nmol ME	61.2 (41.6–89.9)[2]

[1] The ED_{50} value is significantly less than the value of morphiceptin alone.
[2] The ED_{50} is significantly greater than the value of morphiceptin alone.

DPDPE (0.7 nmol) and DPLPE (0.3 nmol) caused a parallel shift to the left of the morphine dose response curve, significantly reducing the ED_{50} of morphine by approximately 4-fold (table III). In contrast, ME (66.2 nmol) and DAMA (0.16 nmol) caused a significant parallel shift of the morphine dose response curve to the right increasing the ED_{50} of morphine by approximately 4-fold. DAPA (0.002 nmol), morphiceptin (1.92 nmol), DMPA (0.003 nmol) and DAGO (0.002 nmol) did not significantly alter the analgesic potency of morphine.

The potentiation or antagonism of analgesia by the enkephalins or their analogs is not restricted to morphine-induced analgesia. For example, LE (68.3 nmol) significantly enhanced the analgesic potency of morphiceptin while ME (66.2 nmol) significantly reduced the potency of morphiceptin (table IV).

The Analgesic and Modulatory Effects of Various Opioid Peptides in Acute and Chronic Morphine-Tolerant Mice

Administration of 100 mg/kg (s.c.) morphine sulfate 3.5 h (acute tolerance) or a 75-mg morphine pellet 72 h (chronic tolerance) prior to the assessment of morphine analgesia resulted in a 6.5- and 10-fold shift (respectively) to the right of the morphine dose-response curve indicating the development of tolerance (table V). In both tolerant states, the analgesic potency of DSTLE, DAMA, DAPA, morphiceptin and DTTLE were significantly reduced indicating analgesic cross-tolerance to morphine (table V; fig. 1). Thus, all of the peptides tested demonstrated cross-tolerance to morphine regardless of their receptor specificity.

Table V. Analgesic effects of morphine, DSTLE and DAMA in morphine naive and morphine-tolerant mice

Compound	Naive	Acute[1]	Chronic[1]
Morphine	3.6 (1.5–8.3)[2, 3]	23.2 (14.3–37.8)[4]	35.8 (21.9–58.4)[4]
DSTLE	1.2 (0.56–2.65)[2, 3]	5.63 (3.30–9.59)[4]	7.50 (5.16–10.89)
[4] DAMA	3.4 (1.9–5.9)[2, 3]	11.1 (6.70–18.4)[4]	10.1 (6.50–15.79)[4]

[1] Acute and chronic morphine tolerance was produced as described in the text.
[2] Values represent ED_{50} and 95% confidence limits in morphine naive mice after i.c.v. administration of compound in the mouse tail-flick assay.
[3] Morphine, DSTLE and DAMA ED_{50} values are representative values of several ED_{50} determinations. On each day, a control ED_{50} was performed to which treatment groups were compared. ED_{50} values in control mice did not differ significantly from day to day.
[4] Value is significantly greater than that obtained in morphine naive mice, $p<0.05$.

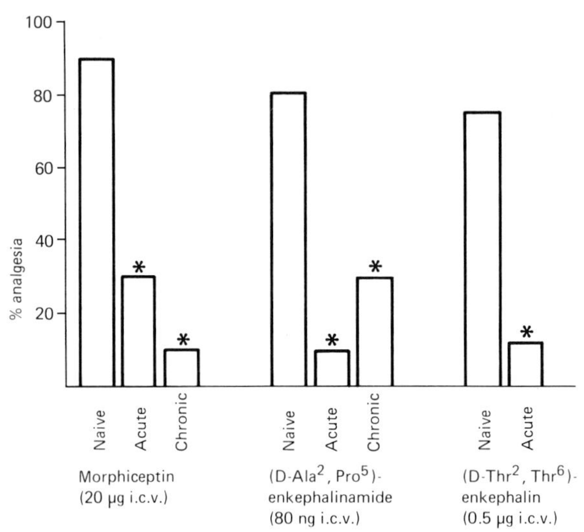

Fig. 1. The analgesic potency of morphiception, DAPA and DTTLE in morphine naive and tolerant mice. Tolerance to morphine was produced as described in the text. The number of animals analgesic is significantly less than that in morphine naive mice (n = 10–15). Fisher's exact probability, $p<0.05$.

Table VI. Modulatory effects of various opioid peptides on morphine-induced analgesia in morphine naive and morphine-tolerant mice

Compound	Naive	Acute[1]	Chronic[1]
Morphine	3.6 (1.5–8.3)[2, 3]	23.2 (14.3–37.8)[3]	35.8 (21.9–58.4)[3]
+LE[4]	0.86 (0.44–1.7)[5]	11.6 (7.4–18.2)[5]	16.4 (10.0–26.8)[5]
+DSTLE[4]	0.9 (0.3–2.3)[5]	6.9 (3.4–13.5)[5]	8.8 (5.2–14.8)[5]
+ME[4]	13.3 (6.8–25.9)	24.0 (12.1–47.6)	40.8 (23.9–69.4)
+DAME[4]	16.8 (10.1–28.1)[6]	86.0 (50.4–146.5)[6]	93.8 (62.6–140.3)[6]

[1] Acute and chronic morphine tolerance was produced as described in the text.
[2] Values represent ED_{50}'s and 95% confidence limits obtained in the mouse tail-flick assay following by i.c.v. administration of drug.
[3] ED_{50} values are representative of several determinations. On each day an ED_{50} 'control' is run to which treatment groups are compared. ED_{50} values did not differ significantly from day to day.
[4] The dose of peptide used in combination with morphine can be found in the text. The same dose was used in morphine naive and acute and chronic morphine-tolerant mice.
[5] Value is significantly less than that of the corresponding control, $p<0.05$.
[6] Value is significantly greater than that of the corresponding control, $p<0.05$.

However, despite tolerance to the analgesic effects of these peptides in a morphine tolerant animal, there was no tolerance to their modulatory effects. Doses of LE (68.3 nmol) and DSTLE (0.073 nmol), which significantly potentiated morphine-induced analgesia in morphine naive mice, were equally effective in acute and chronic morphine-tolerant mice (table VI). The dose of DAMA (0.16 nmol) which significantly antagonized morphine-induced analgesia in morphine naive mice was also equieffective in morphine-tolerant mice (table VI). In contrast to its effect on morphine-naive mice, ME did not significantly alter the potency of morphine in acute morphine-tolerant mice (table VI).

Conclusions

The biochemical documentation of μ- and δ-opioid receptors in rat brain membranes [Chang and Cuatrecasas, 1979] has raised the question of whether one or both of these receptors mediates the analgesic action of opioids. Obviously, several factors other than receptor affinity determine a drug's pharmacological potency (i.e. bioavailability, intrinsic activity, spare receptors). Nonetheless, the activity of a given ligand in in vitro binding or bioassay systems has been consistently demonstrated to be highly correlative to and predictive of analgesic potency [Audigier et al., 1980; Creese and Snyder, 1975]. Based on such correlations, Audigier et al., [1980] and others [Chaillet et al., 1984; Chang et al., 1982; Gasel et al., 1981] have suggested that when heat is used as the nociceptive stimulus, opioid-induced analgesia is mediated by μ-receptor occupancy. The results of analgesic potency determinations of the opioid peptides used in this study appear to disagree with this hypothesis. All of the δ-receptor-preferring peptides, as well as the highly selective δ-receptor agonists DPLPE and DPDPE [Mosberg et al., 1983], were at least as potent, and, in several instances, considerably more potent than the μ-selective peptide morphiceptin or morphine. Discounting the involvement of uncontrollable variables previously alluded to (pharmacokinetic factors; intrinsic activity) which could compromise the extrapolation from in vitro determined receptor affinity to analgesic potency, it seems clear that analgesic potency, per se, is not reflective of μ- or δ-receptor specificity.

Although analgesic potency alone does not distinguish between μ- and δ-receptor selective peptides, a marked difference is apparent when a subanalgesic dose of peptide is coadministered with morphine. Only the purported δ-receptor-selective peptides were capable of modifying the antinociceptive potency of morphine. Quite unexpectedly, the majority potentiated morphine-induced analgesia while met-enkephalin and its stable analog DAMA *antagonized* morphine analgesia. The μ-selective peptides morphiceptin, DAPA, DMPA and DAGO were without modulatory effects (table VII). Additionally, modulation by δ-receptor selective peptides was not unique to morphine, since morphiceptin-induced analgesia was also affected. This has been demonstrated with a variety of other ligands [Vaught and Takemori, 1979b; Vaught et al., 1982] and much of it confirmed in different models by other laboratories [Lee et al., 1980; Sarne et al., 1980; Hylden and Wilcox, 1983; Naranjo and Del Rio, 1982; Chapman et al., 1980; Porreca, personal commun.]. Although there is no absolute method of determining the receptor acted upon in vivo by these peptides, it is likely that at low subanalgesic doses, they act preferentially at the receptor for which they have the greatest affinity. Because the modulation of opioid analgesia appears to be a unique property of δ-specific (preferring) peptides, we have proposed that although analgesia may be predominantly

Table VII. Summary peptide of interactions

Potentiating peptides (receptor preference)		Antagonizing peptides (receptor preference)
LE (δ)	DTTLE (δ)	ME (δ)
DADLE (δ)	DPLPE (δ)	DAMA (δ)
DSTLE (δ)	DPDPE (δ)	
Metkephamide (δ)		
	Non-Modulatory Peptides (Receptor Preference)	
	Morphiceptin (μ)	
	DMPA (μ)	
	DAPA (μ)	
	DAGO (μ)	

mediated by μ-receptor occupancy, the δ-receptor may play an indirect modulatory role in analgesia [Vaught et al., 1982; Barrett and Vaught, 1982]. In addition, we have shown that LE is capable of enhancing, while ME antagonizes the development of tolerance and physical dependence produced by morphine [Vaught and Takemori, 1979a; unpubl. observations]. The enhancement or antagonism of morphine-induced analgesia, tolerance and dependence by opioid peptides suggests that the initial event, in each instance, produced by morphine is affected in a similar manner by subanalgesic doses of these peptides. The fact that these peptides display a high affinity for opioid receptors, that their structure-activity relationship for modulatory activity parallels that for maximum binding affinity [Vaught and Takemori, 1979a] and that naloxone blocks the potentiating effects of LE [Vaught and Takemori, 1979a] supports an opioid receptor-mediated modulatory event.

Utilizing the concepts of selective tolerance development demonstrated in in vitro bioassay tests [Schulz et al., 1980, 1981a, b] we have attempted to further strengthen our hypothesis by examining the ability of LE, ME, DSTLE and DAMA to modify morphine analgesia in morphine tolerant mice. Both tolerance-inducing methods employed produced tolerance to morphine and, as would be expected, cross-tolerance to the μ-selective peptides DAPA and morphiceptin. However, DSTLE, DAMA and DTTLE, delta receptor preferring peptides, similarly displayed analgesic cross-tolerance to morphine. These data suggest that all opioids, regardless of receptor preference, act via a common receptor to produce analgesia. Based on the specificity of morphine for μ-receptors this receptor is presumed to be μ.

Despite a reduction in the analgesic effects of DSTLE and DAMA in morphine tolerant mice, there was no tolerance to their modulatory effects. The same doses of LE and DSTLE which potentiated and of DAMA which antagonized morphine-induced analgesia in morphine naive mice were equally effective in morphine tolerant mice. The only notable exception was ME, for which we have no plausible explanation. This lack of tolerance to the modulatory effects in morphine tolerant animals has also been confirmed using the highly selective δ analogs DPLPE and DPDPE [Porreca, personal commun.]. The fact that in a morphine tolerant animal there exists analgesic cross-tolerance to DSTLE and DAMA with no tolerance to their modulatory effects supports our hypothesis that modulation of analgesia is mediated via a mechanism distinct from that directly associated with the production of analgesia.

Although the proposition of the δ-receptor playing a modulatory role in analgesia based on these in vivo data can be viewed as tentative due to the previously discussed limitations of utilizing an in vivo method with relatively non-selective ligands to distinguish receptor function, several other pieces of biochemical and pharmacological evidence strengthen our argument. We have previously shown that in vitro striatal slices of rat brain accumulate morphine which can be specifically released by infusion of naloxone and have suggested that this naloxone-releasable morphine represents receptor-bound drug [Kitano and Takemori, 1979; Vaught et al., 1981]. Utilizing this system as a model for drug-receptor interactions, we have demonstrated that leucine enkephalin was able to modify narcotic ligand binding characteristics in vitro and in vivo [Vaught et al., 1981]. Additionally, Rothman and Westfall [1981, 1982a, b] and Barrett and Vaught [1983] have demonstrated, using classical ligand binding techniques with computerized curve fitting analysis [Munson and Rodbard, 1980], that in several instances the displacement of a μ-receptor ligand by a δ-receptor ligand or vice versa is not consistent with a competitive two-site model. Based upon an extensive series of ligand-binding experiments, Rothman and Westfall [1981, 1982a, b] suggested that some proportion of mu and delta receptors must co-exist in an 'opioid receptor complex'. Utilizing a different experimental approach, Lee and Smith [1980] have similarly suggested the

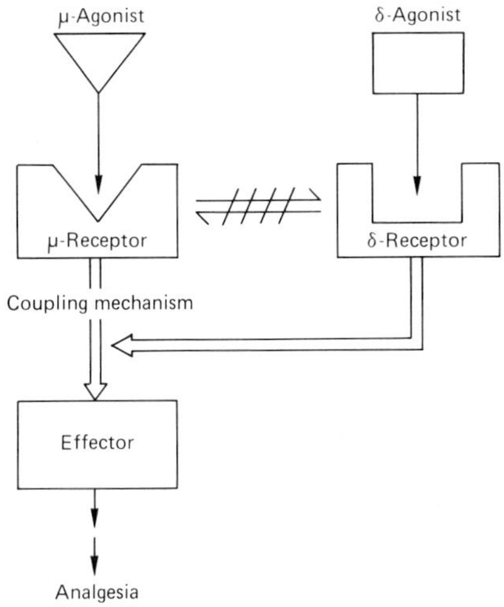

Fig. 2. A model for the opioid receptor complex involved in analgesia.

existence of an opioid receptor complex. Work from our laboratory certainly supports the complex nature of the interactions of μ- and δ-ligands in radioreceptor assays but does not provide any additional evidence for a receptor complex.

The evidence discussed briefly for the co-existence of μ- and δ-receptors in an opioid receptor complex may provide a neurochemical mechanism by which δ-selective ligands can modify opioid-induced analgesia. We feel that the model schematically illustrated in figure 2 represents an initial attempt to integrate the evidence presented here. A detailed explanation of this model can be found elsewhere [Rothman and Westfall, 1981, 1982a, b; Vaught et al., 1982]. In brief, occupation of the μ-receptor by an agonist such as morphine is hypothesized to induce a conformational change in the receptor which allows the coupling of the receptor to its effector, the end result being analgesia. In contrast, occupation of the δ-receptor by a 'potentiating-peptide' (i.e. LE) is hypothesized to induce a conformational change in this receptor which results not in analgesia, but a facilitation of the coupling of the μ-receptor to its effector. This is manifest physiologically as a potentiation of morphine analgesia. Occupation of the δ-receptor by an 'antagonizing-peptide' (i.e. ME) is hypothesized to induce a conformational change in this receptor, which actually inhibits the coupling of the μ-receptor to its effector, manifest physiologically as an antagonism of opioid analgesia.

In addition, an important feature of this model is that it provides a mechanism by which the in vivo analgesic potencies of stable enkephalin analogs may be much greater than predicted by their μ-receptor affinity. Accordingly, an opioid peptide, capable of producing analgesia, can be similarly classified as a 'potentiating' or 'antagonizing' peptide. The ability of a 'potentiating peptide' to bind to the δ-receptor would potentiate the analgesia resulting from its occupation of the μ-receptor. This will result in vivo in an analgesic effect greater than that predicted by its in vitro affinity for the μ-receptor (i.e. DTTLE, DSTLE, DADLE). Based on their relative affinities for μ- and δ-receptor sites, at any given analgesic dose, a certain proportion of drug will be bound to both receptor types. Thus, their analgesic potency is due to their ability to bind to both of these sites. Conversely, an 'antagonizing peptide', capable of producing analgesia, would have an analgesic potency similar to or less than its μ-receptor affinity would indicate.

The underlining question raised by our observations is the role of enkephalins in nociception. The low analgesic efficacy and potency of the enkephalins have been attributed to their rapid in vivo degradation, since enzymic stable analogs produce potent long-lasting analgesic effects. However, Craves et al. [1978] have pointed out that the metabolic degradation of the enkephalins does not solely account for their lack of analgesic potency. Additionally, Bajusz et al., [1978] were unable to find significant correlations between enhanced enzymic stability of various enkephalin analogs and analgesic potency. In fact, Ronai et al., [1981] have suggested that the enkephalin-like nature of the peptide actually inhibits the expression of analgesia. It is our opinion that, based on these data as well as others, the enkephalinergic system is unlikely to be the dominant endogenous 'antinociceptor'. However, leucine and methionine enkephalin are, to be sure, integral components of a complex system. Their selectivity for the δ-opioid-receptor and its potential influence on μ-receptor-mediated events suggests a rather novel role for these endogenous opioids in antinociception. At the neuronal level, LE and ME might be suggested to act as a localized neuronal 'gating' mechanism to facilitate or attenuate endogenous antinociceptive processes, in essence, serve as mediators of sensory homeostasis.

It is interesting to note that the concept of opposing actions of LE and ME is not without precedent in other systems. Gibson et al. [1980] have reported opposing effects of these peptides on plasma corticosterone response to ether stressed mice. Additionally the hypotheses of an opioid receptor complex in which there is an interaction between sites has been used as potential explanation of paradoxical findings in experimental models of shock [Holaday et al., 1985], seizure threshold [Tortella, 1985], and dissimilarities of enkephalin analogs and morphine in morphine-dependent rats [Brady and Holtzman, 1982]. Thus, the potential for modulation between opioid receptor types may play an important role in the overall pharmacology of both exogenous and endogenous opioids.

In conclusion, we do not suggest that this model is complete in all detail. Obviously, the extrapolation of experimental observations from in vitro and in vivo assay conditions meets with great complexity. We do suggest that this represents an important step in addressing the question for the potential of peptide modulation of analgesia. It also provides an alternative way of viewing the role of enkephalins in antinociception and perhaps other physiological processes.

References

Audigier, Y.; Marzarguil, H.; Gout, R.; Cros, J.: Structure-activity relationship of enkephalin analogs at opiate and enkephalin receptors. Correlation with analgesia. Eur. J. Pharmacol. *63:* 35–46 (1980).

Bajusz, S.; Patthy, A.; Kenessey, A.; Graf, L.; Szekely, J.I.; Ronai, A.Z.: Is there correlation between analgesic potency and biodegradation of enkephalin analogs? Biochem. biophys. Res. Commun. *84:* 1045–1053 (1978).

Barker, J.L.; Neale, J.H.; Smith, T.G.; MacDonald, R.L.: Opiate peptide modulation of amino acid responses suggests novel form of neuronal communication. Science *199:* 1451–1453 (1978a).

Barker, J.L.; Smith, T.G.; Neale, J.H.: Multiple membrane actions of enkephalin revealed using cultured spinal neurons. Brain Res. *154:* 153–158 (1978b).

Barrett, R.W.; Vaught, J.L.: The effects of receptor selective opioid peptides on morphine-induced analgesia. Eur. J. Pharmacol. *80:* 427–430 (1982).

Barrett, R.W.; Vaught, J.L.: Evaluation of the interactions of mu and delta selective ligands with [^{3}H]D-Ala^{2}D-Leu^5-Enkephalin binding to mouse brain membranes. Life Sci. *33:* 2439–2448 (1983).

Belluzzi, J.D.; Stein, L.: Enkephalin may mediate euphoria and drive-reduction reward. Nature, Lond. *266:* 556–558 (1977).

Bloom , F. Segal, H.; Ling, H.; Guillemin, R.: Endorphins. Profound behavioral effects in rats suggest new etiological factors in mental illness. Science *194:* 630–632 (1976).

Brady, L.S.; Holtzman, S.: Analgesic effects of intracerebroventricular morphine and enkephalins in non-dependent and morphine-dependent rats. J. Pharmac. exp. Ther. *222:* 190–197 (1982).

Chaillet, P.; Coulaud, A.; ZaJac, J.M.; Fournie-Zaluski, M.C.; Costentin, J.; Roques, B.P.: The μ rather than the δ subtype of opioid receptors appears to be involved in enkephalin-induced analgesia. Eur. J. Pharmacol. *101:* 83–90 (1984).

Chang, K.J.; Cautrecasas, P.: Multiple opiate receptors. Enkephalins and morphine bind to receptors of different specificity. J. biol. Chem. *254:* 2610–2618 (1979).

Chang, K.L.; Cautrecasas, P.; Wei, E.T.; Chang, K.J.: Analgesic activity of intracerebroventricular administration of morphiceptin and β-casomorphins. Correlation with morphine (μ) receptor binding affinity. Life Sci. *30:* 1547–1551 (1982).

Chapman, D.B.; Hu, J.; Way, E.L.: Methionine-enkephalin antagonism and endorphin potentiation of narcotic-induced analgesia. Eur. J. Pharmacol. *65:* 369–377 (1980).

Craves, F.B.; Law, P.Y.; Hunt, C.A.; Loh, H.H.: The metabolic disposition of radiolabeled enkephalins in vitro and in situ. J. Pharmac. exp. Ther. *206:* 492–506 (1978).

Creese, I.; Snyder, S.H.: Receptor binding and pharmacological activity of opiates in the guinea pig intestine. J. Pharmac. exp. Ther. *194:* 205–219 (1975).

Gacel, G.; Fournie-Zaluski, M.C.; Fellion, E.; Roques, B.P.: Evidence for preferential involvement of μ receptors in analgesia using enkephalins highly selective for peripheral μ and δ receptors. J. med. Chem. *24:* 1119–1124 (1981).

Gibson, A.; Hart, S.L.; Shabib, A.: Leucine-enkephalin and methionine-enkephalin produce opposing effects on plasma corticosterone levels in ether stressed mice. Brit. J. Pharmacol. *70:* 509–511 (1980).

Gibson, R.D.; Tingstadt, J.E.: Formulation of a morphine implantation pellet suitable for tolerance-physical dependence studies in mice. J. pharm. Sci. *59:* 426–427 (1970).

Holaday, J.W.; Faden, A.I.: Naloxone reversal of endotoxin-induced hypotension suggests a role of endorphins in shock. Nature, Lond. *275:* 450–451 (1978).

Holaday, J.W.; Long, J.B.; Tortella, F.C.: Evidence for opioid binding site interactions in vivo. Fed. Proc. *44:* (1985).

Hughes, J.: Isolation of an endogenous compound from the brain with pharmacological properties similar to morphine. Brain Res. *88:* 1–14 (1975).

Hughes, J.; Smith, T.; Morgan, B.; Fothergill, L.: Purification and properties of enkephalin the possible endogenous ligand for the opiate receptor; in Goldstein, The opiate narcotics, pp. 1–7 (Pergamon Press, Elmsford, 1975).

Hylden, J.L.K.; Wilson, G.L.: Intrathecal morphine in mice: a new technique. Eur. J. Pharmacol. *67:* 313–315 (1980).

Kitano, T.; Takemori A.E.: Further studies on the enhanced affinity of opioid receptors for naloxone in morphine-dependent mice. J. Pharmac. exp. Ther. *209:* 456–461 (1979).

Laubie, M.; Schmitt, H.; Vincent, M.; Remond, G.: Central cardiovascular effects of morphinimetic peptides in dogs. Eur. J. Pharmacol. *46:* 67–71 (1977).

Lee, N.M.; Leybinl, F.; Chang, J.K.; Loh, H.H.: Opiate and peptide interaction: Effect of enkephalin on morphine analgesia. Eur. J. Pharmacol. *68:* 181–185 (1980a).

Lee, N.M.; Smith, A.P.: A protein-lipid model of the opioid receptor. Life Sci. *26:* 1959–1964 (1980b).

Litchfield, J.T.; Wilcoxon, F.: A simplified method of evaluating dose-effect experiments. J. Pharmac. exp. Ther. *96:* 99–113 (1949).

Mosberg, H.; Hurst, R.; Hruby, V.; Gee, K.; Yamamura, H.; Galligan, J.; Burks, T.: Bis-penicillamine enkephalins possess highly improved specificity toward opioid receptors. Proc. natn. Acad. Sci. USA *80:* 5871–5874 (1983).

Munson, P.J.; Rodbard, D.: Ligand: Versatile approach for characterization of ligand binding systems. Analyt. Biochem. *107:* 735–744 (1980).

Naranjo, J.; Del Rio, J.: Differential effects of D-Ala^2 analogues of enkephalins on substance P-induced analgesia. Eur. J. Pharmacol. *82:* 213–216 (1982).

Olson, G.A.; Olson, R.D.; Kastin, A.J.; Coy, D.H.: Endogenous opiates: 1980. Peptides *2:* 349–369 (1981).

Rigter, H.: Attenuation of amnesia in rats systemically administered enkephalins. Science *200:* 83–85 (1978).

Ronai, A.Z.; Berzetei, I.A.; Szekely, J.I.; Miglesz, E.; Kurgyis, J.; Bajusz, S.: Enkephalin-like character and analgesia. Eur. J. Pharmacol. *69:* 263–271 (1981).

Rothman, R.B.; Westfall, T.C.: Allosteric modulation by leucine enkephalin of [^{3}H] naloxone binding in rat brain. Eur. J. Pharmacol. *72:* 365–368 (1981).

Rothman, R.B.; Westfall, T.C.: Morphine allosterically modulates the binding of [^{3}H]leucine enkephalin to a particulate fraction of rat brain. Molec. Pharmacol. *21:* 538–547 (1982a).

Rothman, R.B.; Westfall, T.C.: Allosteric coupling between morphine and enkephalin receptors in vitro. Molec. Pharmacol. *21:* 548–557 (1982b).

Sarne, Y.; Gothilf, Y.; Weissman, B.A.: Long-term interactions between endogenous and exogenous opiates. Eur. J. Pharmacol. *62:* 117–120 (1980).

Schultz, R.; Wuster, M.; Herz, A.: Differentiation of opiate receptors in the brain by the selective development of tolerance. Pharmac. biochem. Behav. *14:* 75–79 (1981a).

Schultz, R.; Wuster, M.; Krenss, H.; Herz, A.: Lack of cross-tolerance on multiple opiate receptors in the mouse vas deferens. Molec. Pharmacol. *18:* 395–401 (1980).

Schultz, R.; Wuster, M.; Rubini, P.; Herz, A.: Functional opiate receptors in the guinea-pig ileum. Their differentiation by means of selective tolerance development. J. Pharmac. exp. Ther. *219:* 547–550 (1981b).

Tortella, F.C.; Robles, L.; Holaday, J.W.: The anticonvulsant effects of DADLE are primarily mediated by activation of δ opioid receptors. Interaction between and receptor antagonists. Life Sci. *37:* 497–503 (1985).

Vaught, J.L.; Kitano, T.; Takemori, A.E.: Interactions of leucine enkephalin and narcotics with opioid receptors. Molec. Pharmacol. *19:* 236–241 (1981).

Vaught, J.L.; Rothman, R.B.; Westfall, T.C.: Mu and delta receptors. Their role in analgesia and in the differential effects of opioid peptides on analgesia. Life Sci. *30:* 1443–1455 (1982).

Vaught, J.L.; Takemori, A.E.: Characterization of leucine and methionine enkephalin and their interaction with morphine on the guinea pig ileal longitudinal muscle. Res. Commun. chem. Pathol. Pharmacol. *21:* 391–407 (1978).

Vaught, J.L.; Takemori, A.E.: Differential effects of leucine and methionine enkephalin on morphine induced analgesia, acute tolerance and dependence. J. Pharmac. exp. Ther. *208:* 86–90 (1979a).

Vaught, J.L.; Takemori, A.E.: A further characterization of the differential effects of leucine enkephalin, methionine enkephalin and their analogues on morphine-induced analgesia. J. Pharmac. exp. Ther. *211:* 280–283 (1979b).

Dr. Jeffry L. Vaught, Department of Biological Research,
McNeil Pharmaceutical, Spring House, PA 19477 (USA)

Non-Opioid Peptides and Analgesia: Focus on Cholecystokinin

Pain Headache, vol. 9, pp. 194–212 (Karger, Basel 1987)

The Role of Nonopioid Peptides in Pain

Merry Noel[a], *Charles B. Nemeroff*[a, b, 1]

Departments of [a] Psychiatry and [b] Pharmacology, Duke University Medical Center, Durham, N.C., USA

Since the discovery of endogenous opiate-like peptides (opioids), many investigators have focused on biological factors that influence pain perception. In addition to a continually expanding body of knowledge on endogenous opioid peptides, there is also a burgeoning literature on the role of nonopioid peptides in pain physiology and pathophysiology. The neurobiology of nonopioid peptides will be briefly reviewed here, as well as evidence for their involvement in the physiology of chronic pain and their potential utility as novel analgesic agents. These findings were reviewed comprehensively more than two years ago [Luttinger et al., 1984], and are updated here.

More than 40 neuropeptides have now been isolated from the mammalian nervous system (table I). Furthermore, molecular genetic techniques have provided evidence that more than 200 neuropeptides are present in brain. The amino acid sequence of each of the peptides to be discussed in this review is shown in table II. The detailed neuroanatomical localization of these peptides is not included here, except when such localization is particularly relevant to the discussion of pain processes. The interested reader can refer to Hokfelt et al. [1983] and Mense [1983] for these data. Several of the peptides to be discussed have been shown to be released from brain tissue by depolarizing stimuli, to bind to specific, reversible, saturable receptors, and to be degraded by tissue peptidases.

[1] We are grateful to Molly McMullen for preparation of this manuscript. Supported by NIMH MH-39415 and MH-40524. Charles B. Nemeroff is the recipient of a Nanaline Duke Fellowship from the Duke University Medical Center

Also, iontophoretic application of several of these neuropeptides to CNS neurons has been shown to result in changes in neuronal firing rates. Observed changes in the response of laboratory animals and man to painful stimuli has led to the hypothesis that these peptides are involved in the physiology of nociception.

Table I. Nonopioid neuropeptides identified in mammalian brain [Modified from Nemeroff and Bissette, 1986]

Thyrotropin-releasing hormone (TRH)
Gonadotropin-releasing hormone (GnRH, LHRH)
Somatostatin (SRIF, GHIH)
Corticotropin-releasing factor (CRF)
Growth hormone-releasing factor (GRF, GHRH)
Thyroid-stimulating hormone (thyrotropin, TSH)
Adrenocorticotropin (ACTH)
Luteinizing hormone (LH)
Prolactin (PRL)
Growth hormone (GH)
α-Melanocyte-stimulating hormone (α-MSH)
β-Melanocyte-stimulating hormone (β-MSH)
Neurotensin
Substance P
Bombesin
Gastrin
Cholecystokinin
Secretin
Gastrin-releasing peptide
Calcitonin
Carnosine
Insulin
Glucagon
Neuromedin N
Neuropeptide Y
Vasoactive intestinal peptide (VIP)
Vasopressin
Oxytocin
Vasotocin
Delta-sleep inducing peptide (DSIP)
Melanocyte stimulating hormone-release inhibiting factor (MIFI)
Angiotensin
Kallikrein
Bradykinin
Tuftsin

Table II. Peptide sequences

Substance	Peptide sequence	Molecular weight
Bradykinin	Arg-Pro-Pro-Gly-Phe-Ser-Pro-Phe-Arg	1,060.24
Substance P	Arg-Pro-Lys-Pro-Gln-Gln-Phe-Phe-Gly-Leu-Met-NH_2	1,347.80
Met-enkephalin	Tyr-Gly-Gly-Phe-Met	573.75
Leu-enkephalin	Tyr-Gly-Gly-Phe-Leu	555.72
β-Endorphin (β-lipotropin 61-91, human)	Tyr-Gly-Gly-Phe-Met-Thr-Ser-Glu-Lys-Ser-Gln-Thr-Pro-Leu-Val-Thr-Leu-Phe-Lys-Asn-Ala-Ile-Ile-Lsy-Asn-Ala-Tyr-Lys-Lys-Gly-Glu	3,465.63
Dynorphin 1-13 (porcine)	Tyr-Gly-Gly-Phe-Leu-Arg-Arg-Ile-Arg-Pro-Lys-Leu-Lys	1,604.19
Kyotorphin	Tyr-Arg	337.39
(Des,Tyr[1])-endorphin (β-Lipotropin 62-77)	Gly-Gly-Phe-Met-Thr-Ser-Glu-Lys-Ser-Gln-Thr-Pro-Leu-Val-Thr-Leu	1,696.22
Adrenocorticotropic hormone (ACTH, human)	Ser-Tyr-Ser-Met-Glu-His-Phe-Arg-Trp-Gly-Lys-Pro-Val-Gly-Lys-Lys-Arg-Arg-Pro-Val-Lys-Val-Tyr-Pro-Asn Gly-Ala-Glu-Asp-Glu-Ser-Ala-Glu-Ala-Phe-Pro-Leu Gln-Phe	4,541.74
Cholecystokinin (CCK) octapeptide	Asp-Tyr(SO_3)-Met-Gly-Trp-Met-Asp-Phe-NH_2	1,142.31
Somatostatin	Ala-Gly-Cys-Lys-Asn-Phe-Phe-Trp-Lys-Thr-Phe-Thr-Ser-Cys	1,638.12
Neurotensin	pGlu-Leu-Tyr-Glu-Asn-Lys-Pro-Arg-Arg-Pro-Tyr-Ile-Leu	1,084.38
(Arg[8])-vasopressin	Cys-Tyr-Phe-Gln-Asn-Cys-Pro-Arg-Gly-NH_2	1,084.38
Calcitonin (human)	Cys-Gly-Asn-Leu-Ser-Thr-Cys-Met-Leu-Gly-Thr-Tyr-Thr-Gln-Asp-Phe-Asn-Lys-Phe-His-Thr-Phe-Pro-Gln-Thr-Ala-Ile-Gly-Val-Gly-Ala-Pro-NH_2	3,418.41
Bombesin	pGlu-Gln-Arg-Leu-Gly-Asn-Gln-Trp-Ala-Val-Gly-His-Leu-Met-NH_2	1,620.12
Thyrotropin-releasing hormone	pGlu-His-Pro-NH_2	362.42
Melanocyte-stimulating hormone inhibitory factor (MIFI)	Pro-Leu-Gly-NH_2	284.36
α-Melanocyte-stimulating hormone (α-MSH)	Ac-Ser-Tyr-Ser-Met-Glu-His-Phe-Arg-Trp-Gly-Lys-Pro-Val-NH_2	1,665.05
Tuftsin	Thr-Lys-Pro-Arg	500.46

Substance P

Much evidence suggests a role for the undecapeptide substance P as a possible mediator of pain transmission. This evidence has previously been reviewed elsewhere [Luttinger et al., 1984; Sweet, 1979]. Substance P has been demonstrated by immunocytochemical methods to be present in primary afferent fibers innervating the substantia gelatinosa of the spinal cord and trigeminal nerve nucleus [Hokfelt et al., 1977]. These neurons have cell bodies in the spinal dorsal root and trigeminal ganglia, with peripheral distribution in dental pulp and skin.

Several studies suggest a role for substance P in transmitting pain information in unmyelinated primary afferents. For example, electrical stimulation of neonatal rat dorsal root results in release of immunoreactive substance P [Otsuka and Konishi, 1976]. Iontophoretic application of substance P to cat dorsal horn neurons resulted in excitation of approximately half of these cells [Henry et al., 1975], and direct application of substance P to rat spinal cord leads to nociceptive behavior [Seybold et al., 1982]. Release of substance P by capsaicin causes an acute sensation of pain but later a reduction in pain sensitivity once substance P is depleted [Hayes and Tyfers, 1980].

Inhibition of substance P release may be mediated by endogenous opioids. Morphine, the opioid β-endorphin, and a met-enkephalin analogue have been shown to inhibit potassium-induced release of substance P from rat trigeminal nucleus [Jessell and Iversen, 1977]. Moreover, opioid peptide interactions with substance P may differ in various spinal cord regions: Duggan et al. [1979] have described regions of substantia gelatinosa in which met-enkephalin reduced excitation induced by noxious stimuli, but had no effect on the action of substance P.

Substance P has been reported to produce both analgesia and hyperalgesia, depending on the dose and route of administration. Low doses given intraventricularly have been shown to produce analgesia that is naloxone reversible, while hyperalgesia is seen at high doses [Luttinger et al., 1984]. Recently, analogues of substance P have been shown to produce antinociceptive effects after intrathecal administration to rats, but there was a small margin between the effective dose of the peptide and a dose causing irreversible toxic injury to the spinal cord [Post and Paulsson, 1985]. Although considerable data exist on a role for substance P as a neurotransmitter at primary afferent fibers, its role in pain processes remains unclear.

Bradykinin

Bradykinin is a nonapeptide which has been reported to possess algesic (pain-producing) properties. This peptide has been shown to be involved in several pathological processes including inflammation, cardiovascular shock and hypertension, as well as pain [Rocha e Silva, 1970]. Intra-arterial injections of bradykinin to cats and dogs [Guzman et al., 1962] and to guinea pigs [Adachi and Ishii, 1979] elicit algesic behavioral responses. Furthermore, a bradykinin-like substance has been identified by several groups in the superfusate from exposed tooth pulp following noxious stimulation, and data from electrophysiological work gives further evidence for a role of bradykinin in processing noxious stimuli [Luttinger et al., 1984]. In summary, much evidence points toward a role for bradykinin in the periphery as a mediator of pain transmission from damaged tissue; centrally its role in nociception is unclear.

Neurotensin

Neurotensin is a tridecapeptide which was first identified in extracts of bovine hypothalamus by Carraway and Leeman [1973]. The neuroanatomic localization of this peptide has been demonstrated with both radioimmunoassay and immunocytochemical methods. High concentrations of the peptide have been found in many regions of the brain known to be involved in nociception, including the substantia gelatinosa of the spinal cord, periaqueductal gray, amygdala and thalamus [Carraway and Leeman, 1976; Kobayayashi et al., 1977; Uhl et al., 1979; Manberg et al., 1982]. Neurotensin is found in spinal cord interneurons, but not in primary afferent fibers [Leranth et al., 1984]. The distribution of neurotensin has been reported to be strikingly similar to that of enkephalin [Uhl et al., 1977, 1979].

Neurotensin was first shown to induce antinociception by Cline-schmidt and McGuffin [1977]. Intracisternal injection induced antinociception in mice as shown by the hot plate and writhing assays. This finding was confirmed by Nemeroff et al. [1979a]. Of interest was our finding that the potency of neurotensin varies markedly in these assays: a dose as low as 250 pg is effective in the acetic acid writhing assay, while more than 25 ng are needed for the hot plate assay. Localization of this antinociceptive effect has been studied with the use of implanted cannulae [Kalivas et al., 1982a, b], and injection with neurotensin in only a few loci, including the periaqueductal gray and the central nucleus of the amygdala, results in analgesia in the rat.

Neurotensin-induced antinociception has not been found to be reversed by naloxone or affected by drugs that block muscarinic cholinergic, histaminic, β-adrenergic, α-adrenergic or serotonergic receptors [Clineschmidt et al., 1979]. However, electrolytic lesioning of the nucleus raphe magnus abolished the antinociceptive effect of neurotensin injected into the periaqueductal gray [Behbebani and Pert, 1984]. The nucleus raphe magnus is known to be a major source of serotonin-containing neurons. From the latter study and electrophysiological findings, the analgesic effect of neurotensin has been hypothesized to involve excitation of periaqueductal gray, which in turn activates a pain inhibitory system originating in the nucleus raphe magnus.

Structure-activity relations of neurotensin-induced antinociception in the writhing assay have been recently studied by Nicolaides et al., [1985]. Intraventricular neurotensin was a potent analgesic in the mouse, while the C-terminal hexapeptide was three to four times less active; $neurotensin_{6\text{-}8}$ was inactive.

Adrenocorticotropin

Adrenocorticotropin (ACTH) is an anterior pituitary hormone derived from a high molecular weight precursor, pro-opiomelanocortin (POMC). ACTH contains 39 amino acids, and after its release from the pituitary, it stimulates the synthesis and release of glucocorticoids from the adrenal cortex. ACTH has also been found in extrapituitary sites including the CNS [Krieger et al., 1977], adrenal medulla and lymphocytes. Because ACTH is synthesized and released in conjunction with the endogenous opioid, β-endorphin, recent work has examined interactions between these two peptides, as well as effects of ACTH itself on pain systems.

Terenius [1976] reported that ACTH bound to opiate receptors in rat brain in a manner similar to mixed opiate agonist-antagonists. Snell and Snell [1982], using structural analogues of ACTH, obtained evidence that the N-terminal region of ACTH binds to opiate receptors.

Stengaard-Pedersen and Larsson [1981] found similar structural requirements for ACTH binding to receptors for enkephalinamide and naloxone.

Intraventricular administration of ACTH to rats has been reported to produce hyperalgesia in the hot-plate and tail-flick assays [Bertolini et al., 1979]. This effect of ACTH was antagonized by morphine and potentiated by naloxone. In contrast, injection of the ACTH analogue, $ACTH_{4-9}$ (ORG 2766), into the periaqueductal gray of rats has been reported to produce a dose-dependent analgesia in the tail-flick assay; this effect was not altered by naloxone or morphine tolerance [Walker et al., 1981, this vol.].

Interactions between ACTH and morphine have been studied by Amir [1981] using a hot plate procedure in mice. ACTH administered i.v., like morphine, enhanced jumping (an escape response) at low doses, but suppressed this response at higher doses. High dose ACTH modified the hyperalgesic effect of low dose morphine, and low dose ACTH inhibited the analgesic effect of higher dose morphine. These interactions are consistent with the proposed agonist-antagonist effect of ACTH at opiate receptors.

Other studies show an antagonistic effect of ACTH on opiate-induced analgesia when ACTH is administered intraventricularly [Fratta et al., 1981; Smock and Fields, 1981]. Also, ACTH fragments have been reported to inhibit morphine-induced analgesia in rats [Gispen et al., 1976].

Administration of ACTH in low doses into the lumbar subarachnoid space of rats has been shown to reverse the analgesic effect of intrathecal morphine or β-endorphin, while intrathecal ACTH alone produced only small increases in tail-flick latency in about half the animals tested [Belcher et al., 1982]. Taken together, these data are not discordant with the view that ACTH may act at more than one site in the CNS to affect pain processes.

Cholecystokinin

Cholecystokinin (CCK) is a peptide originally discovered in the small intestine and later in brain. Several forms of this peptide are found in the brain and gut. CCK is believed to be involved in the control of appetite and satiety.

Naloxone binding to opiate receptors in rat brain has been shown to be inhibited by the desulfated C-terminal septapeptide of CCK [Schiller et al., 1978]. In mice, subcutaneous administration of CCK octapeptide produced a dose-related, naloxone-reversible antinociception in two paradigms [Zetler, 1980]. This effect was not altered by morphine tolerance. Central administration of CCK to both mice and rats has been studied for antinociceptive effects in the tail-flick assay: although CCK was ineffective in mice in this assay, CCK octapeptide produced significant naloxone-reversible antinociception in rats after direct injection into the periaqueductal gray, caudate nucleus, ventromedial thalamus, cuneiform nucleus, or when injected intrathecally [Jurna and Zetler, 1981].

CCK octapeptide has also been shown to antagonize analgesia produced by electroshock and morphine when given intraperitoneally or perispinally [Faris et al., 1983]. Similarly, intraventricular or intrathecal administration of CCK octapeptide to rats antagonized the effect of morphine analgesia and electroacupuncture analgesia [Han et al., 1985]. Blockade of the effect of endogenous CCK through sequestration by antibodies to CCK resulted in potentiation and prolonged duration of morphine analgesia, but did not affect the rate of development of morphine tolerance [Faris et al., 1984]. Also, antiserum against CCK octapeptide reverses tolerance to electroacupuncture analgesia and changes nonresponders to this analgesia into responders [Han et al., 1985]. Thus, accumulating evidence suggests that CCK may function as an endogenous opiate antagonist, and blocking the effect of this peptide may be one method in which opiate analgesia can be augmented. (The role of CCK in analgesia is discussed in greater detail in the next chapter.)

Somatostatin

Somatostatin is a tetradecapeptide first discovered in extracts of ovine hypothalamus. It is known to inhibit growth hormone secretion, as well as secretion of other pituitary hormones, including thyroid-stimulating hormone (TSH), prolactin and ACTH. In addition, somatostatin appears to act also at extrahypothalamic sites, and substantial concentrations of this peptide are found in several regions of rat brain that have been implicated in nociception, including the midbrain central gray, amygdala and medulla.

Somatostatin also has been shown to be localized in the dorsal horn of the spinal cord [Hokfeldt et al., 1976]. Recent evidence indicates that release of immunoreactive somatostatin from spinal cord occurs in response to noxious thermal, but not mechanical, stimuli [Kuraishi et al., 1985].

Somatostatin appears to interact with opiate receptors. Terenius [1976] reported effects of somatostatin on dihydromorphine and naltrexone binding sites consistent with mixed opiate agonist-antagonist activity. Analogues of somatostatin have been shown by Pugsley and Lippman [1978] to inhibit stereospecific binding to opiate receptors. However, Traber et al. [1977] have studied opiate binding in neuroblastoma-glioma hybrid cells and found somatostatin to have no effect on naloxone binding.

Intraventricular administration of somatostatin has been shown to increase tail-flick latency in rats in a naloxone-reversible manner [Rezek et al., 1978]. Brain tissue incubated in the presence of somatostatin has been shown to competitively inhibit degradation of leu-enkephalin [Barclay and Phillips, 1980], suggesting that somatostatin may induce antinociception by prolonging the effects of endogenous enkephalins.

A recent clinical case report also demonstrates a marked analgesic effect of somatostatin (1–3.2 mg/day) administered by continuous intrathecal infusion to 2 patients (for 25 and 60 days, respectively) with intractable cancer pain [Meynadier et al., 1985]. Moreover, somatostatin has been reported to produce a potent analgesic effect after epidural administration in the treatment of acute postoperative pain and in the treatment of chronic cancer pain [Chrubasik et al., 1984].

Vasopressin

Vasopressin (antidiuretic hormone) is a peptide released from the posterior pituitary in response to changes in plasma osmolarity. In addition, this peptide has also been found in neurons not associated with the hypothalamo-neurohypophysial pathway, including periventricular regions of the diencephalon and mesencephalon implicated in pain transmission [Swanson, 1977; Buijs et al., 1978]. In the spinal cord, vasopressin is concentrated in the substantia gelatinosa, an area implicated in pain processing. However, a recent report indicates that neither a decrease (through lesions) nor an increase (intrathecal injections) in the synaptic activity of spinal cord vasopressin alters nociceptive thresholds in the rat [Millar et al., 1984].

Lysine vasopressin has been shown to produce a potent, dose-dependent antinociceptive response in rats in the tail-flick test after intraventricular or subcutaneous administration [Berntson and Berson, 1980]. This effect was not altered by naloxone. Similarly, vasopression has also been found to produce antinociceptive effects in mice, which are also not reversed by naloxone – this effect was seen after intravenous but not intraventricular administration in the mouse [Berkowitz and Sherman, 1982]. Although i.v. vasopressin injection has been shown to increase β-endorphin concentration in the rat [Anhut et al., 1981], this is not likely to be the mechanism for vasopressin-induced antinociception because of the clear opioid independence of this phenomenon.

Clinical trials with vasopressin have been conducted. Lysine vasopressin given intramuscularly to patients with acute herpes zoster has been reported to reduce both superficial and deep pain [Forssman et al., 1973].

Also of interest is a report of elevated plasma vasopressin concentrations in patients complaining of pain (mostly fractures) in a surgical emergency department compared to pain-free controls [Kendler et al., 1978]. In the latter study, no difference in plasma osmolarity was noted between the two groups. The authors hypothesized that vasopressin may be one of the homeostatic mechanisms that contribute to pain attenuation.

Calcitonin

Calcitonin is a 32 amino acid peptide hormone (table II) that inhibits osteoclastic bone resorption. After peripheral administration, it causes hypocalcemia and hypophosphatemia. Membrane receptors for calcitonin have been demonstrated in rat brain [Fischer et al., 1981]: these receptors are present in high concentrations in the thalamus and brain stem, regions which are thought to be involved in pain transmission.

Preclinical studies have shown that intracerebroventricular injection of calcitonin produces antinociception in rabbits [Braga et al., 1978] and in mice [Yamamoto et al., 1979; Satoh et al., 1979]. Naloxone administration did not alter calcitonin-induced antinociception in these species. Calcium ions given i.c.v. did, however, attentuate the antinociceptive effect of calcitonin in mice [Satoh et al., 1979].

Human trials of calcitonin in certain pain states have provided interesting results. Administration of calcitonin concurrently with increased dietary Ca^{++} has been shown to reduce back pain in osteoporosis [Cohn et al., 1971]. Furthermore, calcitonin has been used for treatment of Paget's disease, and leads to relief of bone pain in this disease, which is often seen prior to biochemical indices of bone disease [Gennari, 1983]. Recently calcitonin has also been shown to be effective for treatment of bone pain in cancer [Hindley et al., 1982; Allan, 1983; Gennari et al., 1985]. Some of these studies have utilized a double-blind design. It is possible, however, that some of the clinical effectiveness of calcitonin is due to effects on the pathophysiology of bone disease rather than to a specific antinociception action.

Tuftsin

Tuftsin is a tetrapeptide (table II) discovered by Najjar and Nishioka [1970]. It is synthesized in the spleen and is also present within the sequence of β-globulin [Nishioka et al., 1972]. In a hot-plate test, intracerebroventricular administration of tuftsin to rats has been reported to produce antinociception that was not blocked by naloxone [Herman et al., 1981].

Bombesin

Bombesin is a tetradecapeptide (table II), first described in frog skin by Anastasi et al., [1971]. This peptide is structurally similar to mammalian gastrin-releasing peptide, which is composed of 27 amino acids [McDonald et al., 1979]. Bombesin-like immunoreactivity has been described in CNS regions implicated in nociception, including the central amygdaloid nucleus, periaqueductal gray, and substantia gelatinosa [Moody et al., 1979]. Pert et al. [1980] have described receptors for bombesin in several rat brain regions, including the periaqueductal gray. Injection of bombesin into rat periaqueductal gray produced antinociceptive responses in the tail-flick and hot-plate tests that were not reversed by naloxone. Intracisternal injection of bombesin in mice did not produce antinociception in the tail immersion test [Nemeroff et al., 1979].

Thyrotropin-Releasing Hormone

Thyrotropin-releasing hormone (TRH) is a tripeptide (table II) that releases thyroid-stimulating hormone and prolactin from the anterior pituitary, but also appears to produce a large number of other direct CNS effects not mediated through the adenohypophysis [see Nemeroff et al., 1979b for a review].

TRH has been reported to exert mixed effects in nociception. No effect on nociception by TRH itself was seen in mice or rats in a variety of tests [Martin et al., 1977; Osbahr et al., 1981]. TRH has been reported to antagonize the antinociceptive effect of neurotensin [Osbahr et al., 1981], but not β-endorphin [Holaday et al., 1978; Osbahr et al., 1981]. Tolerance to opiate-induced analgesia, but not hypothermia, is inhibited by TRH administration to mice [Bhargava, 1981]. Finally, antinociception induced by Δ^9-tetrahydrocannabinol is antagonized by TRH, but not naloxone, in the tail-flick test [Bhargava and Matwyshyn, 1980].

Melanocyte-Simulating Hormone Inhibiting Factor

Melanocyte-stimulating hormone inhibiting factor (MIF-I) is a tripeptide (table II) with a structure identical to the tail of oxytocin. It has been hypothesized to function physiologically as an inhibitor of release of melanocyte-stimulating hormone (MSH) from the pars intermedia of the pituitary. MIF-I has been reported to have a weak antinociceptive effect in a writhing assay after oral administration [Plotnikoff and Kastin, 1974]. MIF-I antagonizes the antinociceptive effect of morphine and enkephalins in the mouse tail-flick assay after intraperitoneal injection but does not affect opioid activity in the vas deferens assay [Kastin et al., 1979]. In contrast, neurotensin-induced antinociception is not altered by intracisternal MIF-I as measured with either the tail immersion or hot-plate tests [Osbahr et al., 1981]. Finally, chronic administration of MIF-I has been shown to block the antinociceptive effect of morphine in the rat tail-flick test, and to attenuate the effect of morphine in the hot-plate test in mice [Dickinson and Slater, 1980]. Thus, MIF-I appears to have mixed effects on nociception.

Melanocyte-Stimulating Hormone

Another peptide observed to have mixed effects on nociception is the tridecapeptide α-melanocyte-stimulating hormone (α-MSH) (table II). This peptide is derived from POMC. Administration of α-MSH with morphine has been reported to reduce tolerance to the opiate [Szekely et al., 1979]. Intracerebroventricular administration of α-MSH has been observed to produce hyperalgesia in rats in a tail-flick test [Sandmand and Kastin, 1981]. In contrast to this report is the finding that α-MSH injected into the periaqueductal gray produced antinociception as well as a behavioral profile similar to that of β-endorphin [Walker et al., 1980]. Thus, the effect of α-MSH on nociception appears to vary according to the site of CNS injection.

Discussion

It appears that a wide variety of nonopioid peptides are capable of influencing the perception and/or response to pain. Few clinical studies on these peptides have been conducted and animal studies have the disadvantage of relying on an assortment of models of acute and chronic pain, none clearly validated.

Several of the peptides discussed here have been reported to induce naloxone-reversible antinociception, including substance P, somatostatin and cholecystokinin. The mechanism by which peptides that clearly do not bind to opiate receptors can modify nociception is uncertain. It is possible that they induce opioid peptide release, or that metabolites of these peptides interact with opiate receptors. Another possibility is that these peptides compete with enzymes that degrade opioid peptides, thus increasing the amount of opioid peptide available at the receptor.

Other peptides produce antinociception that is not reversed by naloxone; these include neurotensin, vasopressin, calcitonin, tuftsin and bombesin. Several peptides appear instead to enhance pain perception, notably bradykinin, substance P and ACTH. Substance P appears able to increase or decrease pain perception, depending on several factors, including dose, species and site of action. Similarly, evidence for TRH, MIF-I and MSH suggests mixed effects on pain processes.

Much work remains to be done before the physiology of nociception is clarified. Hopefully, this growing body of information will lead to the development of novel analgesics as well as providing a better understanding of the pathophysiology of a variety of pain states.

References

Adachi, K.I.; Ishii, K.: Vocalization response to close-arterial injection of bradykinin and other algesic agents in guinea pigs and its application to quantitative assessment of analgesic agents. J. Pharmac. exp. Ther. *209:* 117–124 (1979).

Allan, E.: Calcitonin in the treatment of intractable pain from advanced malignancy. Pharmatherapeutica *3:* 482–486 (1983).

Amir, S.: Effects of ACTH on pain responsiveness in mice. Interaction with morphine. Neuropharmacology *20:* 959–962 (1981).

Anastasi, A.; Erspamer, V.; Bucci, M.: Isolation and structure of bombesin and alytesin, 2 analogous active peptides from the skin of the European amphibians Bombina and Alytes. Experientia *27:* 166–167 (1971).

Anhut, H.; Knepel, W.; Nutto, D.; Hertting, G.: Vasopressin stimulates release of beta-lipotropin and beta-endorphin in concious rats as measured by radioimmunoassay of unextracted plasma. Arch. Pharmacol. *316:* 59–63 (1981).

Barclay, R.K.; Phillips, M.A.: Inhibition of enkephalin-degrading aminopeptidase activity by certain peptides. Biochem. biophys. Res. Commun. *96:* 1732–1738 (1980).

Behbebani, N.N.; Pert, A.: A mechanism for the analgesic effect of neurotensin as revealed by behavioral and electrophysiological techniques. Brain Res. *324:* 35–42 (1984).

Belcher, G.; Smock, T.; Fields, H.L.: Effects of intrathecal ACTH on opiate analgesia in the rat. Brain Res. *247:* 373–377 (1982).

Berkowitz, B.A.; Sherman, S.: Characterization of vasopressin analgesia. J. Pharmac. exp. Ther. *220:* 329–334 (1982).

Berntson, G.G.; Berson, B.S.: Antinociceptive effects of intraventricular or systemic administration of vasopressin in the rat. Life Sci. *26:* 455–459 (1980).

Bertolini, A.; Poggioli, R.; Ferrari, W.: ACTH-induced hyperalgesia in rats. Experientia *35:* 1216–1217 (1979).

Bhargava, H.N.: Dissociation of tolerance to the analgesic and hypothermic effects of morphine by using thyrotropin-releasing hormone. Life Sci. *29:* 1015–1020 (1981).

Bhargava, H.N.; Matwyshyn, G.A.: Influence of thyrotropin-releasing hormone and histidyl-proline diketopiperazine on spontaneous locomotor activity and analgesia induced by delta 9-tetrahydrocannabinol in the mouse. Eur. J. Pharmacol. *68:* 147–154 (1980).

Braga, P.; Ferri, S.; Santagostin, A.; Olgiati, V.R.; Pecile, A.: Lack of opiate receptor involvement in centrally induced calcitonin analgesia. Life Sci. *22:* 971–978 (1978).

Buijs, R.M.; Swaab, D.F.; Dogterom, J.; VanLeeuwen, F.W.: Intra- and extrahypothalamic vasopressin and oxytocin pathways in the rat. Cell Tiss. Res. *186:* 423–433 (1978).

Carraway, R.E.; Leeman, S.E.: The isolation of a new hypothalamic peptide, neurotensin, from bovine hypothalamus. J. biol. Chem. *284:* 6854–6861 (1973).

Carraway, R.; Leeman, S.E.: Characterization of radioimmunoassayable neurotensin in the rat. Its differential distribution in the central nervous system, small intestine, and stomach. J. biol. Chem. *251:* 7045–7052 (1976).

Chrubasik, J.; Meynardier, J.; Blond, S.; Scherpereel, P.; Ackerman, E.; Weinstock, M.; Bonath, K.; Cramer, H.; Wünsch, E.: Somatostatin, a potent analgesic. Lancet *ii:* 1208–1209 (1984).

Clineschmidt, B.V.; McGuffin, J.C.: Neurotensin administered intracisternally inhibits responsiveness of mice to noxious stimuli. Eur. J. Pharmacol. *46:* 395–396 (1977).

Clineschmidt, B.V.; McGuffin, J.C.; Bunting, P.B.: Neurotensin: antinocisponsive action in rodents. Eur. J. Pharmacol. *54:* 129–139 (1979).

Cohn, S.H.; Dombrowski, C.S.; Hauser, W.; Klopper, J.; Atkins, H.L.: Effects of porcine calcitonin on calcium metabolism in osteoporosis. J. clin. Endocr. Metab. *33:* 719–728 (1971).

Dickinson, S.L.; Slater, P.: Opiate receptor antagonism by L-prolyl-L-leucylglycinamide, MIF-I. Peptides *1:* 293–299 (1980).

Duggan, A.W.; Griersmith, B.T.; Headley, P.M.; Hall, J.G.: Lack of effect by substance P at sites in the substantia gelatinosa where Met-enkephalin reduces the transmission of nociceptive impulses. Neurosci. Lett. *12:* 313–317 (1979).

Faris, P.L.; Komisaruk, B.R.; Watkins, L.R.; Mayer, D.J.: Evidence for the neuropeptide cholecystokinin as an antagonist of opiate analgesia. Science *219:* 310–312 (1983).

Faris, P.L.; McLaughlin, C.L.; Baile, C.A.; Olney, J.W.; Komisaruk, B.R.: Morphine analgesia potentiated but tolerance not affected by active immunization against cholecystokinin. Science *226:* 1215–1217 (1984).

Fischer, J.A.; Sagar, S.M.; Martin, J.B.: Characterization of regional distribution of calcitonin binding sites in the rat brain. Life Sci. *29:* 663–671 (1981).

Forssman, O.; Leczinsky, C.G.; Mulder, J.: Synthetic lysine vasopressin in herpetic neuralgia. Acta derm.-vener., Stockh. *53:* 359–362 (1973).

Fratta, W.; Rossetti, Z.L.; Poggioli, R.; Gessa, G.L.: Reciprocal antagonism between ACTH $_{1\text{-}24}$ and beta-endorphin in rats. Neurosci. Lett. *24:* 71–74 (1981).

Gennari, C.: Clinical aspects of calcitonin in pain. Triangle *22:* 157 (1983).

Gennari, C.; Chierichetti, S.M.; Piolini, M.; Vibelli, C.; Agnusdei, D.; Civitelli, R.; Gonnelli, S.: Analgesic activity of salmon and human calcitonin against cancer pain: a double-blind placebo-controlled clinical study. Curr. ther. Res. *38:* 298–308 (1985).

Gispen, W.H.; Buitelaar, J.; Weigant, V.M.; Terenius, L.; deWied, D.: Interaction between ACTH fragments, brain opiate receptors and morphine-induced analgesia. Eur. J. Pharmacol. *39:* 393–397 (1976).

Guzman, F.; Braun, C.; Lim, R.K.S.: Visceral pain and the pseudoaffective response to intra-arterial injection of bradykinin and other algesic agents. Archs. int. Pharmacodyn. Thér. *136:* 353–384 (1962).

Han, J.S.; Ding, X.Z.; Fan, S.G.: Is cholecystokinin octapeptide (CCK-8) a candidate for endogenous anti-opioid substrates? Neuropeptides *5:* 399–402 (1985).

Hayes, A.G.; Tyers, M.B.: Effects of capsaicin on nociceptive heat, pressure and chemical thresholds and on substance P levels in the rat. Brain Res. *189:* 561–564 (1980).

Henry, J.L.; Krnjevic, K.; Morris, M.E.: Substance P and spinal neurones. Can. J. Physiol. Pharmacol. *53:* 423–432 (1975).

Herman, Z.S.; Stachura, Z.; Opielka, L.; Siemion, I.Z.; Nawrocka, E.: Tuftsin and D-Arg3-tuftsin possess analgesic action. Experientia *37:* 76–77 (1981).

Hindley, A.C.; Hill, E.B.; Leyland, M.J.; Wiles, A.E.: A double-blind controlled trial of salmon calcitonin in pain due to malignancy. Cancer Chemother. Pharmacol. *9:* 71–74 (1982).

Hokfelt, T.; Elde, R.; Johansson, O.; Luft, R.; Nilsson, G.; Arimura, A.: Immunohistochemical evidence for separate populations of somatostatin-containing and substance P-containing primary afferent neurons in the rat. Neuroscience *1:* 131–136 (1976).

Hokfelt, T.Q.; Ljungdahl, A.; Terenius, L.; Elde, R.; Nilsson, G.: Immunohistochemical analysis of peptide pathways possibly related to pain and analgesia: enkephalin and substance P. Proc. natn. Acad. Sci. USA *74:* 3081–3085 (1977).

Hokfelt, T.; Skirboll, L.; Lundberg, J.M.; Dalogaard, C-J.; Johansson, O.; Pernow, B.; Jancso, G.: Neuropeptides and pain pathways. Adv. Pain Res. *5:* 227–246 (1983).

Holaday, J.W.; Tseng, L.F.; Loh, H.S.; Li, C.H.: Thyrotropin releasing hormone antagonizes β-endorphin hypothermia and catalepsy. Life Sci. *22:* 1537–1544 (1978).

Jessell, T.M.; Iversen, L.L.: Opiate analgesics inhibit substance P release from rat trigeminal nucleus. Nature, Lond. *268:* 549–551 (1977).

Jurna, I.; Zetler, G.: Antinociceptive effect of centrally administered caerulein and cholecystokinin octapeptide (CCK-8). Eur. J. Pharmacol. *73:* 323–331 (1981).

Kalivas, P.W.; Gau, B.A.; Nemeroff, C.B.; Prange, A.J., Jr.: Antinociception after microinjection of neurotensin into the central amygdaloid nucleus of the rat. Brain Res. *243:* 279–286 (1982a).

Kalivas, P.W.; Jennes, L.; Nemeroff, C.B.; Prange, A.J., Jr.: Neurotensin: topographic distribution of brain sites involved in hypothermia and antinociception. J. comp. Neurol. *210:* 225–238 (1982b).

Kastin, A.J.; Olson, R.D.; Ehrensing, R.H.; Berzas, M.C.; Schally, A.V.; Coy, D.H.: MIF-I's differential actions as an opiate antagonist. Pharmacol. biochem. Behav *11:* 721–723 (1979).

Kendler, K.S.; Weitzman, R.E.; Fisher, D.A.: The effect of pain on plasma arginine vasopressin concentrations in man. Clin. Endocr. *8:* 89–94 (1978).

Kobayayshi, R.M.; Brown, M.; Vale, W.: Regional distribution of neurotensin and somatostatin in rat brain. Brain Res. *126:* 584–588 (1977).

Krieger, D.; Liotta, A.; Suda, T.; Palkovits, M.; Brownstein, M.: Presence of immunoassayable β-lipotrophin in bovine brain and spinal cord: lack of concordance with ACTH concentrations. Biochem. biophys. Res. Commun. *76:* 930–936 (1977).

Kuraishi, Y.; Hirota, N.; Sato, Y.; Hino, Y.; Satoh, M.; Takagi, H.: Evidence that substance P and somatostatin transmit separate information related to pain in the spinal dorsal horn. Brain Res. *325:* 294–298 (1985).

Leranth, C.; Csilik, B.; Knyihar-Csilik, E.: Depletion of substance P and somatostatin in the upper dorsal horn after blockade of axoplasmic transport. Histochemistry *81:* 391–400 (1984).

Luttinger, D.; Hernandez, D.E.; Nemeroff, C.B.; Prange, A.J.: Peptides and nociception. Int. Rev. Neurobiol. *25:* 185–241 (1984).

Manberg, P.J.; Youngblood, W.W.; Nemeroff, C.B.; Rossor, M.N.; Iversen, L.L.; Prange, A.J., Jr.; Kizer, J.S.: Regional distribution of neurotensin in human brain. J. Neurochem. *38:* 1777–1780 (1982).

Martin, B.R.; Dewey, W.L.; Chau-Pham, T.; Prange, A.J., Jr.: Interactions of thyrotropin-releasing hormone and morphine sulfate in rodents. Life Sci. *20:* 715–722 (1977).

McDonald, T.J.; Jornuall, H.; Nilsson, G.; Vagne, M.; Ghatei, M.; Bloom, S.R.; Mutt, V.: Characterization of a gastrin releasing peptide from porcine non-antral gastric tissue. Biochem. biophys. Res. Commun. *90:* 227–233 (1979).

Mense, S.: Basic neurobiologic mechanisms of pain and analgesia. Am. J. Med. *77:* 4–14 (1983).

Meynadier, J.; Chrubasik, J.; Dubar, M.; Wünsch, E.: Intrathecal somatostatin in terminally ill patients. A report of two cases. Pain *23:* 9–12 (1985).

Millan, M.J.; Schmauss, C.; Millan, M.H.; Herz, A.: Vasopressin and oxytocin in the rat spinal cord. Analysis of their role in the control of nociception. Brain Res. *309:* 384–388 (1984).

Moody, T.W.; Pert, C.B.; Jacobowitz, D.M.: Bombesin-like peptides: regional distribution in rat brain; in Gross, Meienhofer, Peptides – structure and biological function, vol. 6, pp. 865–868 (Pierce Chem. Co., Rockford 1979).

Najjar, V.A.; Nishioka, K.: 'Tuftsin'. A natural phagocytosis stimulating peptide. Nature, Lond. *228:* 672–673 (1970).

Nemeroff, C.B.; Osbahr, A.J., III; Manberg, P.J.: Alterations in nociception and body temperature after intracisternally administered neurotensin, β-endorphin, other endogenous peptides and morphine. Proc. natn. Acad. Sci. USA *76:* 5368–5371 (1979a).

Nemeroff, C.B.; Loosen, P.T.; Bissette, G.; Manberg, P.J.; Wilson, I.C.; Lipton, M.A.; Prange, A.J., Jr.: Pharmaco-behavioral effects of hypothalamic peptides in animals and man: focus on thyrotropin-releasing hormone and neurotensin. Psychoneuroendocrinology *3:* 279–310 (1979b).

Nemeroff, C.B.; Bissette, G.: Neuropeptides in psychiatric disorders; in: Berger, Brodie, American handbook of psychiatry, vol. VIII (Basic Books, New York, in press).

Nicolaides, E.D.; Lunney, E.A.; Kaltenbroun, J.S.; Wiley, J.N.; Downs, D.A.: Antiwrithing activity of some peptides related to neurotensin and tuftsin. Int. J. Pept. Prot. Res. *25:* 435–441 (1985).

Nishioka, K.; Constantopoulos, A.; Satoh, P.S.; Najjar, V.A.: The characteristics, isolation and synthesis of the phagocytosis stimulating peptide tuftsin. Biochem. biophys. Res. Commun. *47:* 172–179 (1972).

Osbahr, A.J. III; Nemeroff, C.B.; Luttinger, D.; Mason, G.A.; Prange, A.J., Jr.: Neurotensin-induced antinociception in mice. Antagonism by thyrotropin-releasing hormone. J. Pharmac. exp. Ther. *217:* 645–651 (1981).

Otsuka, M.; Konishi, S.: Substance P: an excitatory transmitter of primary sensory neurons. Cold Spring Harbor Symp. Quant. Biol. *40:* 135–143 (1976).

Pert, A.; Moody, T.W.; Pert, C.B.; DeWald, L.A.; Rivier, J.: Bombesin: receptor distribution in brain and effects on nociception and locomotor activity. Brain Res. *193:* 209–220 (1980).

Plotnikoff, N.P.; Kastin, A.J.: Pharmacological studies with a tripeptide, prolyl-leucyl glycine amide. Archs int. Pharmacodyn. Thér. *211:* 211–224 (1974).

Post, C.; Paulsson, I.: Antinociceptive and neurotoxic actions of substance P analogues in the rat's spinal cord after intrathecal administration. Neurosci. Lett. *57:* 159–164 (1985).

Puglsey, T.A.; Lippman, W.: Effect of somatostatin analogues and 17-alpha-dihydroequilin on rat brain opiate receptors. Res. Commun. chem. Pathol. Pharmacol. *21:* 153–156 (1978).

Rezek, M.; Havlicek, V.; Leybin, L.; LaBella, F.S.; Friesen, H.: Opiate-like naloxone-reversible actions of somatostatin given intracerebrally. Can J. Physiol. Pharmacol. *56:* 227–231 (1978).

Rocha e Silva, M.: Kinin hormones (Thomas, Springfield 1970).

Sandman, C.A.; Kastin, A.J.: Intraventricular administration of MSH induces hyperalgesia in rats. Peptides *1:* 231–233 (1981).

Satoh, M.; Amano, H.; Nakazawa, T.; Takagi, H.: Inhibition by calcium of analgesia induced by intracisternal injection of porcine calcitonin in mice. Res. Commun. chem. Pathol. Pharmacol. *26:* 213–216 (1979).

Schiller, P.W.; Lipton, A.; Horrobin, D.F.; Bodanszky, M.: Unsulfated C-terminal 7-peptide of cholecystokinin. A new ligand of the opiate receptor. Biochem. biophys. Res. Commun. *85:* 1332–1338 (1978).

Seybold, V.S.; Hylden, J.L.K.; Wilcox, G.L.: Intrathecal substance P and somatostatin in rats. Behaviors indicative of nociception. Peptides *3:* 49–54 (1982).

Smock, T.; Fields, H.L.: $ACTH_{1\text{-}24}$ blocks opiate-induced analgesia in the rat. Brain Res. *212:* 202-206 (1981).

Snell, C.R.; Snell, P.H.: A molecular basis for the interactions of corticotropin with opiate receptors. FEBS Lett. *137:* 209–212 (1982).

Stengaard-Pedersen, K.; Larsson, L.I.: Interaction of putative opioid peptides with opiate receptors. Acta pharmac. toxi. *48:* 39–46 (1981).

Swanson, L.W.: Immunohistochemical evidence for a neurophysin-containing autonomic pathway arising in the paraventricular nucleus of the hypothalamus. Brain Res. *128:* 346–353 (1977).

Sweet, W.H.: Neuropeptides and pain. Neurosurgical implications. Clin. Neurosurg. *26:* 657–672 (1979).

Szekely, J.I.; Miglecz, E.; Dunai-Kovacs, Z.; Tarnawa, I.; Ronai, A.Z.; Graf, L.; Bajusz, S.: Attenuation of morphine tolerance and dependence by alpha-melanocyte stimulating hormone (alpha-MSH). Life Sci. *24:* 1931–1938 (1979).

Terenius, L.: Somatostatin and ACTH are peptides with partial antagonist-like selectivity for opiate receptors. Eur. J. Pharmacol. *38:* 211–213 (1976).

Traber, J.; Glaser, T.; Brandt, M.; Klebensberger, W.; Hamprecht, B.: Different receptors for somatostatin and opioids in neuroblastoma X glioma hybrid cells. FEBS Lett. *81:* 351–354 (1977).

Uhl, G.; Kuhar, M.; Snyder, S.H.: Neurotensin immunohistochemical localization in rat central nervous system. Proc. natn. Acad. Sci. USA *74:* 4059–4063 (1977).

Uhl, G.R.; Goodman, R.R.; Snyder, S.H.: Neurotensin-containing cell bodies, fibers and nerve terminals in the brain stem of the rat: immunohistochemical mapping. Brain Res. *167:* 77–91 (1979).

Walker, J.M.; Akil, H.; Watson, S.J.: Evidence for homologous actions of pro-opiomelanocortin products. Science *210:* 1247–1249 (1980).

Walker, J.M.; Berntson, G.G.; Sandman, C.A.; Kastin, A.J.; Akil, H.: Induction of analgesia by central administration of ORG 2766, an analogue of ACTH4–9. Eur. J. Pharmacol. *69:* 71–79 (1981).

Yamamoto, M.; Kumagai, F.; Tachikawa, S.; Maeno, H.: Lack of effect of levallorphan on analgesia induced by intraventricular application of porcine calcitonin in mice. Eur. J. Pharmacol. *55:* 511–513 (1979).

Zetler, G.: Analgesia and ptosis caused by caerulein and cholecystokinin octapeptide (CCK-8). Neuropharmacology *19:* 415–422 (1980).

Charles B. Nemeroff, MD, PhD, Department of Psychiatry, Box 3859,
Duke University Medical Center, Durham, NC 27710 (USA)

Pain Headache, vol. 9, pp. 213–246 (Karger, Basel 1987)

Cholecystokinin and Pain

K.M. Pittaway, R.G. Hill

Parke-Davis Research Unit, New Addenbrookes Hospital Site, Cambridge, UK

Cholecystokinin – General Introduction

Cholecystokinin (CCK), a 33 amino acid peptide was first isolated from the porcine gastrointestinal tract by Mutt and Jorpes [1971] and is known to cause gall bladder contraction, pancreatic enzyme secretion and stimulation of intestinal motility [Morley, 1982]. The presence of CCK in the central nervous system was first detected by Vanderhaegen et al. [1975], who reported that vertebrate brain contained a small peptide which was detectable using gastrin antibodies. Further work by Dockray [1976], Rehfeld [1978] and others showed this gastrin-like material to be a form of CCK which cross-reacted with gastrin antibodies because of a common C-terminal pentapeptide sequence. It is now known that the predominant form in the central nervous system is the sulphated C-terminal octapeptide known as CCK 26-33 or CCK 8, although there are apparently three different forms of sulphated CCK 8 with different net charges [Rehfeld, 1985]. Other forms found to be present in brain were CCK 33 and 39 amino acid residue peptides, CCK 12 like peptides, unsulphated CCK 8 (at concentrations less than 1% of the sulphated form) and CCK 4 or 5 (C-terminal tetra and pentapeptide, respectively). Further details on these fragments and the historical background of CCK is discussed in Beinfeld's [1983] review.

The total amount of CCK in the mammalian central nervous system is higher than that of any other known peptide, and it has a unique distribution with the highest concentrations in the cerebral cortex, striatum, hippocampus, amygdala and parts of the brain stem. Such a distribution has been observed in rats, guinea pigs, cows, monkeys and humans. Radioimmunoassay and histochemical studies have shown CCK to exist in specific neuronal pathways where it may coexist with dopamine in the basal forebrain, with oxytocin and enkephalin in the paraventricular and supraoptic nuclei [Martin et al., 1983], and with substance P in the periaqueductal gray [Skirboll et al., 1983] and in primary afferents [Dalsgaard et al., 1982].

CCK therefore has an extensive but defined distribution within the central nervous system. It is found in high concentrations in synaptosomes [Emson et al., 1980] from which it can be released in a calcium dependent manner on depolarization [Dodd et al., 1980]. The in vivo biosynthesis of CCK in the cerebral cortex has been demonstrated [Goltermann et al., 1980], and mechanisms of enzymatic inactivation have been described [Deschodt-Lanckman et al., 1981]. When applied to the hippocampus, CCK 8 was found to exert excitatory actions on CA1 neurones [Dodd and Kelly, 1981], and a similar action has been noted in the midbrain [Chiodo and Bunney, 1983] and in the dorsal horn of the spinal chord [Jeftinija et al., 1981; Salt and Hill, 1982]. Using ^{125}I-labelled CCK 33, high affinity specific binding sites have been demonstrated both in the pancreas, and in cerebral cortical membrane preparations [Innis and Snyder, 1980; Steigerwalt and illiams, 1984]. CCK 4 is the minimal CCK fragment which binds with nanomolar affinity to central sites, desulphated CCK 8, and gastrin binding with almost the same affinity as CCK 8. In the pancreas, however, the site appears to be much more selective and the unsulphated form of CCK 8, CCK 4, and gastrin bind with a much lower affinity than CCK 8 (sulphated). The binding of CCK to pancreatic receptors can be inhibited by dibutyryl GMP, although this is ineffective at inhibiting CCK binding at central sites [Saito et al., 1980].

Central Effects of Cholecystokinin

CCK 8 has been shown to release growth hormone, ACTH and prolactin and to inhibit the release of thyrotropin and luteinizing hormone, and a modulatory role in regulating the release of other substances has been postulated as reviewed by Morley [1982].

An interaction between CCK and dopaminergic systems is suggested by a number of studies. For example, Katsuura et al. [1985] found that injection of CCK 8 into the nucleus accumbens – an area where dopamine terminals and receptors have been detected and where CCK and dopamine coexist – produced a decrease in locomotor and rearing responses in an open-field situation. CCK 4, however, exerted the opposite effect. Both effects were site specific in that injection into surrounding areas did not produce the same effect. It was suggested by the authors that CCK 4 could exert a negative feedback to modify the effect of CCK 8. CCK 8 has also been shown (after intracerebroventricular (i.c.v.) administration) to reduce the number of ^{3}H-spiperone binding sites and striatal levels of dopamine metabolites in unanaesthetized rats [Mashal et al., 1983] and to potentiate apomorphine-induced inhibition of dopaminergic neurons [Hommer and Skirboll, 1983]. The inter-relationship with the dopaminergic system has led to the suggestion that CCK may have a role in the pathogenesis of schizophrenia and in the therapeutic actions of antipsychotic drugs [Wang et al., 1984]. Post-mortem changes in the CCK content of human brain have been noted in schizophrenia, Parkinson's disease and Huntington's chorea (but not Alzheimer's disease) [Dawbarn and Emson, 1984], but whether such changes are the cause or effect of the disease states is not known.

Reports exist of injection of non-sulphated CCK 8 or CCK 7 into the cerebral ventricles suppressing β-endorphin induced catalepsy [Itoh and Katsuura, 1981], yet, if given subcutaneously, the non-sulphated form is ineffective and CCK 8 produces catalepsy [Kadar et al., 1985]. Subcutaneous administration of CCK 8 has also been reported to inhibit spontaneous rearing activity and prolong hexobarbital sleeping time [Zetler, 1980b]. Antistereotypic effects of CCK 8 have also been noted by Zetler [1985] who concluded that CCK-like peptides could antagonize stereotyped behaviour caused by directly and indirectly acting dopaminergic agonists by a mechanism which differed from that of reference neuroleptics included in his study.

Kadar et al. [1984] found CCK 8 to be anticonvulsant after i.c.v. administration and suggested a role for CCK 8 as an antiepileptic. In another of Zetler's studies [1983b], CCK 8 given subcutaneously was found to antagonize tremors induced by harmine, ibogaine and oxotremorine, independently of its effect on hypothermia and sedation. As a result of this study, Zetler suggested that CCK 8 had a physiological role in the regulation of extrapyramidal motor activity and could be of use to treat patients suffering from Parkinson's disease. Hypothermia after i.c.v. administration of CCK 8 has been reported by Morley et al.

[1981], similar to that produced by neurotensin at equimolar doses, and this effect could be antagonized by TRH.

The reported central effects of CCK 8 are thus many and diverse, which is not unduly surprising considering its wide central distribution and co-localization with other substances. When the role of CCK in pain perception is focused on, then a dichotomy of viewpoints is evident.

The Role of CCK in Pain Perception

One well-described action of CCK 8, when given peripherally or centrally to experimental animals, is the production of antinociception. This effect has been shown by a number of workers under a variety of experimental conditions [Zetler, 1980a; Jurna and Zetler, 1981; Autry et al., 1984; Barbaz et al., 1985], and in some cases the analgesia is reported to be antagonized by naloxone. The CCK-related peptide caerulein and other analogues of similar structure have also been reported to exert antinociceptive actions [Zetler, 1981; Sheehan and DeBelleroche, 1984]. Indeed, caerulein has also been used clinically to relieve pain in biliary colic [Pardo et al., 1984], in patients with arterial insufficiency of the lower extremity (not related to any effects of CCK 8 on blood flow [Basso et al., 1982], and in patients with burns, acute myocardial infarction and malignant tumours [Dolecek et al., 1983]. In all cases where the antinociceptive activity of caerulein has been compared with that of CCK 8, caerulein appears to be the more potent, presumably because of its blocked N-terminal which renders it less susceptible to enzymatic attack. The amino acid sequences of CCK 8, caerulein, Met-enkephalin Arg^6Phe^7 and another CCK-related peptide, FMRF-NH_2 are illustrated in figure 1.

CCK 8 may exert a dual effect in the processing of nociceptive information, as has been suggested previously for substance P and other peptides. CCK 8 is known to be present at all levels of the spinal cord and is concentrated in the dorsal horn [Marley et al., 1982]. Effects of rhizotomies and neonatal capsaicin suggested that all of the CCK 8 present in the dorsal horn was not of primary sensory origin but some was localized to either descending or intrinsic neurons or was present in capsaicin resistant afferent fibres. CCK positive neurons are found in the nucleus raphe magnus and paragigantocellularis and in a population of ventral medullary neurons, suggesting that a CCK-like system orig-

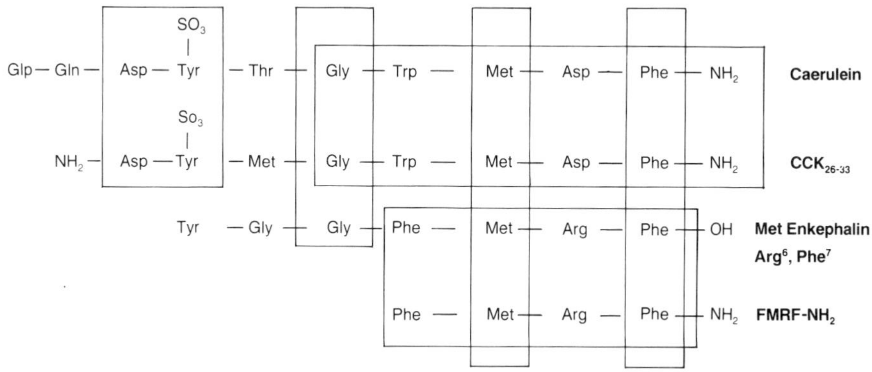

Fig. 1. Amino acid sequences of CCK 8 and related peptides. Caerulein is an amphibian peptide and although an analogue of CCK 8, is generally observed to be more potent because of its protected N-terminal. Met-enkephalin Arg^6,Phe^7 is a proenkephalin-derived enkephalin-related heptapeptide having analgesic properties [Han et al., 1984]. FMRF-NH_2 was initially isolated from molluscan ganglia and exerts potent cardioexcitatory effects in invertebrates. It is present in the central nervous system of vertebrates with high concentrations being found in the spinal cord and hypothalamus, although its physiological role is as yet unknown.

inates in the medulla and projects to the spinal cord. Some of these projecting neurons were found to contain both CCK 8 and 5-HT immunoreactivity [Mantyh and Hunt, 1984]. Coexistence of CCK 8 and substance P in a projection from the periaqueductal gray to the spinal cord and in dorsal root ganglia has been observed [Dalsgaard et al., 1982; Skirboll et al., 1983].

CCK 8 can be released into the lumbar subarachnoid space by electrical stimulation of A delta and C fibres of the sciatic nerve [Yaksh et al., 1982], although it is not clear whether the CCK originates from peripheral afferents, the spinal cord or both. A related observation is that CCK-like peptide levels increase in the dorsal horn of the lumbar region of the spinal cord in polyarthritic rats [Chery-Croze et al., 1985].

CCK 8 excites dorsal horn neurons in vivo and in vitro [Jeftinija et al., 1981; Salt and Hill, 1982], although it was found that both nociceptive and non-nociceptive neurons were equally sensitive to the action of CCK 8. It is noteworthy that, with the exception of solitary tract nucleus neurones [Denavit-Saubie et al., 1985], the actions of CCK on neuronal activity are exclusively excitatory. Many of the above effects are produced by other peptides, such as substance P and bombesin.

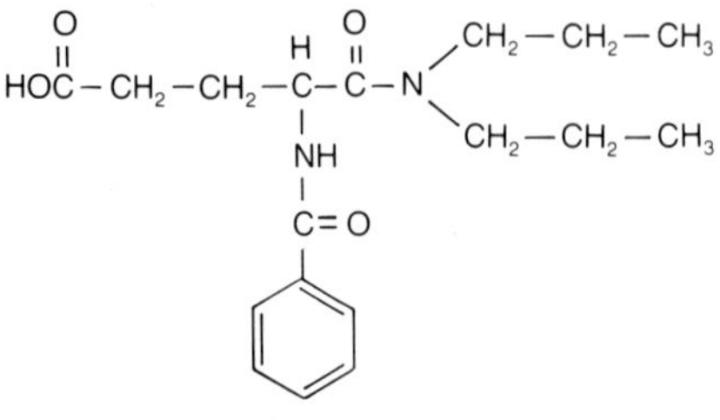

Proglumide

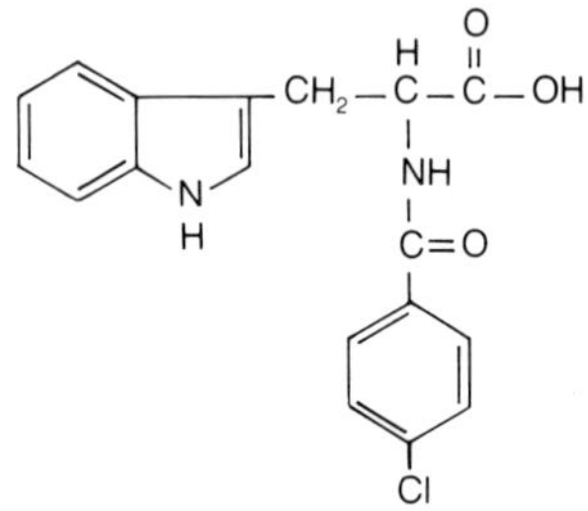

Benzotript

Fig. 2. Chemical structures of the CCK antagonists proglumide and benzotript.

The effect of the putative CCK antagonists proglumide and benzotript (whose structures are shown in figure 2) on the antinociceptive effects of morphine have lead to a number of reports that CCK 8 administered by intracerebroventricular (i.c.v.), intraperitoneal (i.p.) or intrathecal (i.t.h.) routes can antagonize the antinociception produced by morphine and endogenous opioids [Faris et al., 1983; Han et al., 1985; Itoh et al., 1982; Tang et al., 1984a]. Therefore, in addition to the antinociceptive effects reported by Zetler and colleagues, an alternative role for CCK 8 has been suggested by the above workers who postulate that CCK 8 exerts a modulatory effect on the antinociceptive effects of opioids and acts as an endogenous opioid antagonist. Immunization of rats with CCK antiserum or central injection of CCK antiserum has been reported to potentiate and prolong morphine antinociception and partially reverse morphine tolerance [Han et al., 1985; Faris et al., 1984] lending support to this theory. In a similar manner, it has been claimed [Watkins et al., 1984; 1985a, b] that proglumide can enhance opiate analgesia induced by systemic, spinal or i.c.v. morphine, endogenously released opioids or a stable enkephalin analogue DALA (*D*-alanine^{2} methionine enkephalinamide), although biphasic dose response curves were reported with potentiation and attenuation being observed at different doses. Proglumide has also been reported to reverse morphine tolerance [Watkins et al., 1984; Tang et al., 1984a].

A recent report by Tang et al. [1984b] showed that another peptide showing sequence homology with CCK 8, FMRF-NH$_2$ (fig. 1) could also attenuate opioid analgesia, and that an antiserum directed against this peptide induced a long-lasting antinociception which was reversible by naloxone.

Antinociceptive Actions of CCK 8

The antinociceptive effect of CCK 8 has been established in many behavioural tests (table I), and is reported to be 1.5–690 times more potent than morphine on a molar basis depending on the route of administration and test employed [Zetler, 1980a; Jurna and Zetler, 1981]. However, CCK 8 antinociception is not a universal finding, even when similar experimental conditions are used [Hill et al., 1987; Zetler, 1980a; Itoh et al., 1982; Tang et al., 1984a]. Variations in species and strain of animals and conditions in which they are housed and tested explain inconsistencies in reported results from different laboratories, and direct comparisons between antinociceptive effects of CCK 8 cannot often be made because such information is often lacking from published reports.

The hot plate test has been widely used to study CCK 8 antinociception, although variations in results depending on the end point used are encountered. When Woolfe and Macdonald [1944] were characterizing the hot plate test, they noted that mice displayed a characteristic behavioural pattern, exhibiting front paw licking or blowing, then hind paw licking or fanning followed by jumping with increasing time that they were left in contact with the plate. Hind paw licking was recommended as the end point of choice to avoid misinterpretation with respect to grooming responses. Zetler [1983a] found that the jumping response was more readily altered by CCK 8 and morphine than licking, as reflected by lower ED_{50} values when the jumping end point was used. A similar observation has been made by Frederickson et al. [1977] in a study on opioids, and these authors also noted that jump latencies showed a diurnal rhythm but lick latencies did not. Both Frederickson and Zetler discussed the possibility that the licking reaction represented perception of the noxious stimulus, whereas jumping was an indication of the affective properties of pain. Zetler [1983a] showed that apomorphine abolished the effects of morphine antinociception when jumping but not licking latencies were recorded, suggesting that different neural systems were involved in each response.

Whichever end point is used, erroneous interpretations of the antinociceptive efficacy of injected compounds could occur if they had muscle relaxant or sedative properties. O'Neill et al. [1983] developed a system for recording latencies to the first and tenth jump and have used this to provide a gross estimate of possible motor depressant effects of injected compounds. Animals can be observed in the cage before placing

Table I. CCK 8 antinociception

	Dose and route of CCK 8 injection	Antinociception evident?	References
Mouse hot plate (53–54 °C)			
Jump	0.05–1 mg/kg s.c.	yes	Zetler [1980a]
	0.17–1 mg/kg s.c.	yes	Barbaz et al. [1985]
	0.3–3 μg i.c.v.	yes	Barbaz et al. [1985]
	0.05–0.50 mg/kg s.c.	no	Hill et al. [1987]
Lick	0.12 mg/kg s.c.	yes	Zetler [1983a]
	10.0 mg/kg s.c.	yes	Hill et al. [1987]
	3–30 μg i.c.v.	yes	Hill et al. [1987]
Rat hot plate (57 °C), jump	0.8–3.2 μg/kg i.c.v..	no	Itoh et al. [1982]
Rat tail-flick	10–40 ng PAG or i.t.h.	yes	Jurna and Zetler [1981]
	2 μg i.t.h.	no	Tang et al. [1984b]
Mouse tail-flick	1 μg i.c.	yes	Kubota et al. [1985]
	0.05–0.75 mg/kg s.c.	no	Zetler [1980a]
Rat tail immersion	0.05–0.50 mg/kg s.c.	yes	Hill et al. [1987]
Rat paw pressure	3.33 mg/kg s.c.	yes	Hill et al. [1987]
Mouse writhing			
Acetic acid	0.75–2 mg/kg s.c.	yes	Zetler [1980a]
Phenylquinone	3–10 mg/kg s.c.	yes	Barbaz et al. [1985]
	4.5–10 μg i.c.v.	yes	Barbaz et al. [1985]
Acetylcholine	0.37–10 mg/kg s.c.	yes	Hill et al. [1987]

Testing occurred between 15 and 20 min after CCK 8 injection, exceptions being phenylquinone writhing test (5 min) and paw pressure test (30 min).

on the hot plate to check for any obvious sedative effects or motor defects can be tested, for example by noting the number of times an animal crosses a photocell beam when restricted to a certain area [Zetler, 1980a] or in a test such as the rota rod as developed by Jones and Roberts [1968].

Zetler [1980a] reports that CCK 8 given subcutaneously (s.c.) over the dose range 0.05–2.00 mg/kg produced dose dependent sedation, although Barbaz et al. [1985] using similar experimental conditions and dose range did not observe this effect. In our experience, however, with doses of 0.05–10 mg/kg CCK 8 s.c. or 0.003–30 μg i.c.v. sedative effects of CCK 8 were not seen when mice were observed before being placed on the hot plate. Use of the rota rod test in selected experiments also showed no impairment of locomotor performance [Hill et al., 1987]. Sedation has *not* been observed when CCK 8 is administered intrathecally [Zetler, 1980a; Jurna and Zetler, 1981].

Changes in cutaneous blood flow could possibly affect hot plate responses and give 'false positives', exaggerating the antinociceptive potency of compounds tested. However, in human studies [Lynn, 1980; Croze et al., 1977], it has been found that the temperature at which heat pain starts and at which it becomes intolerable were independent of mean skin and body temperatures and the effect of stopping blood flow for up to 17 min. Thermoregulatory effects of peptides have been noted [Yehuda and Kastin, 1980], and CCK 8 has been shown to increase intestinal blood flow [Rozsa and Varro, 1985], an effect thought to be mediated by an action on intrinsic neuronal elements of the intestine rather than a direct effect on vascular smooth muscle. There have been no reports to date of CCK 8 affecting skin blood flow, and it seems unlikely that its antinociceptive effects could be explained by such an action.

Using the jumping end point in the hot plate test, CCK 8 has been shown to be antinociceptive after subcutaneous administration over the range 0.05–1 mg/kg [Zetler, 1980a; Barbaz et al., 1985]. However, our attempt to repeat this experiment [Hill et al., 1987] was unsuccessful, although the jumping latencies recorded for saline treated animals were comparable (71 s). The only obvious difference in experimental conditions used was the strain of mice, and indeed strain differences in mice with respect to the number of naloxone binding sites and sensitivity to exogenous opioids have been reported [Baran et al., 1975; Nilsen, 1961]. Although morphine was tested and found to be antinociceptive by both Zetler and ourselves, available data do not allow a comparison of the relative sensitivity to opioid analgesics of the different strains used. The antinociceptive properties of CCK 8 reported by Zetler were accompanied by sedation, although this was not observed in our experiments.

CCK 8 has been shown to be antinociceptive in the hot plate test after subcutaneous administration when the latency to lick is used as the end point. However, our effective antinociceptive dose [Hill et al., 1987] was approximately 100-fold greater than that reported by Zetler [1983a] when tested under similar conditions.

After intracerebroventricular administration (fig. 3) CCK 8 produced significant antinociception over the range 3–30 μg [Hill et al., 1987]. This range is tenfold greater than that reported by Barbaz et al.

[1985], although in both experiments antinociception was tested for 10 to 15 min after CCK 8 injection. Strain differences again could account for this observation, although the use of jumping rather than licking latencies by Barbaz et al. [1985] may be another explanation. In our experience the licking end point was found to be more reliable.

Itoh et al. [1982], using intracerebroventricular doses of CCK 8 shown to be antinociceptive by others, could not show this effect in the rat hot plate jump test (table I). Whether this lack of effect is due to a different species being employed or a different hot plate temperature (57 °C) is not clear. Species differences in the proportion of opiate receptor subtypes have been documented [Robson et al., 1985]. It is possible that if CCK 8 produced a small antinociceptive effect it would not be easily distinguished from controls, since the temperature of the hot plate and the magnitude of the response are inversely related [Woolfe and Macdonald, 1944] and different temperature plates have actually been used to separate the antinociceptive effects of narcotic agonists and antagonists [O'Callaghan and Holtzman, 1975].

CCK 8 was found to be antinociceptive in the rat tail immersion test over the dose range 0.05–0.50 mg/kg when given subcutaneously and tested 15 min after administration [Hill et al., 1987]. Lack of other reports of CCK 8 antinociception in this test make comparison difficult, although this test has been shown to detect the antinociceptive effects of neurotensin when given intracisternally [Osbahr et al., 1981]. The tail immersion test involves a noxious thermal stimulus like the hot plate and tail flick tests and the end point is thought to be predominantly spinally controlled, being similar to the flexion reflex in man, although vocalization often accompanies removal of the tail from the water (51–52 °C) indicating involvement of higher centres.

The tail flick test [D'Amour and Smith, 1941] is also thought to have a large spinal component in its response, since Irwin et al. [1951] showed the tail flick response to persist in spinalized animals. The tail flick test has been used in rats and mice (table I) and CCK 8 appears to produce a significant antinociceptive effect after injection intrathecally or into the periaqueductal gray [Jurna and Zetler, 1981] or after intracisternal administration [Kubota et al., 1985]. It should be noted that intrathecal administration in this test does not always appear to be effective [Tang et al., 1984a]. Similarly, CCK 8 does not appear to alter mouse tail flick latencies after subcutaneous injection [Zetler, 1980a], even at dose ranges which produced significant antinociception in the hot plate test.

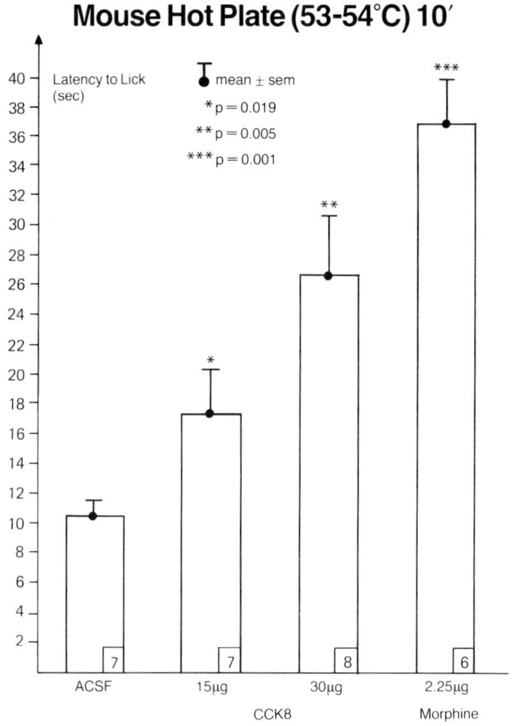

Fig. 3. Effect of CCK 8 in the mouse hot plate test [Pittaway and Hill, unpubl.]. Groups of 6–8 mice were injected i.c.v. with either artificial cerebrospinal fluid (ACSF), CCK 8 (15 or 30 μg) or morphine (2.25 μg); all injections were made in a 10-μl volume and compounds were dissolved in ACSF. Mice were tested on a 53–54 °C hot plate 10 min after injection and latency (s) to licking of forepaws recorded. Values shown are means ± SEM. Latencies associated with drug-treated animals were compared to ACSF controls using the Mann-Whitney U test. The level of significance is denoted as follows: * $p = 0.019$; ** $p = 0.005$; *** $p = 0.001$.

Table I also shows CCK 8 to be antinociceptive in a test employing a noxious mechanical stimulus, the rat paw pressure test. Although CCK 8 was tested over the range 50 μg to 3.33 mg/kg s.c., only the top dose produced a significant effect [Hill et al., 1987]. Since a 30-min period elapsed between injecting and testing in this experiment, it is likely that most of the CCK 8 had been metabolised before testing occurred, and that if testing had been performed 10–15 min after injection a greater antinociceptive effect would have been observed. Sheehan and De Belleroche [1984], using the CCK-related peptide caerulein in the paw pressure test and testing 15 min after injection, found the minimum effective does to be 0.003 mg/kg which was an order or magnitude lower than that needed for the hot plate test. Caerulein however, is known to be more potent than CCK 8 in antinociceptive test [Zetler, 1980a].

Writhing tests are commonly used as a primary screen for potential antinociceptive compounds, since it has been found that all drugs known to be analgesic in man are effective in this test and close correlations are observed in rank order of analgesic potency in mice and men after oral and parenteral administration [Collier et al., 1968]. The writhing test will detect the antinociceptive activity of both centrally and peripherally acting narcotic and non-narcotic analgesics, as well as the activity of narcotic antagonist analgesics [Blumberg et al., 1965]. However, the test can also produce false positives in that writhing agent antagonists and muscle relaxants will also inhibit writhing.

CCK 8 has been shown to produce antinociception in writhing tests after subcutaneous or intracerebroventricular administration when three different writhing agents have been used [Zetler, 1980a; Barbaz et al., 1985; Hill et al., 1987] and effective dose ranges are similar in all reports (table I). It is interesting to note that CCK 8 (and CCK 4) [Pittaway and Hill, unpubl.] appear to be more effective in the writhing than in the hot plate test. This tendency is in contrast to the report by Barbaz et al. [1985], although the experiments are not strictly comparable, since a different end point was used in the hot plate test and a different writhing agent was used, the choice of which is known to influence results [Collier et al., 1964]. The relative order of potencies of bombesin, neurotensin and CCK 8 have also been found to differ in writhing and hot plate tests [Autry et al., 1984], and each of these peptides was found to be more effective than met-enkephalin or morphine tested under the same conditions.

The actual mechanism whereby CCK 8 exerts its antinociceptive effect is unknown at present. Proposals must account for the fact that it is effective after subcutaneous, intracisternal, intrathecal, intracerebroventricular and periaqueductal gray injection (table I). Jurna and Zetler [1981] report that CCK 8 is more effective after intrathecal than periaqueductal gray injection when tested under the same conditions, but such comparative studies are rare. It is interesting to note that caerulein is reported to be ineffective when given intrathecally [Sheehan and De Belleroche, 1984].

The observation that CCK 8 is antinociceptive after subcutaneous administration implies that sufficient amounts must cross the blood-brain barrier if a central site of action is proposed. Oldendorf [1981], using ^{125}I-labelled CCK injected into the carotid artery, found a low extraction after a single circulatory passage and suggested that it was unlikely that significant levels of CCK occurred in the brain after peripheral administration by a direct penetration of the blood-brain barrier. Such observations do not necessarily preclude a central site of action after peripheral administration, since CCK 8 could exert an effect on vascular peptide receptors with release of a second messenger. Other possibilities include an action on receptors in the circumventricular organs or gaining access to cerebrospinal fluid via the choroid plexus.

When the intracerebroventricular route was used [Hill et al., 1987], CCK 8 was targeted towards the cerebral aqueduct and fourth ventricle, areas known to be critical in the production of morphine antinociception [Herz et al., 1970]. It is possible that leakage through the needle track followed by absorption from the subarachnoid space into the systemic circulation occurred, and that even after i.c.v. administration, the effect was produced peripherally. However, the dosage differential between effective antinociceptive doses subcutaneously and intracerebroventricularly together with the large dilution of peptide which would occur when entering the bloodstream makes this an unlikely prospect.

The central antinociceptive action of CCK 8 could be exerted by activating descending inhibitory pathways, as suggested for neurotensin [Behbehani and Pert, 1984] and bombesin [Pert et al., 1980], and indeed CCK 8 has been found to coexist with 5-HT in such a pathway [Mantyh and Hunt, 1984]. CCK 8 is unlikely to exert a direct inhibitory effect, however, since all electrophysiological studies to date, with exception of the study by Denavit-Saubie et al. [1985], show it to have an excitatory effect on neurones tested. The presence of CCK 8 in spinal cord must also be explained and as discussed previously, it may have a dual action, acting as a transmitter of nociceptive information as well as producing antinociception, these effects being elicited at different sites. An alternative possibility is that the CCK 8 present in spinal cord is involved in modulating other functions and is not necessarily involved in nociception.

The observation that CCK 8 is active in tests where noxious thermal, mechanical, or chemical stimuli have been used suggests that it may interact with different neurotransmitter systems, for example, 5-HT is thought to be involved in processing of noxious thermal, and noradrenaline in noxious mechanical information [Kuraishi et al., 1983]. A direct action of CCK 8 on opioid receptors has been proposed as a mode of action and indeed other peptides, e.g. somatostatin and ACTH are reported to have agonist/partial agonist activity on opioid receptors [Terenius et al., 1976]. However, CCK 8 does not appear to bind to opiate receptors, although the unsulphated CCK 7 is reported to have such activity [Schiller et al., 1978].

CCK 8 may exert its antinociceptive effects by causing the release of endogenous opioid peptides [Jurna and Zetler, 1981], and the sulphated but not the unsulphated form of CCK 8 has been shown to release metenkephalin-like immunoreactivity from spinal cord slices [Cesselin et al., 1984]. Naloxone antagonism of CCK 8 antinociception has been reported in the rat tail flick, mouse hot plate, and mouse acetic acid writhing tests [Jurna and Zetler, 1981; Barbaz et al., 1985; Zetler, 1980a], although naloxone appeared to potentiate the antinociceptive effect of caerulein twofold in the rat hot plate and paw pressure tests [Sheehan and De Belleroche, 1984]. Antagonism of peptide induced antinociception by opioid antagonists has also been reported for TRH [Webster et al., 1983] and alpha-MSH [Ohkubo et al., 1985], although most peptides, e.g. bombesin [Pert et al., 1980], vasopressin [Kordower and Bodnar, 1984], calcitonin [Braga et al., 1978], neurotensin [Clineschmidt et al., 1979] and CCK 8 in the phenylquinone writhing test [Barbaz et al., 1985], are not antagonized by naloxone, demonstrating that opioid-independent mechanisms are also involved in the production of antinociception.

The observation by Bates et al. [1983], that salmon calcitonin did not exert a reduction of acetic acid induced writhes in animals treated with parachlorophenylalanine and that 5-HT and calcitonin when given together caused a greater reduction in writhes compared to either agent alone, suggests that 5-HT systems may be involved in eliciting antinociceptive effects of peptides in writhing tests. Atropine has been reported to block the antinociceptive effect of the GABA agonists THIP and baclofen, whereas haloperidol enhanced the effect of baclofen only [Vaught et al., 1985] and TRH has been shown to antagonize the antinociceptive effect of neurotensin [Obsbahr et al., 1981]. It appears, therefore, that peptides may interact with a multitude of neurochemical systems, apart from endogenous opioids, to exert their antinociceptive effects.

CCK 8 has been shown to release substances other than endogenous opioids, e.g. somatostatin [Soll et al., 1985], acetylcholine [Magnani et al., 1984], GABA [Sheehan and De Belleroche, 1983] and substance P [Chang et al., 1984], and it is possible that these substances are involved in cases of non-naloxone-sensitive CCK 8 antinociception. The observation that the antinociceptive effect of CCK 8 can be reversed by benzodiazepine agonists and antagonists [Kubota et al., 1985] raises the possibility that benzodiazepines inhibit the action of CCK 8 through a direct interaction independent of GABA.

Zetler [1980a] showed that in the hot plate jumping test, the antinociceptive effect of CCK 8 had an onset time of 5 min, was maximal at 15 min and lasted 45 min after injection, and this compares closely with our own results [Hill et al., 1987]. CCK 8 was effective 15 and 30 min after injection in both the writhing and tail immersion tests, although in the latter test no effect was observed 45 min after administration.

In comparison with other peptides, the antinociceptive effect of CCK 8 seems relatively short lived, although this may be an apparent effect due to different doses and injection sites used by different workers. Clineschmidt and McGuffin [1977] reported that the antinociceptive effect of 2.5 μg of neurotensin given intracisternally lasted for 2 h. A 2-hour time course for the antinociceptive effect of calcitonin given intracerebroventricularly has also been noted [Braga et al., 1978; Pecile et al., 1975], and it was observed to have a rapid onset although the maximum effect was not observed until 90 min after injection. Bombesin, however, appears to resemble CCK 8 in that it produces its maximum effect 10 min after injection into the periaqueductal gray with antinociceptive activity declining gradually over the course of an hour [Pert et al., 1980].

The relatively short time course of the analgesic effect of CCK 8 raises the possibility that it is rapidly metabolised, and that if this could be prevented, an increased antinociceptive effect should be observed. Indeed, results of in vitro and HPLC studies [Pittaway and Hill, unpubl.] show that CCK 8 is degraded when in contact with brain tissue over the time course during which its antinociceptive effects are evident.

In vitro experiments on CCK 8 metabolism [Deschodt-Lanckman et al., 1983, 1984] suggested that CCK 8 is degraded by a number of enzymes. These were aminopeptidase (cleaving the N-terminal aspartate residue), two aminopeptidases of broad specificity (which sequentially hydrolyse N-terminal residues), a thiol protease and an endopeptidase.

A number of substances are now available to inhibit these enzymes. Bestatin is an aminopeptidase inhibitor and thiorphan an inhibitor of endopeptidase 24,11. Although the effect of thiorphan has not been tested on CCK 8 metabolism, phosphoramidon, another endopeptidase inhibitor, has been shown to be effective, as has bestatin [Deschodt-Lanckman et al., 1984]. Captopril, an angiotensin-converting enzyme (ACE) inhibitor has been used to protect peptides from metabolism, as it was considered at one time that ACE and endopeptidase were the same enzyme, although this matter now seems to be resolved [Swerts et al., 1979].

When thiorphan, captopril and bestatin together with CCK 8 were coinjected intracerebroventricularly, significant potentiation of the antinociceptive effect of CCK 8 was found [Hill et al., 1987]. CCK 8 has been postulated to exert its antinociceptive action through the release of endogenous opioids [Jurna and Zetler, 1981], and it has been noted that both enkephalins and CCK related peptides may share common degrading enzymes [Deschodt-Lanckman and Bui, 1981]. Studies on the metabolism of enkephalins in vitro [McKnight et al., 1983] showed that thiorphan, captopril and bestatin alone could provide protection against degradation, but the greatest effect observed was when a cocktail of all three inhibitors was used, and it was hypothesised that such a cocktail could protect endogenously released enkephalins from breakdown also.

The potentiation of the CCK effect seen in our experiments could therefore be due to inhibition of metabolism of CCK 8 or endogenous opioids, and further experiments would be needed to address this question. Captopril, bestatin and thiorphan have also been shown to possess antinociceptive properties themselves and to potentiate forms of endogenous and exogenous opioid analgesia [Han et al., 1984; Chaillet et al., 1983; Roques et al., 1980], although such effects are not universal findings [Reggiani et al., 1984] and were not evident in our experiments. The antinociceptive effect of enzyme inhibitors is thought to be due to their action in inhibiting the metabolism of opioid peptides although a recent study [Scott et al., 1985], suggests that factors other than enkephalinase inhibition may be important for the antinociceptive activity of thiorphan.

The instability of CCK 8, in contact with brain tissue, raises the possibility that its antinociceptive effects could be due to a fragment rather than the whole octapeptide sequence. A possible candidate is the C-terminal heptapeptide (CCK 7 or CCK 27–33) which has been shown to bind to central CCK-binding sites with approximately the same affinity as CCK 8 [Clark et al., 1984]. The unsulphated form of CCK 7 has also been shown to bind to opiate receptors [Schiller et al., 1978]. However, there are no reports of CCK 7 being antinociceptive in any behavioural studies. The C-terminal tetrapeptide (CCK 4 or CCK 30–33) could also cause the antinociceptive effect of CCK 8 and has also been shown to bind to central CCK binding sites, although less potently than CCK 8 or CCK 7 [Clark et al., 1984]. In our experiments [Hill et al., 1987] CCK 4 has been shown to be antinociceptive in the writhing test although to a lesser extent than CCK 8. CCK 8 and CCK 4 have also been shown to exert opposite behavioural effects on locomotor and rearing activity [Katsuura et al., 1985], although both are known to be potent excitors of pyramidal cells in the hippocampus [Dodd and Kelly, 1981] and of sensory neurones in the trigeminal nucleus [Salt and Hill, 1982]. The possibility that the antinociceptive activity of CCK 8 is mediated by the unsulphated form or the C-terminal pentapeptide is unlikely, since Zetler [1980a, 1981] found unsulphated CCK 8 and pentagastrin to be ineffective in his experimental paradigm. The effect of CCK 8 on antinociception therefore is likely to be mediated through either the whole octapeptide or possibly to a lesser extent by CCK 4.

CCK Antagonists: Effects on CCK 8 and Opioid Antinociception

If CCK 8 exerts its antinociceptive effect via CCK receptors, it would be expected that its action would be diminished or abolished in the presence of CCK antagonists. Experiments that we have performed [Graham et al., 1985] to test this approach using the CCK antagonist proglumide, however, showed this not to be the case (table II; fig. 4). In the mouse hot plate test under conditions where antinociception was evident, proglumide had no effect on hot plate latency itself and did not affect latencies recorded for CCK 8 when coadministered.

The results are in direct contrast to those reported by Barbaz et al. [1985], who noted that the antinociceptive effect of 0.3 μg CCK 8 given intracerebroventricularly in the mouse hot plate tested 15 min after injection could be antagonized by 3 and 10 μg proglumide i.c.v. or 150 mg/kg i.p. A higher dose of proglumide (10 μg i.c.v.) was required to reduce the effect of 3 μg CCK 8. Proglumide was also shown to antago-

Table II. Effect of proglumide on CCK 8 antinociception

	Dose and route of proglumide injection	Dose and route of CCK 8 injection	
Barbaz et al. [1985]			
Hot plate (53 °C), jump	3–10 μg i.c.v.	0.3–3 μg i.c.v.	significant attenuation
	50–150 mg/kg i.p.	0.3 μg i.c.v.	significant attenuation
		0.17–1 mg/kg s.c.	
Phenylquinone writhing	10–100 μg i.c.v.	10 μg i.c.v.	significant attenuation
Graham et al. [1985]			
Hot plate (53–54 °C), lick	5 μg i.c.v.	3–30 μg i.c.v.	no effect

Mice were used in all experiments.
Compounds were coinjected for i.c.v. administration.
Testing occurred 10–15 min after CCK 8 injection.

nize the effect of CCK 8 (10 μg i.c.v.) in the phenylquinone writhing test, although higher doses of proglumide were required. Proglumide given intraperitoneally also antagonized the antinociceptive effect of CCK 8 after subcutaneous injection. The reasons for the discrepancy could be many, although the fact that Barbaz et al. [1985] observed antinociception at doses which we found to be ineffective suggests that possibly the strain of mice they used were more sensitive to the effects of CCK 8 and CCK antagonists. Other factors, however, include different hot plate temperatures (1 °C difference) and use of jumping rather than licking endpoints. It should be noted that a lack of proglumide effect on antinociception produced by caerulein has also been reported [Sheehan and De Belleroche, 1984] using proglumide at doses up to 150 mg/kg given subcutaneously in rat paw pressure and hot plate tests. It is interesting to note that peptide-induced behavioural responses for substance P [Rodriguez et al., 1983] and bombesin [Cowan et al., 1985] appear to be unaffected by their presumed peptide antagonists.

The effects of proglumide have also been studied on morphine antinociception, and potentiation of its effect after parenteral or central administration has been noted for proglumide after intraperitoneal, intrathecal or periaqueductal gray injection when tested in the tail-flick test (table III) [Tang et al., 1984a; Watkins et al., 1984, 1985b; Kinscheck et al., 1983]. Proglumide has also been reported to potentiate morphine analgesia in human subjects [Yang et al., 1984].The potentiation pro-

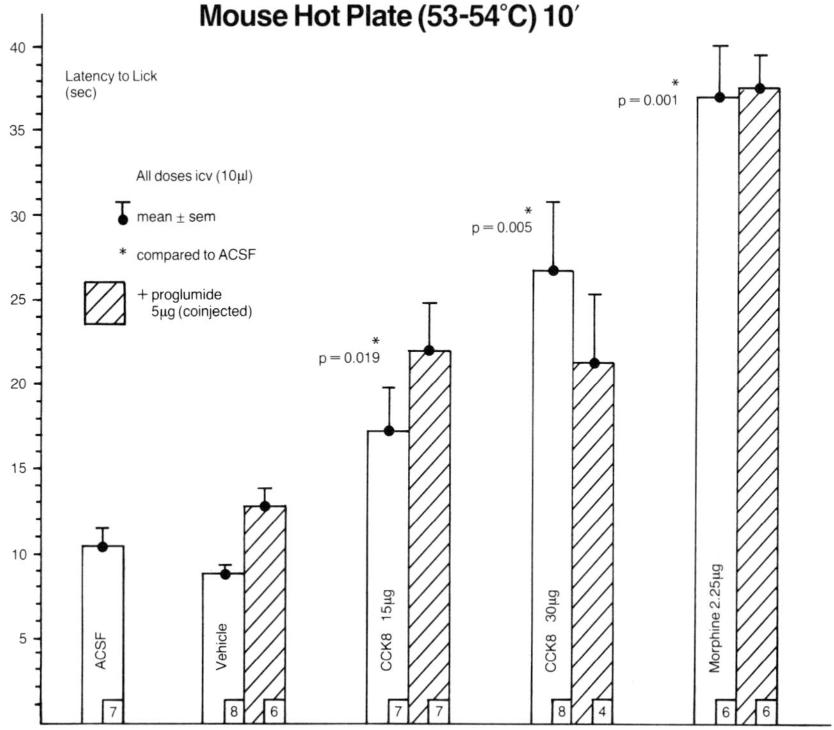

Fig. 4. Effect of proglumide on CCK 8 and morphine antinociception [Pittaway and Hill, unpubl.]. Groups of 4–8 mice were injected i.c.v. with either ACSF, proglumide vehicle (2.5 m*M* $NaHCO_3$, proglumide (5 μg), CCK 8 (15 or 30 μg) or morphine (2.25 μg). CCK 8 and morphine were also each coadministered with proglumide, the 10-μl injection volume being maintained. Mice were tested on a 53–54 °C hot plate 10 min after injection and latency (s) to licking of forepaws recorded. Values shown are mean ± SEM. Latencies associated with CCK 8- and morphine-treated animals were compared to ACSF controls using the Mann-Whitney U test and the level of significance obtained was as follows: p = 0.019; p – 0.005; p – 0.001.

duced by proglumide is not only observed with morphine, but has also been noted for other forms of opioid induced antinociception (table IV) such as front paw foot shock induced analgesia (FPSIA) [Kinscheck et al., 1983; Watkins et al., 1984, 1985a], intrathecal injection of a stable enkephalin analogue *D*-alanine-²methionine enkephalinamide [Watkins et al., 1984, 1985a; Suberg et al., 1984] and intracerebroventricular administration of β-endorphin [Katsuura and Itoh, 1985].

Table III. Effect of proglumide on morphine antinociception

	Dose and route of injection of morphine	Dose and route of injection of proglumide	Attenuation	No effect	Potentiation	References
Rat tail-flick	4 mg/kg s.c.	15 mg/kg i.p.			×	Tang et al. [1984a]
	3 mg/kg i.p.	0.002–0.02 mg/kg i.p.			×	Watkins et al. [1984, 1985b]
		0.20 mg/kg i.p.		×		
		10.0 mg/kg i.p.			×	
	1 μg i.t.h.	0.01 μg i.t.h.			×	Kinscheck et al. [1983]
		5.0 μg i.t.h.	×			
		0.001–0.01 μg i.t.h.			×	Watkins et al. [1984, 1985b]
		0.10 μg i.t.h.		×		
		1–10 μg i.t.h	×			
	3 μg PAG	0.01 μg i.t.h.			×	Kinscheck et al. [1983] Watkins et al. [1985b]
		0.01 μg PAG			×	Watkins et al. [1984, 1985a]
Rat tail immersion	5 mg/kg s.c.	25 mg/kg s.c.		×		Pittaway [1985]
Mouse hot plate	2.5 μg i.c.v.	5 μg i.c.v.		×		Pittaway [1985]

Dosage schedule:
i.c.v. and PAG injections – compounds are coinjected.
Watkins et al. – proglumide given 10 min and immediately prior to morphine (therefore, doses are double those quoted above).
Pittaway – proglumide given 5 min prior to morphine.
Tang et al., Kinscheck et al. – details not available.

Testing schedule:
Testing generally occurred at time points 10→180 min after proglumide injection.
'Potentiation' = Increased onset, peak effect and duration of morphine antinociception.

Our experiments (fig. 4; table III) show [Pittaway, 1985] that proglumide had no effect after subcutaneous or intracerebroventricular administration on antinociception produced by morphine given by the same route, even at doses which other workers had found to be effective. The conditions of the test situation were such that attenuation or potentiation of the morphine effect could have been observed.

Although the usual methodological differences could account for the lack of effect, it should be borne in mind that the observations on which the idea of CCK 8 acting as an endogenous opioid antagonist are based suffer from a number of problems. For example, lack of full experimental details does not allow the methods to be reproduced exactly and vehicle controls are not always included. The reason for the dosage schedule chosen by Watkins et al. in their studies where proglumide is given 10 min and again immediately prior to morphine is unclear, since there are no reports of rapid degradation of proglumide, and Chiodo and Bunney [1983] detected no decrease in its antagonistic effect on neuronal activity evoked by CCK 8 for at least 20 min after intravenous injection. In many studies, proglumide is shown to enhance opioid antinociception, and the antagonistic effect on CCK 8 is inferred without showing that under the same conditions proglumide directly antagonizes the effect of CCK 8. When this control is included in studies [Graham et al., 1985], proglumide appears to have no effect on CCK 8 antinociception, as illustrated in table II and figure 4.

The bulk of work reported on proglumide effects on opioid antinociception have been performed in the rat tail-flick test, although in the mouse the antinociceptive effect of CCK 8 cannot be detected when this test is used [Zetler, 1980a]. The choice of tests used by us (rat tail immersion and mouse hot plate) is unlikely to be the sole reason for failing to observe a potentiating effect of proglumide, since this has been observed in both the rat formalin and hot plate tests [Suberg et al., 1984; Katsuura and Itoh, 1985].

It must be emphasised that potentiation of opioid antinociception by proglumide does not invariably occur and complex dose response relationships have been reported [Watkins et al., 1984, 1985a; Kinscheck et al., 1983]. Potentiation appears to occur with lower doses tested, no effect is seen with intermediate doses and attenuation occurs with higher doses, although the reason for this is not clear, and this biphasic dose-effect relationship of proglumide is often overlooked. Proglumide is also reported to produce attenuation, no effect or potentiation of forms of

Table IV. Effect of proglumide on opioid antinociception (excluding morphine)

Dose and route of administration of opioid	Test	Dose and route of administration of proglumide	Attenu-ation	Poten-tiation	References
FPSIA (front paw shock-induced analgesia)	rat tail-flick	0.001–0.01 μg i.t.h.		×	Kinscheck et al. [1983] Watkins et al [1984, 1985a]
DALA					
3 μg i.t.h.	rat tail-flick	0.01 μg i.t.h.		×	Watkins et al. [1984, 1985a] Suberg et al. [1984]
3 μg i.t.h.	rat formalin test	0.02 μg i.t.h.	×		
1.5 μg i.t.h.	rat formalin test	0.02 μg i.t.h.		×	Suberg et al. [1984]
9.0 μg i.t.h.	tail shock vocalization	0.01 μg i.t.h.	×		
β-Endorphin 2.5 μg i.c.v.	rat hot plate (57 °C)	2 μg i.c.v		×	Katsuura and Itoh [1985]

Dosage schedule:
Watkins et al., Suberg et al. – proglumide given 10 min and immediately prior to opioid (therefore, doses are double those quoted above).
Kinscheck et al. – details not available.
Katsuura and Itoh – coinjected.

Testing schedule:
Testing generally occurred at time points 5–185 min after proglumide injection.
'Potentiation' = Increased onset, peak effect and duration of opioid antinociception.

non-opioid analgesia, such as hind paw foot shock induced analgesia or intrathecal injection of noradrenaline [Watkins et al., 1985a], although the doses required for potentiation are 1–10,000 times higher than those needed to potentiate opioid antinociception.

It is worthwhile to review the evidence that is presently available to support the idea that proglumide is a selective CCK antagonist within the central nervous system. In the periphery, both proglumide and benzotript have been shown to be specific and competitive antagonists of CCK 8 on gastric, ileal, and gall bladder musculature [Collins and Gardner, 1982; Gardner and Jensen, 1984], although anticholinergic and antihistaminergic actions have been reported [Hahne et al., 1981]. CCK 8-induced satiety has been antagonized by proglumide, although higher doses also inhibited bombesin-induced satiety [Collins et al., 1983].

Apparent specific effects of proglumide centrally include antagonism of CCK 8 effects on exploratory behaviour [Katsuura et al., 1984] and at the cellular level. Chiodo and Bunney [1983] found it to selectively block CCK-induced excitation of dopaminergic neurons in the ventral tegmental area, having no effect on activity elicited by neurotensin, met-enkephalin, substance P or histamine. However, Willetts et al. [1985] found that, although proglumide could antagonize CCK 8 excitation of rat dorsal horn neurons, it also produced direct excitatory effects in some neurons tested. It is interesting to note that proglumide does not inhibit the binding of ^{3}H-pentagastrin to rat striatal membranes [Gaudreau et al., 1983], although benzotript appears to displace tritiated Boc-β-Ala CCK 30–33 from cerebral cortical membranes of mice [Clark et al., 1985], and Vigna et al. [1985] report that proglumide can displace [^{125}I] CCK 33 from mouse cerebral cortical membranes although it appears to be a weak antagonist.

Proglumide is known to be anticholinergic [Hahne et al., 1981], and it is possible that the interaction between it and morphine is not related to any mutual CCK involvement, since morphine appears to have cholinergic properties – at least on Renshaw cells [Duggan and Curtis, 1972]. Deschodt-Lanckman and Bui [1982] have shown that proglumide can prevent the degradation of CCK 8 and leu-enkephalin by brain enzymes, thereby shedding doubts on its specificity of action. Until a potent, selective, centrally acting CCK antagonist is available, the question of whether CCK 8 exerts its effects via CCK receptors that are identical to the well described binding sites or whether it acts as an endogenous opioid antagonist is unlikely to be answered satisfactorily.The results of experiments using proglumide should consequently be interpreted with some caution.

The reported potentiation of opioid antinociception by proglumide implies that CCK 8 itself should exert an antagonistic effect on opioid analgesia. A number of studies have been made to test this theory and the ones using morphine are summarised in table V. It can be seen that the antinociceptive effect of morphine in the rat tail-flick test after intraperitoneal, subcutaneous or intrathecal administration can be attenuated by CCK 8 given by the intraperitoneal, intrathecal or intracerebroventricular routes [Faris et al., 1983; Han et al., 1984, 1985; Tang et al., 1984a]. Attenuation of opioid antinociception by FMRF-NH_2 (a peptide showing structural homology with CCK 8; fig. 1) has also been reported [Tang et al., 1984b], and this peptide too has been proposed to act as an endogenous opioid antagonist. Opioid antinociception has also been shown to be potentiated and prolonged in rats immunized against CCK 8 [Faris et al., 1984] or FMRF-NH_2 [Tang et al., 1984b].

Significant attentuation of other forms of opioid antinociception after intraperitoneal, intracerebroventricular or intrathecal injection of CCK 8 have been described. In the rat tail-flick test, Faris et al. [1983] antagonized the antinociceptive effect of front paw shock-induced analgesia (FPSIA), and Han et al. [1984, 1985] reversed the effect of electroacupuncture analgesia. Using the rat hot plate jumping test, Itoh et al. [1982] significantly attenuated the antinociception produced by intracerebroventricular injection of β-endorphin.

The results of our experiments [Pittaway, 1985] on the interaction of CCK 8 and opioids are also summarized in table V. In the rat tail immersion test, when CCK 8 was given before morphine (data not shown), no consistent effect was observed, although Zetler [1982] found that caerulein could enhance the antinociceptive effect of morphine only if given before but not after it. When morphine was given 30 min before CCK 8 in our experiment, a significant increase in the measured effect of a standard dose of morphine was observed.

When both CCK 8 and morphine were coinjected intracerebroventricularly in the hot plate studies (table V; fig. 5) significant increases in hot plate latencies were observed, the potentiation being apparently greater than that expressed from a purely additive effect, implying that different mechanisms of producing antinociception could be involved. Potentiation of opioid analgesia has been noted for a wide range of compounds, including adenosine [Chatterjee et al., 1984] and calcitonin

Table V. Effect of CCK 8 on morphine antinociception

References	Test	Dose and route of morphine injection	Dose and route of CCK 8 injection		
Faris et al. [1983]	tail-flick	10 mg/kg i.p.	5 µg/kg i.p.	apparent attenuation	(a)
Han et al. [1984, 1985]	tail-flick	5 mg/kg s.c.	0.25–4 ng i.t.h. or i.c.v.	dose-dependent attenuation	(b)
Tang et al. [1984]	tail-flick	10 µg i.t.h.	2 µg i.t.h.	significant attenuation	?
Pittaway [1985]	tail immersion	5 mg/kg s.c.	0.50 mg/kg s.c.	significant potentiation	(b)
	hot plate – mouse (53–54 °C)	2.5 µg i.c.v.	3 ng–30 µg i.c.v.	significant dose-dependent potentiation	(c)

All experiments were performed in rats unless stated otherwise.
(a) = CCK given 20 min before morphine; (b) = morphine given 25–30 min before CCK 8; (c) morphine and CCK 8 coinjected. Testing occurred between 5 and 20 min after CCK 8 injection.

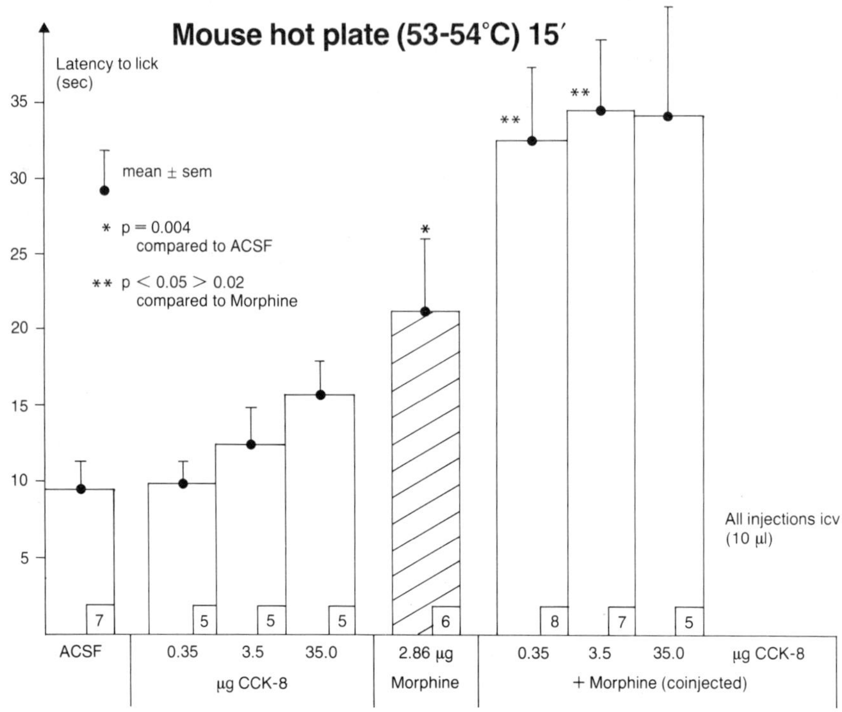

Fig. 5. Effect of CCK 8 on morphine antinociception [Pittaway and Hill, unpubl.]. Groups of 5–8 mice were injected i.c.v. (10-μl volume) with either ACSF, CCK 8 (0.35, 3.5, 35 μg) or morphine (2.86 μg). Three groups of mice received an injection of morphine with each of the doses of CCK 8. Mice were tested on a 53–54 °C hot plate 15 min after injection and latency to licking of forepaws recorded for each animal. Values quoted are mean ± SEM. Latencies associated with different treatments were compared to either ACSF (* $p = 0.04$) or morphine controls (** $p < 0.05 > 0.02$) using the Mann-Whitney U test.

[Braga et al., 1984]. The failure to observe attenuation of morphine analgesia in our experiments is unlikely to be due to the use of the hot plate test or the coadministration procedure, since Itoh et al. [1982] used similar experimental conditions. The differences in results obtained could be due to the injection technique, since although the intracerebroventricular route was used in our experiments, it is noteworthy that in the majority of cases where attenuation has been observed, the animals have been chronically implanted with intracerebroventricular or intrathecal cannulae, rather than injected by direct intracerebral puncture as in our experiments.

It should also be noted that CCK 8 attenuation of opioid antinociception has not so far been reported in mice and that in cases where CCK 8 has caused attenuation, the dose of CCK 8 used was not noted to be directly antinociceptive. It is possible therefore that low (sub-antinociceptive) doses of CCK 8 attenuate morphine antinociception, yet higher doses of CCK 8 which themselves are antinociceptive then potentiate the effect of morphine. A similar interaction has been reported by Chatterjee et al. [1984] for the effect of adenosine agonists on opioid analgesia, and these authors suggested that low doses of adenosine caused attenuation by acting at some hypothetical high affinity site and that higher doses caused potentiation by acting at both high and low affinity sites. However, it seems premature to attempt to interpret CCK/opioid interaction in this way.

In summary, CCK 8 has been shown to be antinociceptive in both rats and mice after central or peripheral administration and when tested against noxious thermal, mechanical or chemical stimuli. The underlying mechanism(s) are unclear and evidence to date suggests the involvement of both opioid and non-opioid systems. Use of the CCK antagonist proglumide has led to the alternative suggestion that CCK 8 may act *in vivo* as an endogenous opioid antagonist. The finding that CCK 8 can potentiate morphine antinociception [Pittaway, 1985] serves to demonstrate that there is a lack of agreement by workers in the field with respect to the effect of CCK 8 on opioid antinociception. Many of the interpretive problems raised can be attributed to experiments which are predominantly of a behavioural nature. It is hoped that more carefully controlled behavioural experiments or a move towards more easily interpreted experiments in vitro will provide data to clarify this rather muddy but interesting field.

References

Autry, W.L.; Barbaz, B.S.; Ambrose, F.G.; Liebman, J.M.: Analgesic profile of CCK-8. Differentiation from opiates and non-opioid peptides. Soc. Neurosci. Abstr. *10:* 54.4 (1984).

Baran, A.; Shuster, L.; Eleftheriou, B.E.; Bailey, D.W.: Opiate receptors in mice: genetic differences. Life Sci. *17:* 633–640 (1975).

Barbaz, B.S.; Autry, W.L.; Ambrose, F.G.; Gerber, R.; Liebman, J.M.: Antagonism of cholecystokinin by naloxone and proglumide in mice. Ann. N.Y. Acad. Sci. *448:* 573–574 (1985).

Basso, N.; Pona, V.; Dintinosante, V.; Fiocca, F.; Desantis, F.; Bagarani, M.; Gizzonio,

D.; Ponzielli, F.: Effect of ceruletide on rest pain in patients with arterial insufficiency of the lower extremity. Eur. J. clin. Pharmacol. *22:* 531–533 (1982).

Bates, R.F.L.; Buckley, G.A.; Eglen, R.M.; McArdle, C.A.: Calcitonin antinociception and serotonergic transmission. Br. J. Pharmacol. *80:* 518P (1983).

Behbehani, M.M., Pert, A.: A mechanism for the analgesic effect of neurotensin as revealed by behavioural and electrophysiological techniques. Brain Res. *324:* 35–42 (1984).

Beinfeld, M.C.: Cholecystokinin in the central nervous system. A minireview. Neuropeptides *3:* 411–427 (1983).

Blumberg, H.; Wolf, P.S.; Dayton, H.B.: Use of writhing test for evaluating analgesic activity of narcotic antagonists. Proc. Soc. exp. Biol. Med. *118:* 763–766 (1965).

Braga, P.; Ferri, S.; Santagostino, A.; Olgiati, V.R.; Pecile, A.: Lack of opiate receptor involvement in centrally induced calcitonin analgesia. Life Sci. *22:* 971–978 (1978).

Cesselin, F.; Bourgoin, S.; Artaud, F.; Hamon, M.: Basic and regulatory mechanisms of in vitro release of Met-enkephalin from the dorsal zone of the rat spinal cord. J. Neurochem. *43:* 763–774 (1984).

Chaillet, P.; Marcais-Collado, H.; Costentin, J.; Yi, C.C.; DeLaBaume, S.; Schwartz, J.C.: Inhibition of enkephalin metabolism by, and antinociceptive activity of Bestatin, an aminopeptidase inhibitor. Eur. J. Pharmacol. *86:* 329–336 (1983).

Chang, R.S.L.; Lotti, V.J.; Chen, T.B.: Further evidence that substance P partly mediates the action of cholecystokinin octapeptide (CCK 8) on the guinea pig ileum but not gall bladder. Studies with a substance P antagonist. Neurosci. Lett. *46:* 71–75 (1984).

Chatterjee, T.K.; Chatterjee, P.; Gebhart, G.F.: Spinal purinergic modulation of morphine-induced antinociception. Soc. Neurosci. Abstr. *10:* 32.8 (1984).

Chery-Croze, S.; Kocher, L.; Bernard, L.; Chayvialle, J.A.: Substance P, somatostatin, vasoactive intestinal polypeptide and cholecystokinin-like levels in spinal cord. of polyarthritic rats. Brain Res. *339:* 183–185 (1985).

Chiodo, L.A.; Bunney, B.S.: Proglumide. Selective antagonism of excitatory effects of cholecystokinin in central nervous system. Science *219:* 1449–1451 (1983).

Clark, C.R.; Daum, P.; Graham, W.; Hughes, J.: Determination of the specificity of the mouse CNS cholecystokinin receptor using two radiolabelled probes. J. Physiol. *354:* 58P (1984).

Clark, C.R.; Daum, P.; Hughes, J.: Proglumide and cholecystokinin (CCK)- 27-32-amide, two reputed peripheral CCK-receptor antagonists do not compete for central CCK-binding sites. Ann. N.Y. Acad. Sci. *448:* 581–582 (1985).

Clineschmidt, B.V.; McGuffin, J.C.: Neurotensin administered intracisternally inhibits responsiveness of mice to noxious stimuli. Eur. J. Pharmacol. *46:* 395–396 (1977).

Clineschmidt, B.V.; McGuffin, J.C.; Bunting, P.B.: Neurotensin. Antinocisponsive action in rodents. Eur. J. Pharmacol. *54:* 129–139 (1979).

Collier, H.O.J.; Hammond, A.R.; Horwood-Barrett, S.; Schneider, C.: Rapid induction by acetylcholine, bradykinin and potassium of a nociceptive response in mice and its selective antagonism by aspirin. Nature, Lond. *204:* 1316–1318 (1964).

Collier, H.O.J.; Dineen, L.C.; Johnson, C.A.; Schneider, C.: The abdominal constriction response and its suppression by analgesic drugs in the mouse. Br. J. Pharmac. Chemother. *32:* 295–310 (1968).

Collins, S.M.; Gardner, J.D.: Proglumide is a specific antagonist of the action of cholecystokinin on isolated smooth muscle cells from guinea pig stomach. Gastroenterology *82:* 1035 (1982).

Collins, S.; Walker, D.; Forsyth, P.; Belbeck, L.; The effect of proglumide on CCK, bombesin and glucagon-induced satiety in the rat. Life Sci. *32:* 2223–2229 (1983).

Cowan, A.; Khunawat, P.; Zuzhu, X.; Gmerek, D.E.: Effects of bombesin on behaviour. Life Sci. *37:* 135–145 (1985).

Croze, S.; Duclaux, R.; Russek, M.: Constancy of heat pain characteristics to changes in skin and body temperature. Brain Res. *131:* 367–372 (1977).

Dalsgaard, C.J.; Vincent, S.R.; Hokfelt, T.; Lundberg, J.M.; Dahlstrom, A.; Schultzberg, M.; Dockray, G.J.; Cuello, A.C.: Coexistence of cholecystokinin and substance P-like peptides in neurons of the dorsal root ganglia of the rat. Neurosci. Lett. *33:* 159–163 (1982).

D'Amour, F.E.; Smith, D.L.: A method for determining loss of pain sensation. J. Pharmac. exp. Ther. *72:* 74–79 (1941).

Dawbarn, D.; Emson P.C.: Cholecystokinin in neurological disease; in De Belleroche, Dockray, CCK in the nervous system, pp. 59–81 (Ellis Horwood, Chichester, 1984).

Denavit-Saubie, M.; Hurle, M.A.; Morin-Sarun, M.P.; Foutz, A.S.; Champagnat, J.: The effects of cholecystokinin-8 in the nucleus tractus solitarius. Ann. N.Y. Acad. Sci. *448:* 375–384 (1985).

Deschodt-Lanckman, M.; Bui, N.D.: Cholecystokinin octa- and tetrapeptide degradation by synaptic membranes. I. Evidence for competition with enkephalins for in vitro common degradation pathways. Peptides *2:* suppl. 2, pp. 113–118 (1981).

Deschodt-Lanckman, M.; Bui, N.D.; Noyer, M.; Christophe, J.: Degradation of cholecystokinin-like peptides by a crude rat brain synaptosomal fraction. A study by high pressure liquid chromatography. Regul. Peptides *2:* 15–30 (1981).

Deschodt-Lanckman, M.; Bui, N.D.: Proglumide inhibits the degradation of cholecystokinin octapeptide and leucine-enkephalin by brain enzymes. Biomed. Res. *3:* 517–524 (1982).

Deschodt-Lanckman, M.; Bui, N.D.; Koulischer, D.; Paroutaud, P.; Strosberg, A.D: Cholecystokinin octa- and tetrapeptide degradation by synaptic membranes. II. Solubilization and separation of membrane-bound CCK-8 cleaving enzymes. Peptides *4:* 71–78 (1983).

Deschodt-Lanckman, M.; Koulischer, D.; Przedborski, S.; Lauwerys, M.: Cholecystokinin octa- and tetrapeptide degradation by synaptic membranes. III. Inactivation of CCK8 by a phosphoramidon-sensitive endopeptidase. Peptides *5:* 649–651 (1984).

Dockray, G.J.: Immunochemical evidence of cholecystokinin-like peptides in brain. Nature, Lond. *264:* 568–570 (1976).

Dodd, P.R.; Edwardson, J.A.; Dockray, G.J.: The depolarisation-induced release of cholecystokinin C-terminal octapeptide (CCK-8) from rat synaptosomes and brain slices. Regul. Peptides *1:* 17–29 (1980).

Dodd, J.; Kelly, J.S.: The actions of cholecystokinin and related peptides on pyramidal neurones of the mammalian hippocampus. Brain Res. *205:* 337–350 (1981).

Dolecek, R.; Jezek, M.; Adamkova, M.; Polivka, P.; Kubis, M.; Sajnar, J.; Doleckova, D.; Kurkova, J.; Zavada, M.: Endorphin releasers. A new possible approach to the treatment of pain after burns – a preliminary report. Burns *10:* 41–44 (1983).

Duggan, A.W.; Curtis, D.R.: Morphine and the synaptic activation of Renshaw cells. Neuropharmacology *11:* 189–196 (1972). Emson, P.C.; Lee, C.M.; Rehfeld, J.F.: Cholecystokinin octapeptide. Vesicular localization and calcium dependent release from rat brain in vitro. Life Sci. *26:* 2157–2163 (1980).

Faris, P.L.; Komisaruk, B.R.; Watkins, L.R.; Mayer, D.J.: Evidence for the neuropeptide cholecystokinin as an antagonist of opiate analgesia. Science *219:* 310–312 (1983).

Faris, P.L.; McLaughlin, C.L.; Baile, C.A.; Olney, J.W.: Morphine analgesia potentiated but tolerance not affected by active immunization against cholecystokinin. Science *226:* 1215–1217 (1984).

Frederickson, R.C.A.; Burgis, V.; Edwards, J.D.: Hyperalgesia induced by naloxone follows diurnal rhythm in responsivity to painful stimuli. Science *198:* 756–758 (1977).

Gardner, J.D.; Jensen, R.T.: Cholecystokinin receptor antagonists. Am. J. Physiol. G. *246:* G471–G476 (1984).

Gaudreau, P.; Quirion, R.; St. Pierre, S.; Pert, C.: Characterisation and visualisation of cholecystokinin receptors in rat brain using H pentagastrin. Peptides *4:* 755–762 (1983).

Golterman, N.R.; Rehfeld, J.F.; Roigaard-Petersen, H.: In vivo biosynthesis of cholecystokinin in rat cerebral cortex. J. biol. Chem. *225:* 6181–6185 (1980).

Graham, W.C.; Meecham, K.G.; Pittaway, K.M.: The antinociceptive effect of cholecystokinin (26-33) in the mouse. Br. J. Pharm. *86:* 409 P (1985).

Hahne, W.F.; Jensen, R.T.; Lemp, G.F.; Gardner, J.D.: Proglumide and benzotript. Members of a different class of cholecystokinin receptor antagonists. Proc. natn. Acad. Sci. USA *78:* 6304–6308 (1981).

Han, J.S.; Fei, H.; Zhou, Z.F.: Met-enkephalin -arg^6-Phe^7-like immunoreactive substances mediate electroacupuncture analgesia in the periaqueductal gray of the rabbit. Brain Res. *322:* 289–296 (1984).

Han, J.S.; Ding, X.Z.; Fan, S.G.; Is cholecystokinin octapeptide (CCK-8) a candidate for endogenous anti-opioid substrates? Neuropeptides *5:* 399–402 (1985).

Herz, A.; Albus, K.; Metys, J.; Schubert, P.; Teschemacher, H.J.: On the central sites for the antinociceptive action of morphine and fentanyl. Neuropharmacology *9:* 539–551 (1970).

Hill R.G.; Hughes, J.; Pittaway, K.M.: Antinociceptive action of cholecystokinin octapeptide (CCK-8) and related peptides in rats and mice: Effects of naloxone and peptidase inhibitors. Neuropharmacology (in press, 1987).

Hommer, D.W.; Skirboll, L.R.: Cholecystokinin-like peptides potentiate apomorphine-induced inhibition of dopamine neurons. Eur. J. Pharmacol. *91:* 151–152 (1983).

Innis, R.B.; Snyder, S.H.: Cholecystokinin receptor binding in brain and pancreas. Regulation of pancreatic binding by cyclic and acyclic guanine nucleotides. Eur. J. Pharmacol. *65:* 123–124 (1980).

Irwin, S.; Houde, R.W.; Bennett, D.R.; Hendershot, L.C.; Seevers, M.H.: The effects of morphine, methadone and meperidine on some reflex responses of spinal animals to nociceptive stimulation. J. Pharmac. exp.Ther. *101:* 132–143 (1951).

Itoh, S.; Katsuura, G.: Suppressive effect of cholecystokinin and its related peptides on a B-endorphin induced catalepsy in rats. Eur. J. Pharmacol. *74:* 381–384 (1981).

Itoh, S.; Katsuura, G.; Maeda, Y.: Caerulein and cholecystokinin suppress beta-endorphin-induced analgesia in the rat. Eur. J. Pharmacol. *80:* 421–425 (1982).

Jeftinija, S.; Miletic, V.; Randic, M.: Cholecystokinin octapeptide excites dorsal horn neurones both in vivo and in vitro. Brain Res. *213:* 231–236 (1981).

Jones, B.J.; Roberts, D.J.: The qualitative measurement of motor inco-ordination in naive mice using an accelerating rotarod. J. Pharm. Pharmac. *20:* 302–304 (1968).

Jurna, I.; Zetler, G.: Antinociceptive effect of centrally administered caerulein and cholecystokinin octapeptide (CCK-8). Eur. J. Pharmacol. *73:* 323–331 (1981).

Kadar, T.; Pesti, A.; Penke, B.; Telegdy, G.: Inhibition of seizures induced by picrotoxin and electroshock by cholecystokinin octapeptides and their fragments in rats after intracerebroventricular administration. Neuropharmacology *23:* 955–961 (1984).

Kadar, T.; Borda, L.; Penke, B.; Kovacs, K.; Telegdy, G.: Cataleptogenic and anticataleptic activity produced by cholecystokinin octapeptides in mice. Neuropeptides *6:* 259–268 (1985).

Katsuura, G.; Hsiao, S.; Itoh, S.: Blocking of cholecystokinin octapeptide behavioural effects by proglumide. Peptides *5:* 529–534 (1984).

Katsuura, G.; Itoh, S.; Hsiao, S.: Specificity of nucleus accumbens to activities related to cholecystokinin in rats. Peptides *6:* 91–96 (1985).

Katsuura, G.; Itoh, S.: Potentiation of B-endorphin effects by proglumide in rats. Eur. J. Pharmacol. *107:* 363–366 (1985).

Kinscheck, I.B.; Watkins, L.R.; Kaufman, E.; Mayer, D.J.: Evidence for a cholecystokinin (CCK)-like endogenous opiate antagonist. Soc. Neurosci. Abstr. *9:* 232.10 (1983).

Kordower, J.H.; Bodnar, R.J.: Vasopressin analgesia. Specificity of action and non-opioid effects. Peptides *5:* 747–756 (1984).

Kubota, K.; Sugaya, K.; Matsuda, I.; Matsuoka, Y.; Terawaki, Y.: Reversal of the antinociceptive effect of cholecystokinin by benzodiazepines and a benzodiazepine antagonist RO 15-1788. Jap. J. Pharmac. *37:* 101–105 (1985).

Kuraishi, Y.; Harada, Y.; Aratani, S.; Satoh, M.; Takagi, H.: Separate involvement of the spinal noradrenergic and serotenergic systems in morphine analgesia. The differences in mechanical and thermal analgesic tests. Brain Res. *273:* 245–252 (1983).

Lynn, B.: Heat pain sensitivity of human skin after mild heat injury and its lack of dependence on the local blood flow. Pain *8:* 189–196 (1980).

Magnani, M.; Mantovani, P.; Pepeu, G.: Effects of cholecystokinin octapeptide and ceruletide on release of acetylcholine from cerebral cortex of the rat in vivo. Neuropharmacology *23:* 1305–1309 (1984).

Mantyh, P.W.; Hunt, S.P.: Evidence for cholecystokinin like immunoreactive neurons in the rat medulla oblongata which project to the spinal cord. Brain Res. *291:* 49–54 (1984).

Marley, P.D.; Nagy, J.I.; Emson, P.C.; Rehfeld, J.F.: Cholecystokinin in the rat spinal cord: Distribution and lack of effect of neonatal capsaicin treatment and rhizotomy. Brain Res. *238:* 494–498 (1982).

Martin, R.; Geis, R.; Holl, R.; Schafer, M.; Voigt, K.H.: Co-existence of unrelated peptides in oxytocin and vasopressin terminals of rat neurohypophyses: immunoreactive methionine[5]-enkephalin, leucine[5]-enkephalin, and cholecystokinin like substances. Neuroscience *8:* 213–227 (1983).

Mashal, R.D.; Owen, F.; Deakin, J.F.W.; Poulter, M.: The effect of cholecystokinin on dopaminergic mechanisms in rat striatum. Brain Res. *277:* 375–376 (1983).

McKnight, A.T.; Corbett, A.D.; Kosterlitz, H.W.: Increase in potencies of opioid peptides after peptidase inhibition. Eur. J. Pharmacol. *86:* 393–402 (1983).

Morley, J.E.; Levine, A.S.; Lindblad, S.: Intraventricular cholecystokinin-octapeptide produces hypothermia in rats. Eur. J. Pharmacol. *74:* 249–251 (1981).

Morley, J.E.: The ascent of cholecystokinin (CCK) from gut to brain. Life Sci. *30:* 479–493 (1982).

Mutt, V.; Jorpes, J.G.: Hormonal polypeptides of the upper intestine. Biochem. *125:* 57P–58P (1971).

Nilsen, P.L.: Studies on algesimetry by electrical stimulation of the mouse tail. Acta Pharmac. tox. *18:* 10–22 (1961).

O'Callaghan, J.P.; Holtzman, S.G.: Quantification of the analgesic activity of narcotic antagonists by a modified hot plate procedure. J. Pharmac. exp. Ther. *192:* 497–505 (1975).

Ohkubo, T.; Shibata, M.; Takahashi, H.; Naruse, S.: Naloxone prevents the analgesic action of alpha-MSH in mice. Experientia *41:* 627–628 (1985).

Oldendorf, W.H.: Blood-brain barrier permeability to peptides. Pitfalls in measurement. Peptides *2:* suppl. 2, pp. 109–111 (1981).

O'Neill, K.A.; Courtney, C.; Rankin, R.; Weissman, A.: An automated, high capacity method for measuring jumping latencies in a hot plate. J. Pharmac. Meth. *10:* 13–18 (1983).

Osbahr, A.J.; Nemeroff, C.B.; Luttinger, D.; Mason, G.A.; Prange, A.J.: Neurotensin-induced antinociception in mice. Antagonism by thyrotropin-releasing hormone. J. Pharmac. exp. Ther. *217:* 645–651 (1981).

Pardo, A.; Celotti, F.; De Paolis, C.: Ceruletide analgesia in biliary colic. Clin. Pharmac. Ther. *36:* 510–514 (1984).

Pecile, A.; Ferri, S.; Braga, P.C.; Olgiati, V.R.: Effects of intracerebroventricular calcitonin in the conscious rabbit. Experientia *31:* 332–333 (1975).

Pert, A.; Moody, T.W.; Pert, C.B.; Dewald, L.A.; Rivier, J.: Bombesin: Receptor distribution in brain and effects on nociception and locomotor activity. Brain Res. *193:* 209–220 (1980).

Pittaway, K.M.: Studies on the antinociceptive properties of CCK 26-33 and related peptides. M. Phil. Thesis, University of Cambridge (1985).

Reggiani, A.; Carenzi, A.; Frigeni, V.; Della Bera, D.: Effect of bestatin and thiorphan on [Met^5] enkephalin-Arg^6 Phe^7 induced analgesia. Eur. J. Pharmacol. *105:* 361–364 (1984).

Rehfeld, J.F.: Immunochemical studies on cholecystokinin. II. Distribution and molecular heterogeneity in the central nervous system and small intestine of man and hog. J. biol. Chem. *253:* 4022–4030 (1978).

Rehfeld, J.F.: Neuronal cholecystokinin. One or two multiple transmitters? J. Neurochem. *44:* 1–10 (1985).

Robson, L.E.; Gillan, M.G.E.; Kosterlitz, H.W.: Species differences in the concentrations and distributions of opioid binding sites. Eur. J. Pharmacol. *112:* 65–71 (1985).

Rodriguez, R.E.; Salt, T.E.; Cahusac, P.M.B.; Hill, R.G.: The behavioural effects of intrathecally administered [D-Pro^2, D-$Trp^{7,7}$]-substance P, an analogue with presumed antagonist actions, in the rat. Neuropharmacology *22:* 173–176 (1983).

Roques, B.P.; Fournie-Zaluski, M.C.; Sorocca, E.; Lecomte, J.M.; Malfroy, B.; Llorens, C.; Schwartz, J.C.: The enkephalinase inhibitor thiorphan shows antinociceptive activity in mice. Nature, Lond. *288:* 286–288 (1980).

Rosza, Z.; Varro, V.: Mechanism of action of cholecystokinin on intestinal blood flow; interactions with opioid peptides and vasoactive intestinal peptide. Neuropeptides *6:* 71–81 (1985).

Saito, A.; Sankaran, H.; Goldfine, I.D.; Williams, J.A.: Cholecystokinin receptors in the brain. Characterization and distribution. Science *208:* 1155–1156 (1980).

Salt, T.E.; Hill, R.G.: The effects of C-terminal fragments of cholecystokinin on the firing of single neurones in the caudal trigeminal nucleus of the rat. Neuropeptides *2:* 301–306 (1982).

Schiller, P.W.; Lipton, A.; Horrobin, D.F.; Bodanszky, M.: Unsulphated C-terminal 7-Peptide of cholecystokinin. A new ligand of the opiate receptor. Biochem. biophys. Res. Commun. *85:* 1332–1338 (1978).

Scott, W.L.; Mendelssohn, L.E.; Cohen, M.L.; Evans, D.A.; Frederickson, R.C.A.: Enantiomers of [R,S]-thiorphan. Dissociation of analgesia from enkephalinase A inhibition. Life Sci. *36:* 1307–1313 (1985).

Sheehan, M.J.; Belleroche, J. de: Facilitation of GABA release by cholecystokinin and caerulein in rat cerebral cortex. Neuropeptides *3:* 429–434 (1983).

Sheehan, M.J.; Belleroche, J. de: Central actions of cholecystokinin: behavioural and release studies; in De Belleroche, Dockray, CCK in the nervous system, pp. 110–127 (Ellis Horwood, Chichester, 1984).

Skirboll, L.; Hokfelt, T.; Dockray, G.; Rehfeld, J.; Brownstein, M.; Cuello, A.C.: Evidence for periaqueductal cholecystokinin-substance P neurons projecting to the spinal cord. J. Neurosci. *3:* 1151–1157 (1983).

Soll, A.H.; Amirian, D.A.; Park, J.; Elashoff, J.D.; Yamada, T.: Cholecystokinin potently releases somatostatin from canine fundic mucosal cells in short-term culture. Am. J. Physiol. *248:* G569–G573 (1985).

Steigerwalt, R.W.; Williams, J.A.: Binding specificity of the mouse cerebral cortex receptor for small cholecystokinin peptides. Regul. Peptides *8:* 51–59 (1984).

Suberg, S.N.; Culhane, E.S.; Carstens, E.; Watkins, L.R.: Behavioural and electrophysiological investigations of opiate/cholecystokinin interactions. Adv. Pain Res. Ther. *9:* 541–553 (1985).

Swerts, J.P.; Perdrisot, R.; Patey, G.; Baume, S. de la; Schwartz, J.C.: 'Enkephalinase' is distinct from brain 'angiotensin converting enzyme'. Eur. J. Pharmacol. *57:* 279–281 (1979).

Tang, J.; Chou, J.; Iadarola, M.; Yang, H.Y.T.; Costa, E.: Proglumide prevents and curtails acute tolerance to morphine in rats. Neuropharmacology *23:* 715–718 (1984a).

Tang, J.; Yang, H.Y.T.; Costa, E.: Inhibition of spontaneous and opiate modified nociception by an edogenous neuropeptide with Phe-Met-Arg-Phe-NH_2-like immunoreactivity. Proc. natn. Acad. Sci. USA *81:* 5002–5005 (1984b).

Terenius, L.: Somatostatin and ACTH are peptides with partial antagonist-like selectivity for opiate receptors. Eur. J. Pharmacol. *38:* 211–213 (1976).

Vanderhaegen, J.J.; Signeau, J.C.; Gepts, W.: New peptide in the vertebrate CNS reacting with antigastrin antibodies. (Nature, Lond. *257:* 604–605 (1975).

Vaught, J.L.; Peley, K.; Costa, L.C.; Setler, P.; Enna, S.J.: A comparison of the antinociceptive responses to the GABA-receptor agonists THIP and Baclofen. Neuropharmacology *24:* 211–216 (1985).

Vigna, S.R.; Szecowka, J.; Williams, J.A.: Do antagonists of pancreatic cholecystokinin receptors interact with central nervous system cholecystokinin receptors? Brain Res. *343:* 394–397 (1985).

Wang, R.Y.; White, F.J.; Voigt, M.M.: Cholecystokinin, dopamine and schizophrenia. TIPS *Oct.:* 436–438 (1984).

Watkins, L.R.; Kinscheck, I.B.; Kaufman, E.F.S.; Miller, J.; Frenk, H.; Meyer, D.J.: Cholecystokinin antagonists selectively potentiate analgesia induced by endogenous opiates. Brain Res. *327:* 181–190 (1985a).

Watkins, L.R.; Kinscheck, I.B.; Mayer, D.J.: Potentiation of opiate analgesia and apparent reversal of morphine tolerance by proglumide. Science *224:* 395–396 (1984).

Watkins, L.R.; Kinscheck, I.B.; Mayer, D.J.: Potentiation of morphine analgesia by the cholecystokinin antagonist proglumide. Brain Res. *327:* 169–180 (1985b).

Webster, V.A.D.; Griffiths, E.L.; Slater, P.: Antinociceptive effects of thyrotrophin-releasing hormone and its analogues in the rat periaqueductal grey region. Neurosci. Lett. *42:* 67–70 (1983).

Willetts, J.; Urban, L.; Murase, K.; Randic, M.: Actions of cholecystokinin octapeptide on rat spinal dorsal horn neurones. Ann N. Y. Acad. Sci. *448:* 385–402 (1985).

Woolfe, G.; Macdonald, A.D.: The evaluation of the analgesic action of pethidine hydrochloride (Demerol). J. Pharmac. exp. Ther. *80:* 300–307 (1944).

Yaksh, T.L.; Abay, E.O., Go, V.L.W.: Studies on location and release of cholecystokinin and vasoactive intestinal peptide in rat and cat spinal cord. Brain Res. *242:* 279–290 (1982).

Yang, H.Y.T.; Tang, J.; Costa, E.: Modulation of opioid analgesia by endogenous peptides. Int. Endocoids, Fort Worth. 1984, Abstr. 71, p. 467.

Yehuda, S.; Kastin, A.J.: Peptides and thermoregulation. Neurosci. biobehav. Rev. *4:* 459–471 (1980).

Zetler, G.: Analgesia and ptosis caused by caerulein and cholecystokinin octapeptide (CCK-8). Neuropharmacology *19:* 415–422 (1980a).

Zetler, G.: Effects of cholecystokinin-like peptides on rearing activity and hexobarbitol induced sleep. Eur. J. Pharmacol. *66:* 137–139 (1980b).

Zetler, G.: Central effects of ceruletide analogues. Peptides *2:* suppl. 2, pp. 65–69 (1981).

Zetler, G.: Caerulein and morphine. An attempt to differentiate their antinociceptive effects. Pharmacology *25:* 121–129 (1982).

Zetler, G.: Apomorphine separates the antinociceptive effects of cholecystokinin octapeptide and ceruletide from those of morphine. Eur. J. Pharmacol. *92:* 151–154. (1983a).

Zetler, G.: Cholecystokinin octapeptide (CCK-8), ceruletide and analogues of ceruletide. Effects on tremors induced by oxotremorine, harmine and ibogaine. A comparison with prolyl-leucylglycine amide (MIF), anti-parkinsonian drugs and clonazepam. Neuropharmacology *22:* 757–766 (1983b).

Zetler, G.: Antistereotypic effects of cholecystokinin octapeptide (CCK-8), ceruletide and related peptides on apomorphine-induced gnawing in sensitised mice. Neuropharmacology *24:* 251–259 (1985).

R.G. Hill, PhD, Parke-Davis Research Unit, New Addenbrookes Hospital Site, Hills Road, Cambridge CB2 2QB (UK)

Pain Headache, vol. 9, pp. 247–265 (Karger, Basel, 1987)

Interaction of Cholecystokinin and Opioids in Pain Modulation

S.N. Suberg, L.R. Watkins [1]

Department of Animal Physiology, University of California, Davis, Calif. USA

A variety of peptides which are classically associated with the gastrointestinal tract have been identified in the vertebrate [Vanderhaeghen, et al., 1975], porcine [Rehfeld, 1978], amphibian [Beinfeld et al., 1983], mollusk [Osborne et al., 1982], and human [Sakamoto, 1984] central nervous systems (CNS). One of these gastrointestinal hormones, which is of interest to our laboratory, is the sulfated octapeptide variant of cholecystokinin (CCK). Vanderhaeghen et al. [1975] described a new peptide extracted from the brains of several vertebrates. The peptide was described as 'gastrin-like', due to its reactivity with antiserum produced against the heptadecapeptide, gastrin. A few years later Dockray et al. [1978] isolated two CCK peptides from sheep brain, reporting that the new brain peptide resembled the carboxyl terminal of the octapeptide variant of CCK. Subsequent studies using a variety of techniques have confirmed that the new peptide described by Vanderhaeghen closely resembles sulfated CCK-8 and not gastrin [Rehfeld, 1978; Ryder et al., 1981; Larsson and Rehfeld, 1979; Micevych et al., 1982].

The distribution of CCK in the brain has its largest concentration in the cerebral cortex [Beinfeld et al., 1981; Jones, 1975]. Within the cerebral cortex of the rat, the highest concentrations are found in the cingulate, pyriform and entorhinal areas [Beinfeld et al., 1981]. Outside the cortex the highest concentration is in the caudate nucleus [Beinfeld et al., 1981]. Many investigators have shown that CCK may be localized in the

[1] The authors gratefully thank Ms. Elizabeth Culhane and Ms. Cindy Thurston for their invaluable technical and editorial assistance and Ms. Paula Buchignani for her excellent preparation of the manuscript.

cells, fibers and terminals of the cerebral cortex, amygdala, hippocampus, hypothalamus, periaqueductal gray, dorsal raphe, and dorsal horn of the spinal cord. CCK containing neurons have also been found to project from the periaqueductal gray [Skirboll et al., 1983] and medulla oblongata [Mantyn et al., 1984] to the spinal cord. Zarbin and coworkers [Zarbin et al., 1984] have compiled an extensive report on the localization of CCK receptors, with receptors being concentrated in the olfactory bulb, cortex and pretectal region. A substantial number of sites were also found in the basal ganglia, amygdala, hippocampus, and dorsal horn of the spinal cord.

The physiological actions attributed to CCK are as diverse as its distribution. The physiological action of CCK in the gastrointestinal tract is excitatory. It is a potent activator of digestion, pancreatic secretion, and gall-bladder contraction, as well as increasing gut motility. In the CNS, the physiological actions attributed to CCK include regulation of satiation [Stein and Woods, 1981], modulation of catecholaminergic activity [Fuxe et al., 1980] and regulation of hypothalamic peptides [Itoh et al., 1979]. After central administration, CCK has been reported to produce a variety of effects. These include hyperglycemia [Morley and Levine, 1981], hypothermia [Katsuura et al., 1981], depression of certain types of emotional behavior [Zetler, 1981], rotational syndrome [Mann et al., 1980] and, in high doses, analgesia [Zetler, 1980]. On a cellular level, electrophysiological data have shown both excitatory and inhibitory actions of CCK in the CNS. Excitatory effects have been observed with application of CCK to spinal [Phillis and Kirkpatrick, 1980; Jeftinija et al., 1981] and medullary dorsal horn [Salt and Hill, 1982] neurons. In slice preparations of immature rat spinal dorsal horn neurons, CCK induced a slow, reversible, dose-dependent depolarization [Willitts et al., 1985].

Faris et al. [1973] noted that several of the physiological actions of CCK are opposite to those of opiates. Some examples noted include: (1) Short-term food intake decreases after intraperitoneally or intracerebroventricularly administered CCK [Della-Fera and Baile, 1979], in contrast to increases in food intake after morphine or opioid administration [Morley and Levine, 1983]. (2) Stress-induced feeding is mediated by endogenous opiates [Morley and Levine, 1980] and blocked by CCK [Nemeroff et al., 1978].

Based on these and other examples, Faris et al., [1983] hypothesized that CCK may physiologically function as an endogenous opiate antagonist. Since publication of their paper, additional evidence has been noted, thus increasing and supporting the likelihood that CCK may interact with opiate/opioid systems. As an example, it is now known that CCK will physiologically increase respiration [Gillis et al., 1983], whereas a well-known side effect of morphine is respiratory depression. Also, importantly, many of the CCK-containing loci are areas observed to contain high concentrations of endogenous opioids, with many of these areas implicated in pain modulation. Results from this study suggest that the distribution of enkephalins and CCK are strikingly similar. Therefore, a functional relationship may exist between them, though sulfated CCK-8 does not bind to opiate receptors in the brain [Stengaard-Pederson and Larsson, 1981].

In support of the hypothesis that CCK can function as a physiological opiate antagonist, Faris et al., [1983] found that CCK did indeed antagonize opiate analgesia produced by morphine administration as well as analgesia produced by the release of endogenous opiates; that is induced by front paw foot-shock and classical conditioning [Watkins et al., 1982]. CCK has also been found to antagonize analgesia produced by administration of β-endorphin [Itoh, 1982]. In contrast, when CCK was administered prior to nonopiate analgesia (induced by hind paw foot-shock), there was no attenuation of analgesia. Thus, these findings indicate that CCK may be selectively modulating the analgesic effects of opiates and endogenous opioids.

How might CCK be modulating the effects of opiates? CCK may be modulating opiates through a negative feedback mechanism, such that administration or presence of opiates/opioids induces endogenous release of CCK. The release of CCK may occur to return the animal toward a state of basal pain sensitivity. This phenomenon may be homeostatically valuable in pain of acute origin to maintain awareness of tissue-damaging stimuli. In the case of chronic pain it might only add to the suffering. Thus, preventing CCK from interacting with opiates/opioids may prolong the action of opiates, thereby enhancing the degree and duration of pain relief.

This hypothesis has recently been tested using proglumide, a putative CCK receptor antagonist [Hahn et al., 1981; Rovati, 1976]. Proglumide, a glutaramic acid derivate, has been used for the treatment of ulcers in Europe [Rovati, 1976] and has been shown to be an antagonist of CCK receptors in the pancreas. Though a recent study suggests that proglumide does not bind to cortical CCK receptors [Clark et al., 1985], there is overwhelming behavioral and electrophysiological evidence that proglumide blocks the effects of CCK in the CNS [Willits et al., 1985; Chiodo and Bunney, 1983; Gillis et al., 1983, Kinscheck et al., 1983; Tang et al., 1983; Watkins et al., 1985]. Although there is a discrepancy between the receptor studies and electrophysiological/behavioral studies, all of the in vivo and cell culture studies clearly show that proglumide does selectively and profoundly block CCK-induced effects in the CNS.

If endogenous CCK acts as an endogenous opiate antagonist, and proglumide blocks CCK, then proglumide administration would be expected to potentiate opiate analgesia. In support of this hypothesis, Watkins et al., [1985] reported that proglumide potentiates the onset, peak, and duration of opiate analgesia [Watkins et al., 1984, 1985; Suberg et al., 1985] as measured by the tail-flick test. Potentiation was observed when proglumide was combined with either intrathecal (IT) morphine or endogenous opiates (It-*D*-alanine-methionine enkephalinamide, DALA; induced release by front paw footshock). Importantly, Watkins et al., [1985] observed that proglumide neither produced analgesia when injected in the absence of opiates nor potentiated nonopiate forms of analgesia. These results imply both that proglumide is selectively interacting with opiate systems in order to produce these effects and that CCK is not tonically released but rather induced to release by the presence of opiates.

The studies of Faris et al., [1983] and Watkins et al., [1984, 1985] assessed analgesia only by using the tail-flick test. Thus, in our first series of studies, we wanted to test whether proglumide's ability to potentiate opiate/opioid analgesia was specific to the type of noxious stimulus applied. That is, whether proglumide would potentiate the effects of opiates when assessing analgesia using a variety of painful stimuli [Suberg, 1985a]. Secondly, we wanted to examine electrophysiologically the effect of proglumide on morphine induced suppression of responses from single dorsal horn neurons to noxious heating of their receptive fields [Suberg et al., 1985b]. The results from these studies stimulated our investigations to test whether CCK modulated the activation of only certain types of opiate receptors; that is, whether proglumide would potentiate analgesia induced by a variety of opioid receptor agonists.

The methodology and results of our first series of studies have been published elsewhere [Suberg et al., 1985a], but our findings can be summarized as follows. Briefly, adult male rats were implanted with

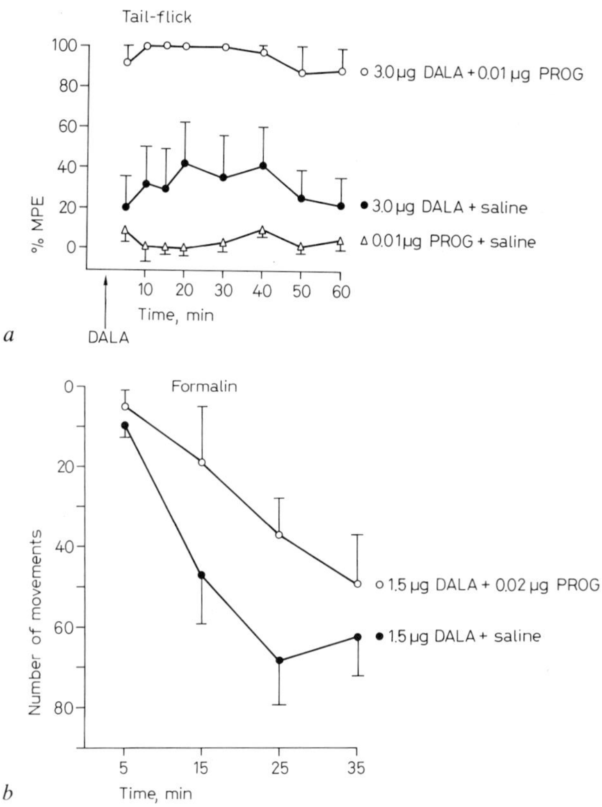

Fig. 1. a Potentiation of lumbosacral intrathecal (IT) DALA analgesia by IT proglumide (PROG) in the tail flick test. PROG significantly ($p < 0.001$) potentiates DALA analgesia in the tail flick test. Graph shows mean percent of maximal effect ± SE, across time, after 3.0 μg DALA IT (●, n = 7), 3.0 μg DALA + 0.01 μg PROG IT (○, n = 7) or 0.01 μg PROG IT (△, n = 5). *b* Increase in DALA analgesia by PROG in the formalin test. This graph plots the number of leg flinches across time, after 1.5 μg DALA IT (●, n = 14) or 1.5 μg DALA + 0.02 μg PROG IT (○, n = 5). PROG significantly ($p < 0.005$) increased the degree of DALA analgesia in the formalin test as compared to analgesia produced by DALA alone. From Suberg et al., [1985a].

intrathecal (IT) catheters which terminated at the lumbosacral enlargement [Watkins and Mayer, 1982]. After rats were habituated to the restraining cylinders used to hold them during behavioral experiments, they were injected IT with either proglumide or equivolume saline at 10 min before and then again just before DALA. DALA was chosen based on the previous behavioral studies of Watkins et al. [1985] showing that proglumide was highly effective in potentiating analgesia produced by this opiate. Behavioral assessments performed were modifications of the tail-flick [D'Amour and Smith, 1941], hot plate [Woolfe and MacDonald, 1944], formalin [Dubuisson and Denis, 1977], tail-shock vocalization [Charpentier, 1968] and tail-pinch [Collier, 1962] tests, as well as assessments of general motor and sensory functions.

In the tail-flick and formalin tests, increasing dosages of DALA produced an increasing degree of analgesia. A potentiation of DALA analgesia by proglumide occurred only in these same tests (fig. 1). Since general motor (posture; locomotion; righting; grasping and corneal reflexes) and sensory (light touch/pressure; startle) functions were not impaired, potentiation of DALA analgesia by proglumide cannot be explained by generalized impairment of either motor or sensory systems. Furthermore, in the hot-plate, tail-shock vocalization, and tail-pinch tests, no clear dose-response relationship was observed with DALA. Interestingly, when proglumide was administered along with DALA, no potentiation of the analgesic effects of DALA was evident (fig. 2). We postulated that, in tests where DALA did not produce dose-dependent analgesia, DALA may not have been the appropriate opiate for blocking the noxious stimuli associated with those tests. This postulate arose from earlier observations that various types of pain may be modulated by different neural systems [Abbott et al., 1981, 1982; Dennis and Melzack, 1980]. Concurrent with the work of Schmauss and co-workers [Schmauss et al., 1983; Schmauss, 1984], our laboratory recognized the possibility that different opiate receptors modulate different pain modalities. Multiple opiate receptor types [Martin et al., 1976] have been found in the rat spinal cord [Traynor et al., 1982], thus supporting this possibility. The reason that DALA does not produce dose-dependent analgesia in the hot-plate, tail-shock and pinch tests may be because DALA is not specific for the neural modulating systems involved.

Knowing that DALA is a mu/delta receptor agonist [Lord et al., 1977], it appears that mu and/or delta receptors are involved in modulating painful stimuli arising during the tail-flick and formalin tests. Since proglumide can potentiate these DALA analgesias, proglumide may interact with mu and/or delta receptors.

In the electrophysiological studies, proglumide markedly potentiated the morphine induced suppression of nociceptive transmission in a naloxone reversible manner (fig. 3). These studies demonstrate at a cellular level that proglumide enhances the inhibitory effects of opiates

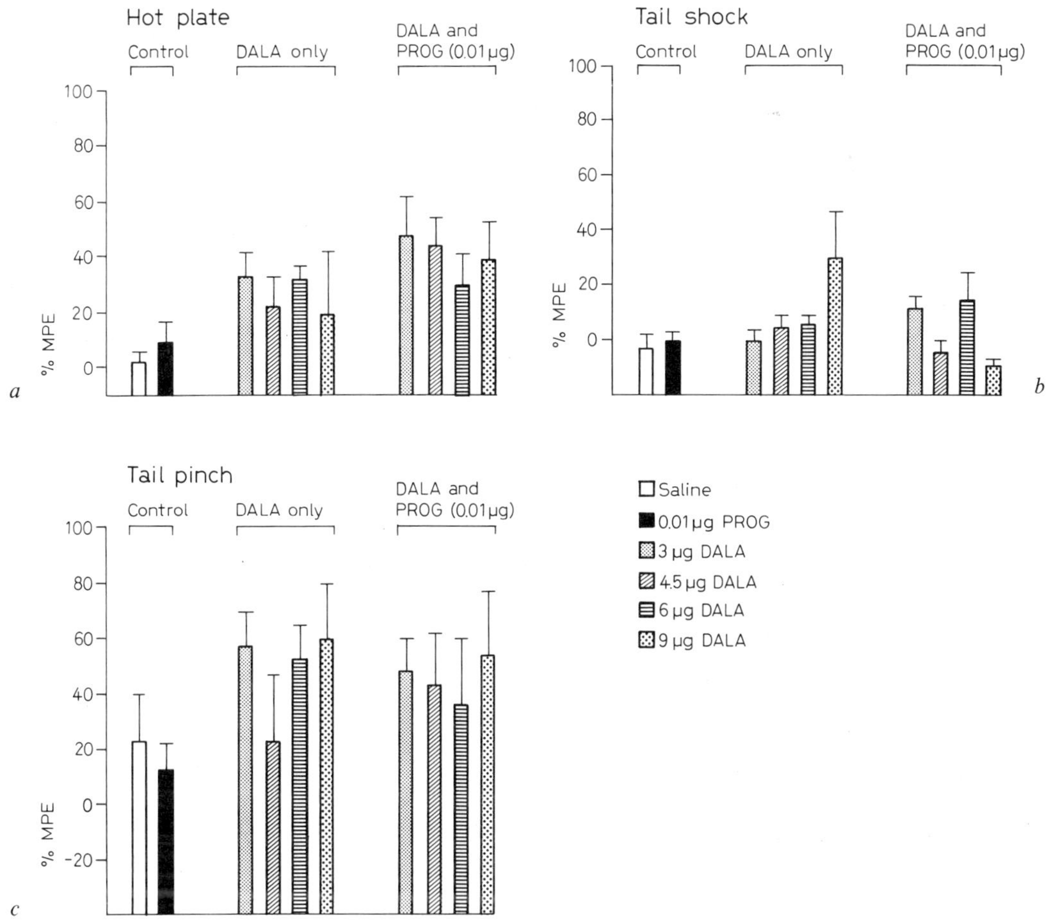

Fig. 2. Effect of IT DALA (3.0, 4.5, 6.0 or 9.0 μg, n = 10 for each dose) combined either with IT saline or proglumide (PROG, 0.01 μg, n = 10). Data are expressed as mean percent of maximal effect ± SE for the 3 pain tests. *a* Hot plate test: No clear dose-response relationship occurred with increasing doses of DALA and there was no significant potentiation of DALA analgesia with PROG. *b* Tail-shock vocalization test: No clear dose-response relationship occurred with increasing doses of DALA, though with 9.0 μg there was a significant ($p < 0.005$) increase in DALA analgesia. PROG produced no significant potentiation of DALA analgesia and at 9.0 μg DALA + 0.01 μg PROG there was an attenuation of DALA analgesia. *c* Tail-pinch test: No dose-response relationship occurred with increasing doses of DALA and there was no significant potentiation of DALA analgesia with PROG. From Suberg et al., [1985a].

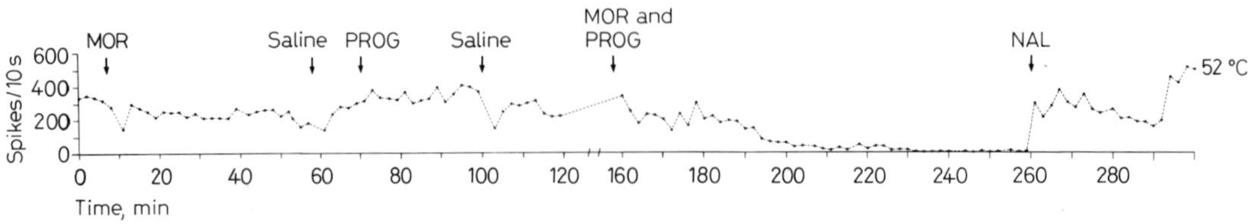

Fig. 3. Proglumide markedly potentiates morphine-induced inhibition of spinal nociceptive transmission. Time course of effect of morphine (MOR, 0.1 μ) and proglumide (PROG, 0.04 ng) applied directly onto the dorsal aspect of the lumbosacral spinal cord, and of systemic naloxone (NAL, 10 mg/kg i.v.), during noxious heating of the rat's hind foot pad. Graph plots the number of action potentials recorded from a single unit during the 10-second heat period, against time. After application of morphine, a 50% reduction in the unit's response to 52 °C was not observed until 50 min. After application of PROG alone, the unit's firing rate remained within ± 10% of predrug level. Following application of MOR + PROG, the unit's firing rate was more rapidly reduced to 50% than by morphine alone (26 vs. 50 min) and to 97% at 100 min postdrug application. Naloxone immediately reversed the reduced firing rate to predrug levels. From Suberg et al. [1985a].

without causing any inhibition when applied alone. The potentiation of morphine inhibition implies that there is a proglumide/morphine interaction that is occurring at the level of the spinal cord. Since morphine is a mu receptor agonist [Lord et al., 1977], these results confirm and extend our behavioral conclusions that the proglumide/opiate interaction may occur at the mu receptor.

To test further whether proglumide was selectively interacting with the mu receptor, we studied the effect of proglumide on analgesia induced by a variety of opioid receptor agonists, as assessed by a variety of painful stimuli. The effects of morphine, *D*-alanine-leucine-enkephalin (DADL), and U50-488H were tested alone and in combination with proglumide in general motor and sensory function tests and in the tail-flick, tail-shock vocalization, hot-plate, pinch, and formalin tests. These compounds were chosen for their selectivity towards mu, delta and kappa opioid receptors, respectively. All compounds were administered IT using the same methodology as mentioned above.

Proglumide (0.01 μg) potentiated the analgesia induced by morphine (1.0, 3.0, 4.5, 6.0 and 9.0 μg), DADL (0.1, 1.0 and 10.0 μg), and U50-488H (50 and 100 μg) in a very selective manner (table I). In contrast to our observations using DALA, enhancement of analgesia was not seen in the tail-flick test when proglumide was combined with either morphine, DADL or U50-488H. Though Watkins et al., [1985] did see only a very

Table 1. Effect of Proglumide on Analgesia Induced by a Variety of Opioid Agonists

Test	DALA μ/δ	DADL δ	Morphine μ	U50-488H κ
Tail Flick	+	0	+[c]	0
Tail Shock	0	0	+	0
Hot Plate	0	+[a]	0	0
Pinch	0	+[b]	+	0
Formalin	+	0	−[d]	−[e]

\+ Potentiation ($p < 0.05$)
0 No effect
− Attenuation ($p < 0.05$)
a Trend towards potentiation at 0.1 μg dose only
b Potentiation ($p < 0.01$) at 0.1 μg dose only
c See references 45 and 62
d Attenuation at 1.5 μg dose only
e Attenuation at 50 μg dose only

small enhancement of analgesia with 1 μg of morphine co-administered with 0.01 μg proglumide, statistical significance was reached using a trend analysis over time. There was no significant potentiation of morphine analgesia with proglumide at any single point in time. In the present series of studies, potentiation did occur in the following instances. In the tail-shock test, potentiation by proglumide was seen only with one dose of morphine (4.5 μg), in the pinch test with 0.1 μg DADL and 100 μg of U50-488H. In the hot-plate and tail-shock tests, proglumide did consistently enhance the analgesic effects of 0.1 μg DADL. In the formalin test, proglumide did not significantly potentiate the effects of the opioid agonists at the doses tested. Proglumide when combined with 1.5 μg morphine or 50 μg U50-488H significantly attenuated analgesia induced by these opioids. From these results, it appears that proglumide most consistently potentiates analgesia when mu, delta or mu/delta agonists are used. Results of our electrophysiology experiments also strongly support a proglumid/mu receptor interaction.

In addition to CCK modulation of acute activation of opiate systems, data is accumulating which strongly suggests involvement of CCK in the phenomenon of morphine tolerance. Morphine tolerance has been implicated to occur largely through the mu receptor [De Lander, 1984], and proglumide has been noted to both prevent and reverse morphine tolerance. Tang et al. [1983] showed that proglumide given concurrently with injections of morphine reduced morphine tolerance and also delayed the onset of tolerance. Maier and co-workers [unpubl. observations] and Rovati et al. [1985] have recently verified and extended these findings, adding support that CCK antagonists can prevent the development of morphine tolerance. Concurrent with the work of Tang et al. [1983], Watkins [1984] demonstrated that proglumide could also dramatically 'reverse' morphine tolerance. They showed that doses of morphine which produced no analgesia when delivered to morphine-tolerant animals produced marked analgesia when combined with proglumide. In an extension of these initial observations, Kinscheck et al. [1983] exposed animals to both short- and long-term tolerance regima. Animals exposed to a longer tolerance regimen required a higher dose of proglumide to induce analgesia when co-administered with morphine. These latter data suggest that the progressive development of morphine tolerance may parallel a progressive enhancement of CCK synthesis/release secondary to repeated opiate administration. Such a hypothesis is consistent with the idea that CCK systems are activated in response to activation of opiate systems in a negative feedback manner. In support of these ideas, we have provided initial evidence that radioimmunoassayable levels of CCK progressively increase, compared to saline-injected controls, following acute morphine administration, a short-term tolerance regimen, and a long-term tolerance regimen [Watkins et al., 1984].

There are several conclusions that may be drawn from our work and that of other investigators regarding cholecystokinin and proglumide and their interaction in pain modulation: (1) At a behavioral level, cholecystokinin has been shown to antagonize analgesia induced by morphine and opioids (either endogenously released or exogenously administered) [Faris et al., 1983; Kinscheck et al., 1983; Watkins et al., 1985]. At an electrophysiological level CCK can antagonize, whereas CCK antibodies can enhance, morphine induced suppression of dorsal horn neuronal response to painful stimuli [Suberg et al., 1985]. (2) Cholecystokinin does not bind to opiate receptors in the brain [Stengaard-Pederson and Larsson, 1981]. Parallel studies have not yet been done in spinal cord, the major confirmed site of opiate-CCK interactions for pain modulation. (3) Proglumide is a cholecystokinin receptor antagonist in the periphery [Hahn et al., 1981; Rovati, 1976], but does not appear to bind to cholecystokinin receptors in the cerebral cortex [Clark et al., 1985]. To date, parallel binding studies have not been done in spinal cord, the major confirmed site of proglumide-opiate interaction for pain modulation. (4) Proglumide does reverse the excitatory effects of cholecystokinin in the central nervous system [Willitts et al., 1985; Chiodo et al., 1983; Gillis et al., 1983; Kinscheck et al., 1983; Tang et al., 1983; Watkins et al., 1985]. (5) Proglumide can potentiate the analgesic effects of mu, delta and mu/delta opioid receptor agonists [Watkins et al., 1984; Suberg et al., 1985a]. Proglumide enhances the onset, degree and duration of opiate analgesia. At an electrophysiological level, proglumide enhances the onset, degree and duration of morphine-induced inhibition of dorsal horn neuronal response to pain. (6) Proglumide prevents and reverses morphine tolerance [Watkins et al., 1984; Tang et al., 1984; Kinscheck et al., 1983]. At present, proglumide does not appear to affect withdrawal (dependence).

Initially, we had thought that proglumide might potentiate analgesia whenever the administered opiate/opioid produced dose-dependent analgesia for a given type of painful stimulus. When we used different opiate receptor agonists, we did observe some dose-dependent relationships which were not associated with any observable potentiation of analgesia by proglumide. Proglumide most consistently potentiated the dose-dependent analgesia produced by mu, delta and mu/delta agonists (morphine, DADL and DALA respectively), indicating both that proglumide only potentiates opiate analgesic effects where the opiate effectively modulates the applied painful stimulus (i.e. heat, shock, formalin) and that the opiate/opioid needs to be of the mu or delta subtype. Potentiation seen in DADL induced analgesia might be explained by DADL's putative capability to bind to the lower affinity (mu-like) delta receptor binding site [Rothman et al., 1985a, b].

There is a wealth of excellent studies which show that CCK antagonizes opiate analgesia, that proglumide antagonizes CCK in the CNS and that proglumide potentiates opiated mediated analgesia. But does proglumide potentiate opiate analgesia by acting as a CCK antagonist? If proglumide's sole action were as a CCK antagonist, one would expect that, in instances where administration of opioids induced release of endogenous CCK, co-administration of proglumide should prevent the antagonistic action of CCK on opiate analgesia. From our initial studies

[Suberg et al., 1985b], proglumide potentiated the analgesic effect of DALA as assessed by the tail-flick and formalin tests. This indicates that DALA both (1) induced the release of CCK, and (2) was associated with the antinociceptive system selective for these types of nociceptive input. Knowing that DALA is a mu/delta receptor agonist, it appears that (1) mu/delta-like agonists induce the release of CCK, and (2) mu/delta-like agonists are selective for cutaneous thermal and irritative chemical types of nociceptive inputs [Schmauss and Yaksh, 1984]. If DALA induces the release of CCK and is selective for modulating noxious stimuli of cutaneous thermal and chemical origin, one would expect that in the hot plate test (a noxious cutaneous thermal stimulus) proglumide would potentiate DALA analgesia. Our studies show that proglumide neither potentiated DALA analgesia, as measured by the hot-plate test, nor did it attenuate it. These results raise the following possibilities: (1) DALA may not induce the release of endogenous CCK. If there is no CCK released, there is no substrate for which proglumide may exert its antagonistic action upon. (2) DALA may not be selective for the noxious stimulus induced by the hot-plate test. Since DALA did not induce a dose response analgesia in this test, it appears that DALA is not selectively modulating nociceptive input from the hot plate. Thus, if administration of DALA did induce endogenous CCK release and administration of proglumide antagonized that CCK release, potentiation of opioid analgesia is not measurable because DALA did not produce analgesia. (3) Proglumide may not act as a CNS CCK antagonist. If DALA induces release of endogenous CCK and proglumide does not antagonize CCK in the CNS, what mechanism is responsible for the observation that proglumide does potentiate DALA analgesia as measured by the tail-flick and formalin test? Proglumide may counteract the DALA induced release of CCK by potentiating the analgesic action of DALA. If DALA induces a clear-cut dose-response analgesia, we can expect to see a potentiation of DALA analgesia by proglumide. This was observed in the tail-flick and formalin tests. Where DALA did not produce a clear-cut dose-response analgesia (hot plate, tail-shock, and pinch tests), there obviously was no analgesia which proglumide could potentiate. Thus, it appears that proglumide may not act solely as a CCK antagonist, but it may indirectly or directly interact with DALA-activated opioid mechanisms.

Are opiate-induced release of endogenous CCK and opiate/CCK interactions selective for specific opiate receptor subtypes? Previous studies using morphine (the prototype for the mu receptor) showed that CCK does antagonize morphine-induced analgesia [Faris et al., 1983], as well as inhibit morphine-induced suppression of spinal nociceptive transmission [Suberg, Culhane, Carstens, Watkins, manuscript in preparation]. Our studies, as well as those of others [Watkins et al., 1984; Rovati et al., 1985], have shown that proglumide also potentiates the pain suppressive effects induced by morphine. Our hypothesis that morphine induces the release of endogenous CCK is fully supported by our preliminary findings that there are increased levels of radioimmunoassayable CCK in the dorsal horn of the spinal cord both in animals receiving a single dose of morphine as well as in animals made tolerant to morphine [Watkins, 1984]. Taken together, these results indicate that mu receptor agonists induce the release of endogenous CCK.

When DADL was used (a known delta receptor agonist), our studies showed that proglumide only potentiated analgesia induced by this ligand in the pinch test and at only one dose (0.1 μg). In a cutaneous thermal test (the hot plate test), proglumide tended towards potentiating the analgesic effect of DADL at only one dose (0.1 μg). These data do not clearly indicate if delta receptor agonists induce the release of CCK.

The kappa receptor agonist, U50-488H (50 and 100 μg) produced no consistent analgesia on any of our tests. Schmauss et al. [1984] and Yaksh [1984] have also reported similar results. In our studies, there was a slight trend towards analgesia as measured by the formalin test. Schmauss et al. [1984] did see similar results with another chemical irritant test (the writhing test). Thus, it appears that the spinal kappa receptor is not involved in modulating nociceptive information of cutaneous thermal or of mechanical origin. When proglumide was co-administered with either 50 or 100 μg of U50-488H, we again did not see any analgesia as measured by our tests. Therefore, it appears that kappa agonists do not induce the release of endogenous CCK at the level of the spinal cord.

In contrast to administration of opioids whose receptor specificities have been well defined, little is known regarding either the identity or receptor specificity of endogenous opioids which are released during various forms of environmentally induced analgesias. Given the differential interaction of proglumide with various opioid receptors, it is plausible that testing for proglumide-opioid interactions could give some clue as to the endogenous opioid released by analgesia-causing environmental stimuli. Taking front paw foot-shock induced analgesia as an example, this form of environmentally induced analgesia [Watkins and Mayer, 1982] has been shown to be opiate-mediated at the level of the spinal cord, antagonized by CCK, and potentiated by proglumide. This information, taken together with the conclusion derived from our present series of experiments, indicates that the endogenous opiates released via front paw foot shock may interact at the mu opiate receptor. In fact, Yaksh and Elde [1981] and recently Cesselin et al. [1985] have shown that methionine enkephaline-like material (methionine enkephalin is a mu receptor agonist) is released from the spinal cord of cats and rats subjected to noxious stimuli.

From our data, a correlation is noted between the type of receptor ligand effective in inducing analgesia and the type of ligand with which proglumide most consistently interacts. Our observations indicate that proglumide does not potentiate all forms of opioid-induced analgesia (table I). Our data also show that proglumide most consistently potentiates analgesia induced by morphine and DALA, mu and mu/delta receptor ligands. This indicates that mu and/or mu/delta receptor ligands induce the release of CCK, or that possibly proglumide may be interacting at the mu or a mu/delta receptor complex. Proglumide may be interacting in an allosteric fashion (as proposed for CCK) [Stengaard-Pederson and Larsson, 1981] at the mu receptor and/or at the mu/delta receptor complex. The action may be one of a neuromodulator, in that proglumide may 'prime' an opiate receptor to accentuate its normal activity. To date, we are not aware of any binding studies indicating whether proglumide does or does not influence binding at mu or delta receptors. Information of this type is necessary to address any of these speculations.

In conclusion, it appears that proglumide may physiologically function in two ways: (1) to act as a CCK antagonist when opiates induce release of endogenous CCK, and/or possibly (2) to influence binding of opiate ligands to their receptors.

The implications of cholecystokinin in pain modulation are that it acts as an endogenous opiate antagonist and that it appears to contribute to the development of morphine tolerance. Therefore, antagonizing CCK may prevent many of the complications associated with administration of narcotics. Proglumide may be acting as a CCK antagonist in the CNS, suppressing the antagonistic activity that CCK has on opiate systems. But as speculated above, proglumide may have neuromodulatory activity at an opiate receptor subtype or receptor complex.

Clinical applications for proglumide have been suggested in that it could be used for prevention and reversal of opiate tolerance [Suberg et al., 1985a; Tang et al., 1983; Kinscheck et al., 1983] and also, due to its ability to potentiate certain opiates, to allow physicians to administer lower dosages of opiates for longer periods of time to manage pain. We have also suggested that proglumide or compounds like proglumide be used in treating chronic pain [Suberg, 1985a], where tonic pain appears to induce release of endogenous opiates. Thus, the potentiation of the effects of opiates by proglumide has broad biomedical implications in terms of managing chronic and acute pain in man.

References

Abbott, F.V.; Franklin, K.B.J.; Ludwick, R.J.; Melzack, R.: Apparent lack of tolerance in the formalin test suggests different mechanisms for morphine analgesia in different types of pain. Pharmacol. Biochem. Behav. *15:* 637–640 (1981).

Abbott, F.V.; Melzack, R.; Samuel, C.: Morphine analgesia in the tail-flick and formalin pain tests is mediated by different neural systems. Expl. Neurol. *75:* 644–651 (1982).

Beinfeld, M.C.; Meyer, D.K.; Eskay, R.L.; Jensen, R.T.; Brownstein, M.J.: The distribution of cholecystokinin immunoreactivity in the central nervous system of the rat as determined by radioimmunoassay. Brain Res. *212:* 51–57 (1981).

Beinfeld, M.C.; Trubatch, J.R.; Brownstein, M.J.: Cholecystokinin peptides in the brain and pituitary of the bullfrog *Rana catesbrana.* Distribution of characterization. Brain Res. *268:* 192–196 (1983).

Cesselin, F.; LeBars, D.; Bourgoin, S.; Artaud, F.; Gozlan, H.; Clot, A.M.; Besson, J.M.; Hamon, M.: Spontaneous and evoked release of methionine-endephalin-like material from rat spinal cord in vivo. Brain Res. *339:* 305–313 (1985).

Charpentier, J.: Analysis and mechanisms of pain in animals: A new concept of pain; in Solvairac, Cahn, Charpentier, Pain, pp. 171–200 (Academic Press, London, 1968).

Chiodo, L.A.; Bunney, B.S.: Proglumide. Selective antagonism of excitatory effects of cholecystokinin in the central nervous system. Science *219:* 1449–1451 (1983).

Clark, C.R.; Daum, P.; Hughes, J.: Lack of competition between two reputed peripheral cholecystokinin receptor antagonists for central cholecystokinin binding sites. Ann. N.Y. Acad. Sci. *448:* 581–582 (1985).

Collier, H.O.J.: Multiple toe-pinch test for potential analgesic drugs; in Keele, Smith, The assessment of pain in man and animals, pp. 262–270 (Latimer, Trend, Plymouth, 1962).

D'Amour, F.E.; Smith, D.L.: A method for determining loss of pain sensation. J. Pharmac. exp. Ther. *72:* 74–79 (1941).

DeLander, G.E.; Portoghese, P.S.; Takemori, A.E.: Role of spinal mu opioid receptors in the development of morphine tolerance and dependence. J. Pharmac. exp. Ther. *231:* 91–96 (1984).

Della-Fera, M.A.; Baile, C.A.: Cholecystokinin octapeptide: continuous picomole injections into the cerebral ventricles of sheep suppress feeding. Science *206:* 471 (1979).

Dennis, S.G.; Melzack, R.: Pain modulation by 5-hydroxytryptaminerigc agents and morphine as measured by three pain tests. Expl. Neurol. *69:* 260–270 (1980).

Dockray, G.J.; Gregory, R.A.; Hutchinson, J.B.: Isolation, structure and biological activity of two cholecystokinin octapeptides from sheep brain. Nature Lond. *274:* 711–713 (1978).

Dubuisson, D.; Dennis, S.G.: The formalin test. A quantitative study of the analgesic effects of morphine, mepiridine and brain-stem stimulation in rats and cats. Pain *4:* 161–174 (1977).

Faris, P.L.; Komisaruk, B.R.; Watkins, L.R.; Mayer, D.J.: Evidence for the neuropeptide cholecystokinin as an antagonist of opiate analgesia. Science *219:* 310–312 (1983).

Fuxe, K.; Andersson, K.; Locatelli, V.; Agnati, L.F.; Hokfelt, T.; Skirboll, L.; Mutt, V.: Cholecystokinin peptides produce marked reduction of dopamine turnover in discrete areas of rat brain following intraventricular injection. Eur. J. Pharmacol. *67:* 329–331 (1986).

Gillis, R.A.; Quest, A.J.; Pagani, F.D.; Dias-Souza, J.; Traveira Da Silva, A.M.; Jensen, R.T.; Garvey, T.Z.; Hamosh, P.: Activation of CNS *CCK* receptors. J. Pharmac. exp. Ther. *224:* 408–414 (1983).

Gillis, R.A.; Quest, J.A.; Pagani, F.D.; Souza, J.D.; DaSilva, A.M.T.; Jensen, R.T.; Garvey, T.Q.; Hamosh, P.: Activation of central nervous system cholecystokinin receptors stimulates respiration in the cat. J. Pharmac. exp. Ther. *224:* 408–414 (1983).

Hahne, W.F.; Jensen, R.T.; Lemp, G.F.; Gardner, J.D.: Proglumide and benzotript: members of a different class of cholecystokinin receptor antagonists. Proc. natn. Acad. Sci. USA *78:* 6304–6308 (1981).

Itoh, S.; Hirota, R.; Katonura, G.; Odaguchi, K.: Adrenocortical stimulation by a cholecystokinin preparation in the rat. Life Sci. *25:* 1725–1730 (1979).

Itoh, S.; Katsuura, G.; Maeda, Y.: Caerulein and cholecystokinin suppress β-endorphin-induced analgesia in the rat. Eur. J. Pharmacol. *80:* 421–425 (1982).

Jeftinija, S.; Miletic, V.; Randic, M.: CCK-8 excites dorsal horn neurons in vivo and in vitro. Brain Res. *213:* 231–236 (1981).

Jones, E.G.: Varieties and distribution of non-pyramidal cells in the somatic sensory cortex of the squirrel monkey. J. comp. Neurol. *160:* 205 (1975).

Katsuura, G.; Hirota, R.; Itoh, S.: Cholecystokinin-induced hypothermia in the rat. Experientia *37:* 60 (1981).

Kinscheck, I.B.; Watkins, L.R.; Kaufman, E.; Mayer, D.J.: Evidence for a cholecystokinin (CCK)-like endogenous opiate antagonist. Soc. Neurosci. Abstr. *9:* 792 (1983).

Larsson, L.I.; Rehfeld, J.F.: Localization and molecular heterogeneity of cholecystokinin in the central and peripheral nervous system. Brain Res. *165:* 201–218 (1979).

Lord, J.A.H.; Waterfield, A.A.; Hughes, J.; Kosterlitz, H.W.: Endogenous opioid peptides. Multiple agonists and receptors. Nature Lond. *267:* 495–499 (1977).

Mann, J.F.E.; Boucher, R.; Schiller, P.W.: Rotational syndrome after central injection of C-terminal 7-peptide of cholecystokinin. Pharmacol. Biochem. Behav. *13:* 125–127 (1980).

Mantyh, P.W.; Hunt, S.P.: Evidence for cholecystokinin-like immunoreactive neurons in the rat medulla oblongata which project to the spinal cord. Brain Res. *291:* 49–54 (1984).

Martin, W.R.; Eades, C.G.; Thompson, J.A.; Huppler, R.E.; Gilbert, P.E.: The effects of morphine- and nalorphine-like drugs in the nondependent and morphine dependent spinal dog. J. Pharmac. exp. Ther. *197:* 517–532 (1976).

Micevych, P.E.; Yaksh, T.L.; Go, V.L.W.: Opiate-mediated inhibition of the release of cholecystokinin and substance P but not neurotensin from cut hypothalamic slices. Brain Res. *250:* 283–289 (1982).

Morley, J.E.; Levine, A.S.: Stress-induced eating is mediated through endogenous opiates. Science *209:* 1259 (1980).

Morley, J.E.; Levine, A.S.: Intraventricular CCK-8 produces hyperglycemia in rats. Life Sci. *28:* 2187 (1981).

Morley, J.E.; Levine, A.S.: Opioid modulation of appetite. Neurosci. Biobehav. Rev. *7:* 281 (1983).

Nemeroff, C.B.; Osbahr, A.J.; Bissette, G.; Jahnke, G.; Lipton, M.A.; Prange, A.J.: Cholecystokinin inhibits tail-pinch induced eating in rats. Science *200:* 793–794 (1978).

Osborne, N.N.; Cuello, A.C.; Dockray, G.J.: Substance P and cholecystokinin-like peptides in helix neurons and cholecystokinin and serotonin in a giant neuron. Science *216:* 409–411 (1982).

Phillis, J.W.; Kirkpatrick, J.R.: The actions of motilin, luteinizing hormone releasing hormone, cholecystokinin, somatostatin, VIP and other peptides on rat cerebral cortical neurons. Can. J. Physiol. Pharmacol. *58:* 612–623 (1980).

Rehfeld, J.F.: Immunochemical studies on cholecystokinin. II. Distribution and molecular heterogeneity in the central nervous system and small intestine of man and hog. J. biol. Chem. *253:* 4022–4030 (1978).

Rothman, R.B.; Danks, J.A.; Herkenham, M.; Jacobsen, A.E.; Burke, T.R., Jr.; Rice, K.C.: Evidence that the delta-selective alkylating agent, FIT, alters the mu-noncompetitive opiate delta binding site. Neuropeptides *6:* 227–237 (1985a).

Rothman, R.B.; Danks, J.A.; Jacobson, A.E., Burke, T.R., Jr.; Rice, K.C.: Leucine enkephalin-noncompetitively inhibits the binding of [^{3}H]naloxone to the opiate mu-recognition site. Evidence for delta-mu binding site interactions in vivo. Neuropeptides *6:* 351–363 (1985b).

Rovati, A.L.: Inhibition of gastric secretions by antigastrinic and H_2 blocking agents. Scand. J. Gastroent. *11:* 113 (1976).

Rovati, C.L.; Sacerbote, P.; Panerai, A.E.: Effects of proglumide on morphine analgesia and tolerance. Ann. N.Y. Acad. Sci. *448:* 630–632 (1985).

Ryder, S.W.; Eng, J.; Shaus, E.; Yalow, R.S.: Extraction and immunochemical characterization of cholecystokinin-like peptides from pig and rat brain. Proc. natn. Acad. Sci. USA *78:* 3891–3896 (1981).

Sakamoto, N.; Takatsuji, K.; Shiosaka, S.; Tateishi, K.; Hashimura, E.; Miura, S.; Hamaoka, T.; Tohyama, M.: Cholecystokinin-8-like immunoreactivity in the pre- and post-central gyri of the human cerebral cortex. Brain Res. *307:* 77–83 (1984).

Salt, T.E.; Hill, R.G.: The effects of C-terminal fragments of cholecystokinin on the firing of single neurones in the caudal trigeminal nucleus of the rat. Neuropeptides *2:* 301–306 (1982).

Schmauss, C.; Yaksh, T.L.: In vivo studies on spinal opiate receptor systems mediating antinociception. II. Pharmacological profiles suggesting a differential association of mu, delta and kappa receptors with visceral chemical and cutaneous thermal stimuli in the rat. J. Pharmac. exp. Ther. *228:* 1–12 (1984).

Schmauss, C.; Yaksh, T.L.; Shimohigashi, Y.; Jensen, T.S.; Harty, G.; Robard, D.: Studies on the pharmacology of the spinal opiate receptors which modify the rats response on cutaneous thermal and visceral chemical evoked responses. Test selectivity and a dis-association of μ, δ and κ-receptors. Soc. Neurosci. Abstr. *9:* 740 (1983).

Skirboll, L.; Hokfelt, T.; Dockray, G.; Rehfeld, J.; Brownstein, M.; Cuello, A.C.: Evidence for periaqueductal cholecystokinin-substance P neurons projecting to the spinal cord. J. Neurosci. *3:* 1151–1157 (1983).

Stein, L.J.; Woods, S.C.: Cholecystokinin and bombesin act independently to decrease food intake in the rat. Peptides *2:* 431 (1981).

Stengaard-Pederson, K.; Larsson, L.I.: Localization and opiate receptor binding of enkephalin, CCK and ACTH/B-endorphin in the rat central nervous system. Peptides *2:* suppl. 1, pp. 3–19 (1981).

Suberg, S.N.; Culhane, E.S.; Carstens, E.; Watkins, L.R.: Behavioral and electrophysiological investigations of opiate/cholecystokinin interactions. Adv. Pain Res. Ther. *9:* 541–553 (1985a).

Suberg, S.N.; Culhane, E.S.; Carstens, E.; Watkins, L.R.: The potentiation of morphine-induced inhibition of spinal transmission by proglumide, a putative cholecystokinin antagonist. Ann. N.Y. Acad. Sci. *448:* 660–662 (1985b).

Suberg, S.N.; Culhane, E.S.; Rosenquist, G.; Carstens, E.; Watkins, L.R.: Effect of anti-cholecystokinin antibody (AB) on morphine (MOR) induced suppression of spinal nociceptive transmission. Soc. Neurosci. Abstr. *11:* 284 (1985c).

Tang, J.; Chou, J.; Iadarola, M.J.; Yang, H.-Y.T.; Costa, E.: Possible participation of endogenous cholecystokinin 8 (CCK) in the tolerance to morphine analgesia in the rat. Soc. Neurosci. Abstr. *9:* 288 (1983).

Traynor, J.R.; Kelly, P.D.; Rance, M.J.: Multiple opiate binding sites in rat spinal cord. Life Sci. *31:* 1377–1380 (1982).

Vanderhaeghen, J.J.; Signeau, J.C.; Gepts, W.: New peptide in the vertebrate CNS reacting with anti-gastrin antibodies. Nature *257:* 604–605 (1975).

Watkins, L.R.; Cobelli, D.A.; Faris, P.; Aceto, M.D.; Mayer, D.J: Opiate vs. nonopiate foot-shock-induced analgesia (FSIA). The body region shocked is a critical factor. Brain Res. *242:* 299–308 (1982).

Watkins, L.R.; Kinscheck, I.B.; Rosenquist, G.; Miller, J.; Wimberg, S.; Frenk, H.; Kaufman, E.; Coghill, R.; Suberg, S.N.; Mayer, D.J.: The enhancement of opiate analgesia and the possible reversal of morphine tolerance by proglumide. Ann. N.Y. Acad. Sci. *448:* 676–677 (1984).

Watkins, L.R.; Kinscheck, I.B.; Mayer, D.J.: Potentiation of opiate analgesia and apparent reversal of morphine tolerance by proglumide. Science *244:* 395–396 (1984).

Watkins, L.R.; Kinscheck, I.B.; Mayer, D.J.: Potentiation of morphine analgesia by the CCK antagonist proglumide. Brain Res. *327:* 169–180 (1985).

Watkins, L.R.; Mayer, D.J.: The neural organization of endogenous opiate and non-opiate pain control systems. Science *216:* 1185 (1982).

Willitts, J.; Urban, L.; Murase, K.; Randic, M.: Actions of cholecystokinin octapeptide on rat spinal dorsal horn neurons. Ann. N.Y. Acad. Sci. *448:* 385–402 (1985).

Woolfe, G.; MacDonald, A.D.: The evaluation of the analgesic action of pethidine hydrochloride (Demerol). J. Pharmac. exp. Ther. *80:* 300–307 (1944).

Yaksh, T.L.: Multiple spinal opiate receptor systems in analgesia. Adv. Pain Res. Ther. *6:* 197–215 (1984).

Yaksh, T.L.; Elde, R.P.: Factors governing release of methionine enkephaline-like immunoreactivity from mesencephalon and spinal cord of the cat in vivo. J. Neurophysiol. *46:* 1056–1075 (1981).

Zarbin, M.A.; Innis, R.B.; Wamsley, J.K.; Snyder, S.H.; Kuhar, M.J.: Autoradiographic localization of cholecystokinin receptors in rodent brain. J. Neurosci. *3:* 877–906 (1983).

Zetler, G.: Analgesia and ptosis caused by caerulein and cholecystokinin octapeptide (CCK-8). Neuropharmacology *19:* 415–422 (1980).

Zetler, G.: Central depressant effects of caerulein and cholecystokinin-octapeptide (CC-8) differ from those of diazepam and haloperidol. Neuropharmacology *20:* 277–283 (1981).

S.N. Suberg, PhD, Preclinical CNS Biology, Bristol-Myers Company,
5 Research Parkway, Wallingford, CT 06492 (USA)

Pain Headache, vol. 9, pp. 266–282 (Karger, Basel 1987)

Neurochemical Mechanisms of Acupuncture Analgesia

K. Tsou

Shanghai Institute of Materia Medica, Department of Pharmacology, Chinese Academy of Sciences, Shanghai, China

Introduction

Many years ago, we demonstrated in rabbits, by means of intracerebral microinjection of morphine, that the most sensitive sites for analgesia were periventricular and periaqueductal gray [Tsou et al., 1962]. Microinjection of the opiate antagonist nalorphine in these two areas blocked the analgesic effect of intravenous morphine [Tsou, et al., 1963]. These results were summarized and published in 1964 [Tsou et al., 1964]. Similar results in primates were obtained later in a more systematic and detailed study by Pert and Yaksh [1974]. With tooth pulp electric stimulation techniques in cats, we showed in 1960 that intravenous morphine did not block the evoked potentials in periaqueductal gray. Evoked potentials in the posterior nuclear group (PO) region were blocked instead. This site, however, was not sensitive to morphine by local microinjection [Tsou, 1960]. A hypothesis was thus put forward that morphine might activate instead of inhibit certain mechanisms in the central gray to block pain transmission in other parts of the central nervous system indirectly. Morphine analgesia and brain stimulation-produced analgesia may have a common mode of action through the periventricular and periaqueductal gray, but the analgesia would be finally achieved at the spinal level by a descending inhibitory mechanism from the brain stem.

A direct spinal mechanism of analgesia should not be neglected. For example, we showed in 1964 in spinal cats that the superficial, high threshold, dorsal horn neurons were highly sensitive to intravenous morphine. Neurons in deeper layers of the dorsal horn were also sensitive [Tsou and Chang, 1964]. In the intact animals, a descending inhibition appeared to be more important, as demonstrated in our recent study [Hu et al., 1984]. Reserpine was first found by Schneider [1954] to antagonize morphine analgesia. We studied the dose-response relations of systemic, intrathecal and intraventricular morphine, compared in normal and reserpin-treated rats with the tail-flick method. Reserpine caused a marked parallel shift to the right of the dose-response curve of systemic morphine, but the antagonism was surmountable. Reserpine had no effect on intrathecal morphine analgesia while the analgesic action of intraventricular morphine was practically eliminated by reserpine (fig. 1). These findings suggest that when a small dose of morphine is used, morphine acts primarily on the supraspinal sites; the direct spinal action becomes apparent only when a high systemic dose is used. These results also pointed out in addition that supraspinal morphine analgesia is mediated by monoamines. A similar conclusion was reached by many investigators with different approaches.

With regard to the descending pathway of supraspinal morphine analgesia, the relative importance of 5-HT versus norepinephrine is still unresolved. However, our results indicate that the descending norepinephrine pathway is more important. This conclusion comes from our study on the interaction of α-methyl-DOPA and reserpine on morphine analgesia [Tsou and Tu, 1963]. α-Methyl-DOPA is an amino acid decarboxylase inhibitor, but its decarboxylase inhibitory effect soon wears off, and it is decarboxylated into α-methyl-dopamine and further becomes α-methyl-norepinephrine, a false transmitter of the noradrenergic neurons. We found that injection of α-methyl-DOPA, prior to or simultaneously with reserpine, blocked the antagonism of morphine analgesia by reserpine. Rat brain stem specimens, processed according to Hillarp and Falck but without monoamine oxidase inhibitor pretreatment, showed that the brain stem noradrenergic neurons had a yellowish-green fluorescence. After reserpine, these neurons lost their fluorescence. α-Methyl-DOPA treatment caused the appearance of a blue-green fluorescence in these neurons, reflecting the presence of the false transmitter, α-methyl-norepinephrine. This blue-green fluorescence was resistant to reserpine, in contrast to the yellowish-green fluorescence of norepinephrine. Therefore, histochemical data support that α-methyl-norepinephrine may mediate morphine analgesia resistant to reserpine. Furthermore, bilateral dorsal lateral funiculus lesions at the third tho-

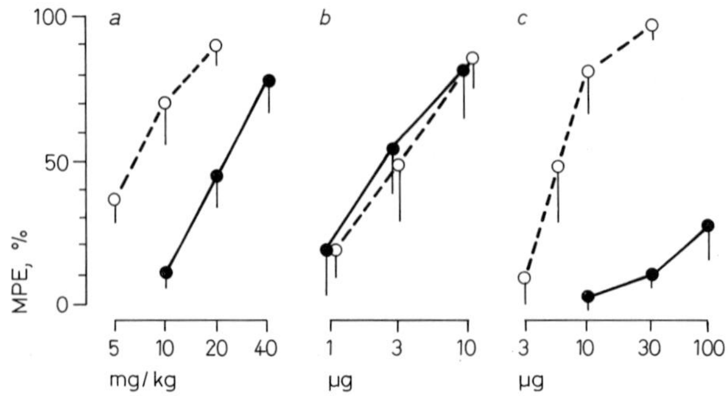

Fig. 1. Effects of reserpine on the dose-response curves of subcutaneous, intrathecal and intraventricular morphine analgesia. *a* Subcutaneous injection. *b* Intrathecal injection. *c* Intraventricular injection. Ordinate: MPE (maximum possible effect); abscissa: dose in log scale. ○ = Control rats; ● = reserpinized rats. Each point represents the mean of MPE ± SEM for 5–9 rats. From Hu et al. [1984].

racic level eliminated the α-methyl-DOPA pretreated reserpine-resistant morphine analgesia, indicating that the norepinephrine descending pathways are located within the dorsal lateral funiculus [Tsou et al., 1980].

The role of norepinephrine in supraspinal morphine analgesia is further strengthened by our recent study with the selective noradrenergic neurotoxin DSP-4 (N-[2-chloroethyl]-N-2-bromobenzylamine) [Zhong et al., 1985]. DSP-4 was injected intrathecally into the lumbar subarachnoid space of the rat. After ten days the tail-flick block dose-response curves of intraperitoneal and intraventricular morphine were determined, and the norepinephrine and 5-HT contents in brain stem and spinal cord were measured. The analgesic effect of 10 or 30 µg of intraventricular morphine and 10 or 30 mg/kg or intraperitoneal morphine were significantly attenuated by DSP-4 treatment. Intrathecal DSP-4 depleted norepinephrine in the spinal cord but not in the brain stem. Both brain stem and spinal cord 5-HT were not affected (fig. 2, 3). The result of this study differs from our previous experiments with reserpine [Hu et al., 1984], in that the intrathecal DSP-4 antagonism of supraspinal morphine analgesia was surmountable. Since all monoamines are depleted by reserpine, while only norepinephrine is depleted by DSP-4, this result may indicate that when a larger dose of morphine is injected intraventricularly, monoaminergic pathways other than norepinephrine are also involved.

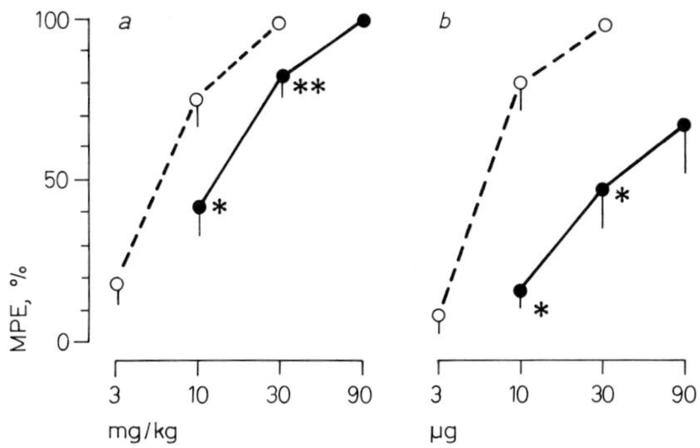

Fig. 2. Effects of intrathecal DSP4 on the dose-response curves of intraperitoneal and intraventricular morphine analgesia. *a* Intraperitoneal injection. *b* Intraventricular injection. ○ = Control rats; ● = DSP4-pretreated rats. Each point represents the mean of MPE ± SEM for 8 rats. * $p<0.02$; ** $p<0.01$. From Zhong et al., [1985].

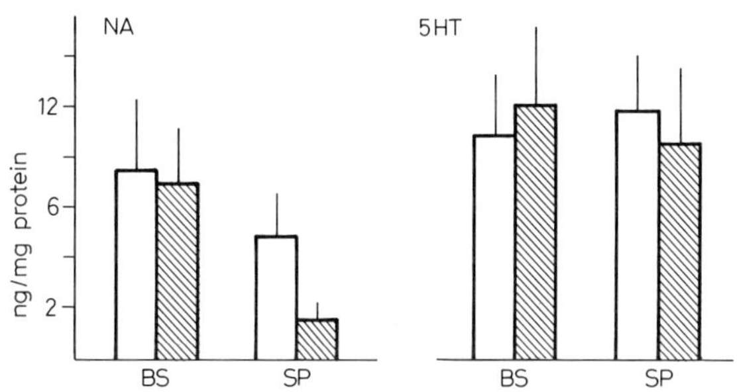

Fig. 3. Effects of intrathecal DSP4 on the brain stem and spinal cord norepinephrine (NA) and 5-hydroxytryptamine (5-HT) contents. Ordinate: Contents expressed in ng/mg protein; abscissa: BS = brain stem; SP = spinal cord; hollow column = control rats 10 days after intrathecal saline; hatched column = rats 10 days after intrathecal DSP4 300 µg; vertical bar on top = SEM of 10 (NA) or 8 (5-HT) samples each containing two brain stem or two spinal cords. From Zhong et al., [1985].

The Role of Serotonin in Acupuncture Analgesia

Our laboratory had been working on morphine analgesia for many years when we embarked on the acupuncture analgesia project in 1973. The first study [Yi et al., 1977a] focused on the role of 5-hydroxytryptamine (5-HT) in acupuncture analgesia in rabbits. Pain sensitivity was measured by potassium iontophoresis on the ear tips through two cotton wick electrodes, one on each ear. The current passing through the ears was increased steadily and the electric current that caused a sudden struggle of the rabbit was recorded as the pain threshold. Animals were implanted with plastic cannulas, one in the lateral ventricle and another in the aqueduct, suspended in cloth slings, and the ventricular system perfused with [^{3}H]-5-HT-containing artificial CSF, during which time [^{3}H]-5-HT was presumably taken up by the serotonergic nerve endings. The perfusion was then continued with [^{3}H]-5-HT free CSF. Under control conditions, the radioactivity in the perfusates collected at 10 min intervals decreased in a time-dependent fashion. If the rabbits were manually acupunctured by twirling needles in the Tsu-san-li points of both hind legs for 20 min, the radioactivity in the perfusates increased concomitantly with the elevation of pain threshold. Intravenous injection of morphine (5 mg/kg), by contrast, produced a prominent and long-lasting elevation of pain threshold without any increase of radioactivity in the perfusates. Therefore, there was a distinct difference between acupuncture and morphine analgesia in regard to 5-HT.

A second experiment determined the effect of chemical lesioning of serotonergic neurons on acupuncture analgesia in rats. The serotonergic neurotoxin 5,6-dihydroxytryptamine was first injected into the lateral ventricle one week prior to assessment of the analgesic effect of electroacupuncture. Pain sensitivity was assessed using the tail-flick test. Electroacupuncture analgesia was induced by application of current through needled electrodes inserted in the Huantiao point of each hind limb. The two electrodes were connected to a G 6805 therapeutic stimulator for 30 min which produced dense and sparse impulses. This procedure caused a conspicuous increase in the tail-flick latency in control rats. Acupuncture analgesia, however, was significantly reduced in 5,6-DHT-treated rats. Histochemical examination of the raphe nucleus neurons in these rats revealed degeneration of the cell bodies and swelling of axons and loss of fluorescence in the nerve terminals [Yi et al., 1977b].

Role of Opioid Peptides in Acupuncture Analgesia

After the discovery of enkephalins by Hughes et al. [1975a, b] we started working on the role of these endogenous opioids in acupuncture analgesia. We were encouraged by the findings of Mayer et al. [1977] that

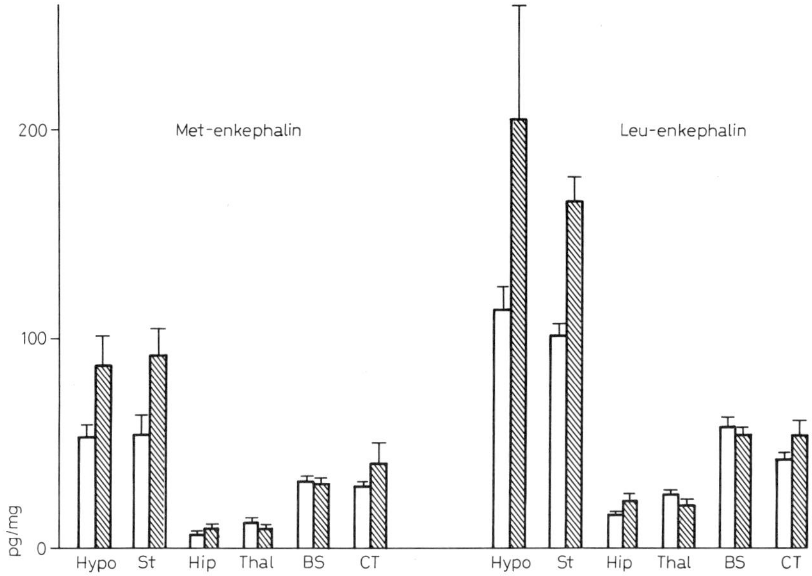

Fig. 4. Effect of acupuncture on rabbit brain regional contents of enkephalins. White column = control group; striated column = acupuncture group; Hypo = hypothalamus; St = striatum; Hip = hippocampus; Thal = thalamus; BS = brain stem; CT = cortex. From Tsou et al. [1980].

naloxone could partially antagonize the acupuncture analgesia and those of Sjolund et al. [1977] indicating that electric acupuncture increased the level of endorphins in the lumbar CSF of patients.

We developed a radioimmunoassay for leu- and met-enkephalin and began to study their roles in acupuncture analgesia. The immunogens were prepared by condensation of polylysyl succinic acid with the N terminals of the two enkephalins. Our leu-enkephalin antiserum could detect 20 pg leu-enkephalin and the met-enkephalin antiserum could detect 150 pg met-enkephalin. The cross-reaction of met-enkephalin with the leu-enkephalin antiserum was 3.3%, while that of leu-enkephalin with met-enkephalin antiserum was less than 1% [Lo et al., 1980].

In experiments using rabbits, animals were subjected, as before, to alternate twirling of the bilateral Tsu-san-li points for 20 min. This procedure caused pain threshold to increase more than 80%, as measured by potassium iontophoresis. After stimulation, the rabbit was decapitated and the brain was put immediately into boiling saline for 5 min. The

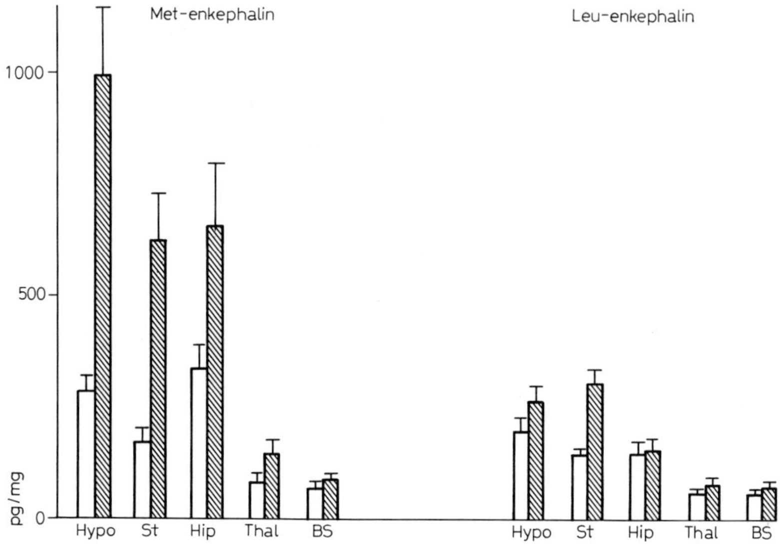

Fig. 5. Effect of acupuncture on rat brain regional contents of enkephalins. White column = control; striated column = acupuncture group; abbreviations, see fig. 4. From Tsou et al. [1980].

brain was then dissected to study the following regions: hypothalamus, striatum, hippocampus, thalamus, brain stem and cortex. Subsequently, met- and leu-enkephalin contents measured by radioimmunoassay.

In the nonaccupuncture control group, enkephalin contents were high in the hypothalamus and striatum, moderate in the brain stem and cortex, and low in the hippocampus and thalamus. After acupuncture, enkephalin content increased markedly in the hypothalamus and the striatum to about 1.6–1.8 times control levels. No significant changes were found in the other four brain regions (fig. 4) [Tsou et al., 1980].

In the rat experiments, pain thresholds, measured 5 s after 30 min electric acupuncture, were increased 92%. Once again, the most prominent changes in the enkephalin levels were found in the hypothalamus and the striatum. Compared with the control group, met-enkephalin increased 3.5 times, while leu-enkephalin increased 1.4–2 times (fig. 5).

It is well known that acupuncture produces only a transient analgesia in animals and that enkephalins degrade rapidly after release. If enkephalins are indeed involved in acupuncture, inhibition of their degradation should prolong the analgesic response. To test this hypoth-

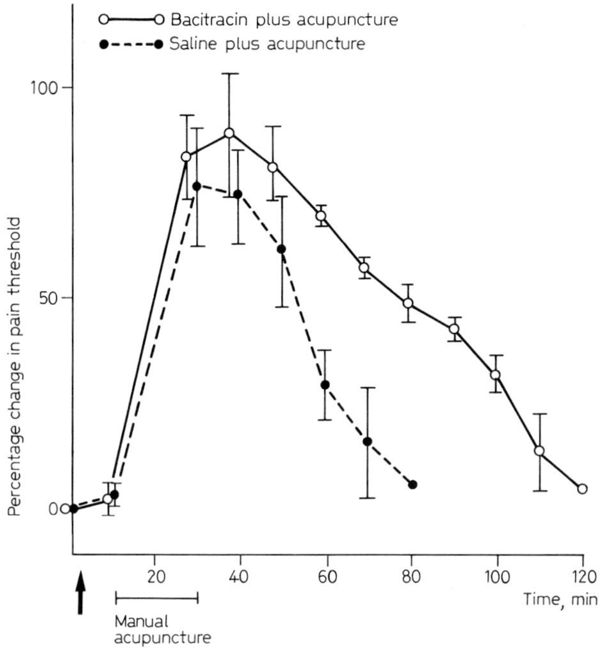

Fig. 6. Enhancement of acupuncture analgesia by intraventricular bacitracin. Arrow designates intraventricular injection of bacitracin. From Tsou et al. [1980].

esis, we injected the peptidase inhibitor bacitracin (50 μg/50 μl) into the lateral ventricle of the rabbit prior to acupuncture. In these animals acupuncture analgesia was conspicuously prolonged (fig. 6). Whereas the acupuncture-induced elevated pain threshold of rabbits injected with saline returned to preacupuncture level at 30–40 min after the cessation of stimulation, the pain threshold in the bacitracin-treated rabbits was still markedly elevated [Tsou et al., 1979]. Moreover, intravenous naloxone was capable of reversing the acupuncture analgesia during the bacitracin prolonged analgesia period, implicating that the prolongation is due to endogenous morphine-like substances. Finally, when rabbits were sacrificed at 30–40 min after the cessation of acupuncture, the enkephalin contents in the hypothalamus and the striatum were found to be highest in the bacitracin plus acupuncture group and lowest in the bacitracin alone controls. The saline plus acupuncture group showed a residual elevation (fig. 7). These results indicated that the enkephalinergic neurons have a low level of activity during the resting state.

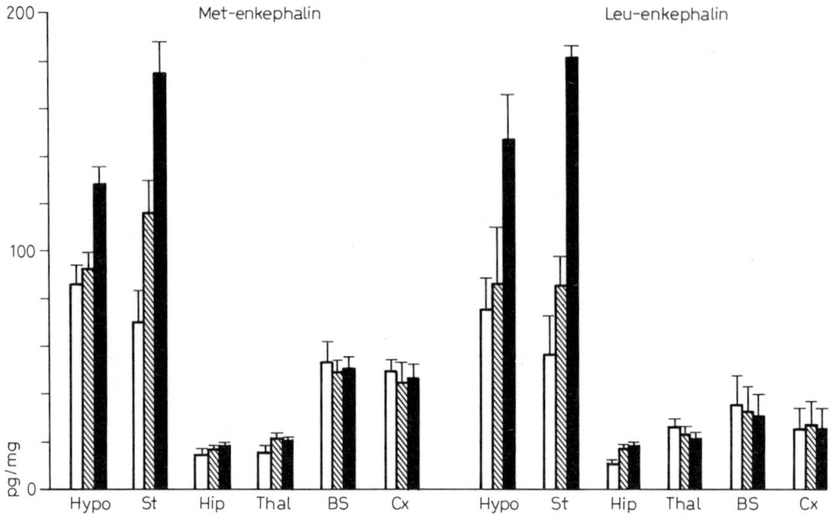

Fig. 7. Effect of intraventricular bacitracin on acupuncture-induced alterations of brain enkephalin contents. White column = intraventricular bacitracin; striated column = intraventricular saline plus acupuncture; dark column = intraventricular bacitracin plus acupuncture. Rabbits were decapitated and dissected at 30 min after the cessation of needling. AC = Nucleus accumbens; for other abbreviations, see fig. 1. From Tsou et al. [1980].

To learn if acupuncture releases brain enkephalins, we measured the met-enkephalin contents in the CSF [Tsou et al., 1979]. Experiments were performed in bacitracin and bacitracin plus acupuncture groups. As before, pain threshold was still highly elevated in the latter group 30 min after the cessation of acupuncture, the time at which cisternal CSF was collected. The CSF content of met-enkephalin was significantly higher in the bacitracin plus acupuncture group, suggesting that enkephalinergic neurons were activated by acupuncture releasing enkephalin into the CSF.

Since it was reported that the protein synthesis inhibitor cycloheximide interfered with the incorporation of tritium-labelled tyrosine into enkephalins in vitro [Hughes et al., 1978], we studied the effect of this drug injected in vivo on acupuncture analgesia. Thirty-eight rats were divided into the following groups: (1) control; (2) electric acupuncture; (3) cycloheximide; and (4) cycloheximide plus acupuncture. The dosage of cycloheximide was 500 pg per rat (i.c.v.). Cycloheximide alone had no discernible effect on basal pain threshold, but it greatly attenuated electric acupuncture analgesia, and decreased met-enkephalin content by 47% in the hypothalamus and 20% in the striatum. Electric acupuncture caused increases in met-enkephalin of 113% in the hypothalamus and 200% in the striatum. In the cycloheximide plus acupuncture group, however, met-enkephalin was only increased 28% and 120% respectively. That the amplitude of met-enkephalin increase caused by acupuncture was greatly reduced by cycloheximide pretreatment suggests an acceleration of enkephalin biosynthesis by acupuncture [Wu et al., 1980]. Overall, the effects of cycloheximide on enkephalin in the striatum was less obvious than in the hypothalamus. This might be because very little cycloheximide reaches the contralateral striatum after unilateral intraventricular injection.

To study further the dynamic changes in the biosynthesis and release of enkephalins by acupuncture, we studied in vitro potassium induced leu-enkephalin release from the striatal slices of control and electric acupunctured rats [Zhu et al., 1985]. Rats were stimulated as described above for 20 min and were decapitated either immediately, 15, 30, or 60 min after the termination of electric acupuncture. Striatal slices, of 150–200 μm thick and weighing 30 mg, were perfused with Krebs solution in a perfusing chamber. Perfusates were collected at 2 min intervals, and used for leu-enkephalin radioimmunoassay. When the striatal slices from control rats were perfused with a Krebs solution containing 30 m*M* potassium, leu-enkephalin release first was observed after a 2-min delay, attained its peak rate in 3 min, decreased gradually thereafter, approaching the basal rate in 7 min. In rats killed immediately after the termination of 30 min of electric acupuncture, the striatal slices were refractory to high potassium stimulation. There was a partial recovery of potassium-induced leu-enkephalin release in the group of rats killed 15 min after the cessation of electric acupuncture, and a full recovery was seen in the 60 min group (fig. 8). In view of the fact that the enkephalin content was increased in the rats killed immediately after the cessation of acupuncture, the demonstration of decreased potassium-induced leu-enkephalin was rather surprising. However, it might be explained if we consider the possibility of enkephalin being stored in two pools, one releasable and the other nonreleasable. Perhaps a long period of electric acupuncture (30 min) depletes the releasable pool, while accelerating the biosynthesis and the processing of the enkephalin precursors. The newly formed enkephalin may take 30–60 min to become releasable [Zhu et al.; 1985].

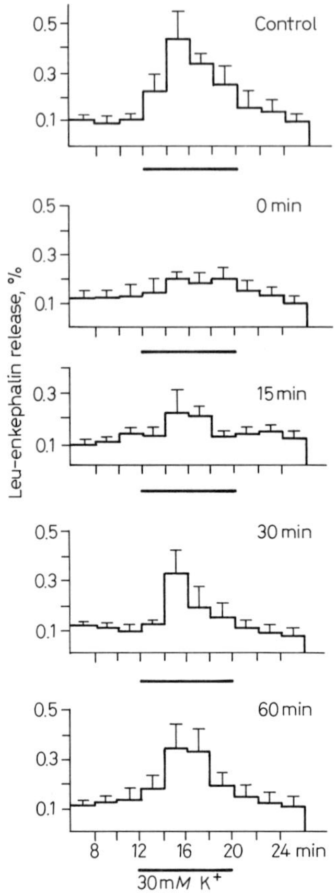

Fig. 8. Decreased release of potassium-induced release of leu-enkephalin from striatal slices of acupunctured rats. Ordinate: Leu-enkephalin release expressed as the percentage of total leu-enkephalin in the slices; abscissa: time scale. Potassium failed to increase the release of leu-enkephalin from the striatal slices of rats killed immediately (0 min) or 15 min after the termination of electroacupuncture. From Zhu et al. [1985].

To substantiate our working hypothesis that acupuncture accelerated enkephalin biosynthesis, the level of high molecular weight leu-enkephalin containing precursors was directly measured after electric acupuncture [Ding et al., 1986]. Leu-enkephalin and its precursors in the pooled striatal extracts of 10 rats were fractionated on a Sephadex G-75 column, according to the methods of Lewis et al. [1978]. The eluates were collected at 2-min intervals, and two peaks, representing large and small molecular weight peptides, were monitored by absorbance at 254 nm.

The first peak was in the void volume while the second peak was in the salt volume. Leu-enkephalin could be detected directly in the eluates corresponding to the second peak. The leu-enkephalin immunoreactivity in the first peak and the intermediate region were detectable only after trypsin and carboxypeptidase B digestion. The eluates collected in the 8th to 24th tubes were pooled and represented the processed intermediates of the precursors. The levels of proenkephalin, intermediates and leu-enkephalin were observed in the following groups, each containing 10 rats: nonacupuncture controls, and groups killed either immediately, 0.25, 0.5, 1, 2, 4, 8, 16, 24, 48 and 96 h after the termination of 30 min of electric acupuncture (fig. 8). Leu-enkephalin levels were very prominently elevated in the 0-hour group and decreased gradually in the 0.25-, 0.5- and 1-hour groups. In the 2- and 4-hour groups, leu-enkephalin level was below control values. From 8 h onward, leu-enkephalin levels again increased steadily but were never as high as in the 0-hour group. The proenkephalin and intermediates, by contrast, showed no change in the 0-, 0.25-, 0.5-hour groups, but started to increase after 1 h. Their levels got progressively higher until 96 h. This trend of increasing proenkephalin and intermediates preceded the second phase of leu-enkepahlin increase (fig. 9).

In the 0-hour group, free leu-enkephalin level was very high, but the proenkephalin and processed intermediates did not change. Since a large amount of leu-enkephalin was released, only an accelerated biosynthesis and processing of the precursors could keep the levels of proenkephalin and the intermediates essentially unchanged. Therefore, the acceleration of biosynthesis and processing of the proenkephalin had already started during 30 min electric acupuncture. This acceleration of biosynthesis and processing outlasted the acupuncture for as long as 96 h.

A Possible Role of β-Endorphin in Acupuncture

Immunohistochemical data of Watson et al. [1978] and Bloom et al. [1978] have shown that β-endorphin-containing neurons are distinct from enkephalin-containing neurons. The β-endorphin cells are located mainly in the basal hypothalamus. In order to study the involvement of β-endorphin in acupuncture, the Halasz knife was used to isolate the basal hypothalamus from the rest of the brain. The effect of such an isolation on acupuncture analgesia and met-enkephalin content in the basal

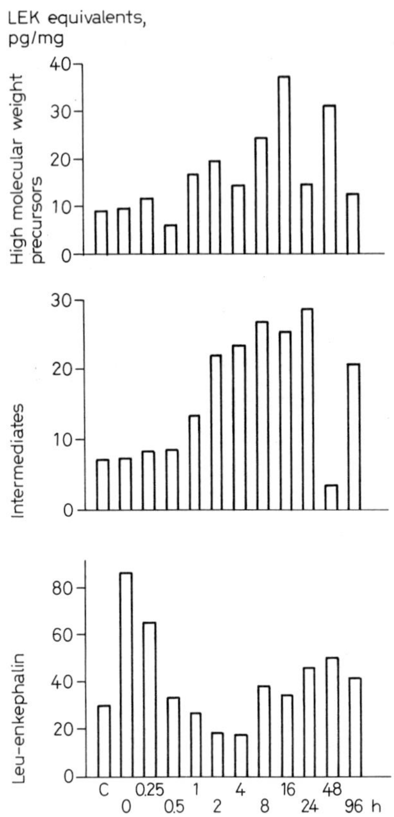

Fig. 9. Effect of electroacupuncture on the levels of rat striatal leu-enkephalin and its high molecular weight precursors and intermediates. Ordinate: Leu-enkephalin equivalents (pg/mg); abscissa: hours after the termination of electroacupuncture. From Ding et al. [1986].

hypothalamus could then be determined. Thirty rats were divided into the following groups: (1) control; (2) hypothalamus isolation; (3) hypothalamus isolation plus electric acupuncture; (4) sham operation; and (5) sham operation plus electric acupuncture. Compared to controls, rats in group 5 showed a prominent increase of pain threshold. This analgesia was reduced in rats with isolated hypothalami (group 3), suggesting the importance of a hypothalamic mechanism in the production of acupuncture analgesia.

The met-enkephalin content in the hypothalamus was about the same in the isolated hypothalamus and those of sham operated rats.

Table I. Effects of morphine and acupuncture on met-enkephalin levels in rabbit brain

	Striatum	Hypothalamus, pg/mg protein	Thalamus	CSF, pg/ml
Control	289 ± 16	218 ± 8	54 ± 2	1,127 ± 38
Morphine	405 ± 15***	268 ± 17*	57 ± 2	1,390 ± 82***
Acupuncture	383 ± 4*	321 ± 16***	57 ± 4	1,479 ± 67***
Morphine and acupuncture	420 ± 37*	380 ± 23***	56 ± 2	1,560 ± 327**

* $p<0.05$; ** $p<0.01$; *** $p<0.005$.

Electric acupuncture, however, failed to elicit the usual increase in the met-enkephalin content in the isolated hypothalamus, indicating that the incoming impulses were essential for the acupuncture-induced enkephalin elevation.

Is the Acupuncture Effect Specific?

To test the specificity of the acupuncture effect, a comparison was made with amphetamine-induced excitation, hot or cold environment, and bone fracture [Wang et al., 1985]. Each group consisted of 8 rats. In the amphetamine group, rats were injected with amphetamine sulfate 10 mg/kg intraperitoneally and grouped together. In the hot temperature group, rats were kept in an incubator at 40 °C. In the cold temperature group, rats were kept at 4 °C. In the fifth group, the right tibia of the rats was fractured. In each case, the treatment lasted for 30 min prior to decapitation. As before, electric acupuncture increased the levels of met-enkephalin and leu-enkephalin in the hypothalamus and the striatum. Amphetamine decreased the levels of the two enkephalins in both the hypothalamus and striatum, exposure to either hot or cold temperature decreased leu-enkephalin in the hypothalamus, and bone fracture increased the two enkephalins in the hypothalamus and thalamus. The heterogeneous nature of these results indicates that the effects of acupuncture are not attributable to the effects of stress per se on brain enkephalin levels.

It should also be noted that the acupuncture methods used in our experiments may be themselves noxious. If the effects of acupuncture on enkephalin content are due to exposure to pain, these should be blocked by morphine pretreatment. To our surprise, we found that in rabbits morphine produced increases of met-enkephalin in the hypothalamus, striatum and the CSF similar to those caused by electric acupuncture (table I) [Wu et al., 1984]. Moreover, when morphine was combined with acupuncture, additive effects, not antagonism, were observed. The mechanism by which morphine produces an increase of enkephalin level is unknown. Similar results were observed earlier by Dewey and his group [1979].

Conclusions

The present results suggest that acupuncture may well be another stimulus for activation of the central gray mechanisms of analgesia, which in turn modulate other pain-sensitive brain sites. Enkephalins released locally in the periventricular gray and periaqueductal gray may be responsible for acupuncture analgesia. Additionally, enkephalins released from striatum into the CSF may also act in the periventricular and periaqueductal gray. Finally, direct and indirect analgetic spinal mechanisms are likely to be activated.

Acupuncture analgesia is a very complicated phenomenon with many transmitters and modulators likely to be involved. Enkephalin participation is a single element in this complex picture. However, acupuncture activation of the enkephalinergic neuronal system can be widely utilized as a means for studying the dynamic changes in the biosynthesis, processing, release and degradation of brain and spinal enkephalinergic systems.

References

Bloom, F.E.; Battenberg, E.; Rossier, J.; Ling, N.; Guillemin, R.: Neurons containing beta-endorphin in rat brain exist separately from those containing enkephalins. Immunocytochemical studies. Proc. Natn. Acad. Sci. USA *75:* 1591–1595 (1978).

Ding, X.-H.; Ji, X.-Q.; Tsou, K.: Dynamic changes in the levels of enkephalin precursors, processed intermediates and leu-enkephalin in rat caudate nucleus following electroacupuncture. Chin. J. Physiol. Sci. *2:* 42–46 (1986).

Fu, T.-C.; Dewey, W.L.: Morphine antinociception. Evidence for the release of endogenous substance(s). Life Sci. *25:* 53–60 (1979).

Hu, G.-Y.; Zhong, F.-X.; Tsou, K.: Dissociation of supraspinal and spinal morphine analgesia by reserpine. Eur. J. Pharmacol. *97:* 129–131 (1984).

Hughes, J.: Isolation of an endogenous compound from the brain with properties similar to morphine. Brain Res. *88:* 295–308 (1975a).

Hughes, J.; Kosterlitz, H.W.; McKnight, A.T.: The incorporation of [^{3}H]tyrosine into enkephalin of striatal slices of guinea pig brain. Br. J. Pharmacol. *65:* 396 (1978).

Hughes, J.; Smith, T.W.; Kosterlitz, H.W.; Fothergill, L.A.; Morgan, B.A.; Morris, H.R.: Identification of two related pentapeptides from brain with potent opiate agonist activity. Nature Lond. *258:* 577–579 (1975b).

Lewis, R.V.; Stein, S.; Gerber, L.D.; Rubinstein, M.; Udenfriend, S.: High molecular weight opioid-containing proteins in striatum. Proc. Natn. Acad. Sci. USA *75:* 4021–4023 (1978).

Lo, E.-S.; Wu, J.-B.; Yi, C.-C.; Wang, F.-S.; Tsou, K.: Radioimmunoassay for enkephalins. Acta biochim. biophys. Sin. *12:* 115–125 (1980).

Mayer, D.J.; Price, D.D.; Rafii, A.: Antagonism of acupuncture analgesia in man by the narcotic antagonist naloxone. Brain Res. *121:* 368–372 (1975).

Pert, A.; Yaksh, T.L.: Site of morphine induced analgesia in the primate brain. Relation to pain pathways. Brain Res. *80:* 135–140 (1974).

Schneider, J.A.: Reserpine antagonism of morphine analgesia in mice. Proc. Soc. exp. Med. *37:* 614–616 (1954).

Sjolund, B.; Erikson, M.: Increased CSF levels of endorphins after acupuncture. Acta physiol. scand. *100:* 382–384 (1977).

Tsou, K.: A study on the site of analgesic action and the effects on higher nervous activity of morphine. Selected papers of Science and Technology in Shanghai 1960, Volume VI, 271–276 (Shanghai Science and Technology Publishers, Shanghai).

Tsou, K.: Antagonism of morphine analgesia by localized intracerebral microinjection of nalorphine. Acta physiol. sin. *26:* 332–338 (1963).

Tsou, K.; Chang, Z.-L.: Morphine effects on the spinal cord electrical activity. Acta physiol. sin. *27:* 369–377 (1964).

Tsou, K.; Hu, G.-Y.; Zhao, D.-D.; Ji, X.-Q.; Yi, C.-C.: The role of noradrenergic neurons in morphine analgesia. Acta pharmacol. sin. *1:* 85–88 (1980).

Tsou, K.; Jang, C.-S.: Analgesic effect of intraventricular and intracerebral microinjection of morphine. Acta physiol. sin. *25:* 119–128 (1962).

Tsou, K.; Jang, C.-S.: Studies on the site of analgesic action of morphine by intracerebral microinjection. Scient. sin. *13:* 1099–1109 (1964).

Tsou, K.; Tu, Z.-H.: Effects of reserpine, alpha-methyl-DOPA and their interactions on morphine analgesia. Acta physiol. sin. *26:* 360–366 (1963).

Tsou, K.; Wu, S.-H.; Wang, F.-S.; Ji, X.-Q.; Chang, T.-S.; Lo, E.-S; Yi, C.-C.: Increased levels of endorphins in the cisternal CSF of rabbits in acupuncture analgesia. Acta physiol. sin. *31:* 371–376 (1979).

Tsou, K.; Yi, C.-C.; Wang, F.-S.; Lo, E.-S.; Zhang, Z.-X.; Wu, S.-H.: Alterations of enkephalin contents in brain areas by acupuncture. Kexue Tongbao *25:* 78–82 (1980).

Tsou, K.; Yi, C.-C.; Wu, S.-H.; Wang, F.-S.; Lo, E.-S.; Ji, X.-Q.; Yu, Y.-K.; Zhao, D.-D.: Enkephalin involvement in acupuncture analgesia-radioimmunoassay. Scient. Sin. *23:* 1197–1220 (1980).

Tsou, K.; Zhao, D.-D.; Wu, S.-H.; Chang, T.-S.; Lo, E.-S.; Wang, F.-S.: Enhancement of acupuncture analgesia with concomitant increase of enkephalins in striatum and hypothalamus by intraventricular bacitracin. Acta physiol. sin. *31:* 377–381 (1979).

Wang, F.-S.; Wu, S.-H.; Zhang, Z.-X.; Tsou, K.: The brain enkephalin contents under four stressful conditions. Acta physiol. sin. *37:* 374–376 (1985).

Watson, S.J.; Akil, H.; Richard, C.N.; Barchas, J.D.: Evidence for two separate opiate peptide neuronal systems and the coexistence of beta-lipotropin, beta-endorphin and ACTH immunoreactivities in the same hypothalamus neurons. Nature, Lond. *275:* 226–228 (1978).

Wu, S.-H.; Chang, T.-S.; Wang, F.-S.; Tsou, K.: Effects of intraventricular cycloheximide on acupuncture analgesia and brain met-enkephalin. Acta physiol. sin. *32:* 79–81 (1980).

Wu, S.-X.; Wang, F.-S.; Zhang, Z.-X.; Tsou, K.: Effects of morphine on met-enkephalin contents in rabbit brain and CSF. Kexue Tongbao *29:* 840–841 (1984).

Yi, C.-C.; Lu, T.-H.; Wu, S.H.; Tsou, K.: A study on the release of [^{3}H]-5-hydroxytryptamine from brain during acupuncture and morphine analgesia. Scient. Sin. *20:* 113–124 (1977a).

Yi, C.-C.; Wu, S.-H.; Yu, Y.-K.; Wang, F.-S.; Ji, X.-Q.; Zhao, D.-D.; Tsou, K.: Effect on acupuncture analgesia by intraventricular injection of 5,6-dihydroxytryptamine. A functional and fluorescence histochemical study. Kexue Tongbao *22:* 43–48 (1977b).

Zhong, F.-X.; Ji, X.-Q.; Tsou, K.: Intrathecal DSP-4 selectively depletes spinal noradrenaline and attenuates morphine analgesia. Eur. J. Pharmacol. *116:* 327–330 (1985).

Zhu, X.-Z.; Wang, F.-S.; Yi, C.C.; Tsou, K.: Decreased potassium induced leu-enkephalin release from striatal slices after electric acupuncture. Chin. J. Physiol. Sci. *2:* 146–150 (1985).

K. Tsou, MD, PhD, Shanghai Institute of Materia Medica, Department of Pharmacology, Chinese Academy of Sciences, Shanghai 200031 (China)

Pain Headache, vol. 9, pp. 283–294 (Karger, Basel 1987)

Analgesic Effects of Electrical Brain Stimulation and Stress

Brain Stem Mechanisms

J. Timothy Cannon[a], *John C. Liebeskind*[b]

[a] Department of Psychology, University of Scranton, Pa., USA; [b] Department of Psychology, University of California at Los Angeles, Calif., USA

It has been 20 years since Melzack and Wall [1965] published their Gate Control Theory of pain. Included in this seminal work was a proposal that the brain could modulate the processing of nociceptive information at the spinal level. At that time, very little was known about the specifics of such a descending system. Particularly during the past decade, however, there has been rapid expansion of information concerning endogenous pain-inhibitory mechanisms. Models of the neural circuitry underlying pain inhibition have kept pace with this expansion by becoming more intricate [Basbaum and Fields, 1978, 1984]. Similarly, there is growing appreciation that responses to noxious stimulation rely upon complex interactions between environement and organism.

This review will focus on the analgesic effects of electrical brain stimulation (stimulation-produced analgesia) and some forms of stress (stress-induced analgesia). Research in these areas has provided some of the first glimpses of the neural and biochemical bases of endogenous pain modulation. Additionally, it has contributed to our understanding of variables that define when these systems are called into play. Researchers in this field hope that such insights will contribute to more effective management of clinical pain states. With this goal in mind it is gratifying that there has already been a rapid transition from basic animal research [Mayer et al., 1971, Reynolds, 1969] to successful clinical trials in which electrical brain stimulation effectively attenuates some chronic pain states in man [Adams, 1976; Hosobuchi et al., 1977; Richardson and Akil, 1977a, b].

Multiple Mechanisms of Pain Inhibition

Ten years ago it was widely accepted that the body possessed a single pain-inhibitory system, and that the recently discovered opioids were an essential component of this system. Early evidence to the contrary sometimes was downplayed or overlooked. Within recent years, however, it has become clear that there exist multiple substrates for pain modulation that, to varying degrees, are distinct both biochemically and anatomically. Not all of these analgesic substrates depend upon an opioid link [Terman et al., 1984; Watson and Mayer, 1982]. Making matters more complex, there appear to be more than one type of opioid analgesic state [Terman et al., 1984], and it is probable that multiple forms of nonopioid analgesia also exist.

Stimulation-Produced Analgesia

Reynolds [1969] first observed stimulation-produced analgesia (SPA) using electrode placements in the periaqueductal gray (PAG) of the rat midbrain. This initial publication provided what is still the most dramatic demonstration of the analgesic potency of brain stimulation. Abdominal surgery was performed on conscious rats using midbrain stimulation as the only analgesic intervention. While the PAG was stimulated, the animals were normally active and responsive to non-noxious stimuli; yet, they were totally unresponsive to noxious stimulation.

Mayer et al. [1971] performed a series of studies defining several important features of SPA. These include: (1) in addition to the PAG, other medial brain stem loci have analgesic influences; (2) SPA shares with opiates the ability to inhibit spinally mediated nociceptive reflexes; (3) brain stimulation can suppress nociception as potently as a large dose of morphine; (4) a rat learned to self-administer brain stimulation only in the presence of noxious stimulation; (5) analgesia can be present in one body region while normal nociception exists elsewhere; (6) brain stimulation can alter nociception selectively, leaving responses to other sensory qualities and motivational states unaffected.

In view of these findings and what was known at that time about opiates' analgesic action, Mayer et al. [1971] proposed the existence of a brain stem system that had pain modulation as its normal role. This system could be activated by either opiates or direct electrical stimulation, and its analgesic effects were exerted by descending projections to the spinal cord. Much recent research supports this hypothesis.

Working with a variety of species [Adams, 1976; Goodman and Holcombe, 1976; Mayer et al., 1971; Oliveras et al., 1971] and analgesic tests [Balagura and Ralph, 1973; Dennis et al., 1980, Mayer et al., 1971], researchers consistently have been able to produce analgesia by stimulating the central gray of the brain stem. The periaqueductal subdivision of the central gray has been the most extensively studied in infrahuman species. The periventricular gray of the third ventricle and adjoining diencephalic regions often have been the stimulation loci of choice in human work, due in part, to the disphoria that can accompany midbrain stimulation [Richardson and Akil, 1977a; Young et al., 1984]. The integrity of the caudal PAG may be necessary for the expression of SPA from more rostral regions of the central gray [Rhodes, 1979].

Other SPA-positive brain regions include several of the raphe nuclei [Cannon et al., 1982; Oliveras et al., 1979; Oliveras et al., 1975], medial forebrain bundle [Balagura and Ralph, 1973], subdivisions of the hypothalamus [Carr and Uysal, 1985], nucleus of the solitary tract [Lewis et al., 1985], and frontal lobe [Hardy, 1985]. A number of these regions have extensive interconnections with the central gray. Consideration of SPA derived from each of these sites is beyond the scope of this review; however, the nucleus raphe magnus (NRM) and the adjoining medullary reticular formation will be discussed briefly. With the exception of the caudal PAG, this subarea of the rostral medulla has been the most extensively examined for SPA. It also serves as a relay to the spinal cord for some analgesic effects of PAG stimulation (see below and [Bausbaum and Fields, 1978; Prieto, 1983]).

Akil et al. [1976] provided the first direct evidence of a link between the analgesic effects of opiates and brain stimulation. Naloxone, a highly specific opiate antagonist, partially antagonized SPA from electrode placements in the caudal periaqueductal gray of the rat. Subsequent investigators have found that naloxone can antagonize SPA in other species, including man [Adams, 1976; Oliveras et al., 1973; Richardson and Akil, 1976], and from other stimulation loci [Lewis et al., 1985; Oliveras et al., 1977; Richardson and Akil, 1977b].

Naloxone antagonism strongly suggested that SPA relies upon an opioid link. Also pointing to the involvement of opioids was the finding that SPA from the PAG exhibits tolerance with repeated stimulation and cross-tolerance with systemically administered morphine [Mayer and Hayes, 1975]. In humans, SPA was found to be associated with a release of opioid-like immunoreactive materials into cerebrospinal fluid [Akil et al., 1978; Hosobuchi et al., 1979].

Despite the array of observations linking SPA and opioids, some problems existed. When considering the PAG, doses of naloxone that are known to completely block the analgesic effects of high doses of morphine at most only partially disrupted SPA [Akil et al., 1976]. Additionally, some investigators found naloxone to be completely without effect [Pert and Walter, 1976; Yaksh et al., 1976]. A partial resolution to this problem was offered by the discovery of naloxone-sensitive and naloxone-insensitive SPA sites in the PAG region [Cannon et al., 1982]. Naloxone reliably reduces, but does not eliminate, the analgesic influence of electrode placements within and just below the dorsal raphe nucleus. In contrast, naloxone is without effect on SPA from more dorsally located placements in the midline PAG. Further suggesting the differential involvement of opioids in SPA derived from these two midbrain areas is the observation that only analgesia from the naloxone-sensitive region exhibits cross-tolerance when animals are repeatedly exposed to opioid forms [Terman et al., 1985]. SPA from the dorsal, naloxone-insensitive region does not exhibit such cross-tolerance.

Paralleling the preceding, serotonergic involvement in SPA from the caudal midbrain appears to depend upon the placement of stimulating electrodes in or near the dorsal raphe nucleus [Akil and Liebeskind, 1975; Akil and Mayer, 1972]. Other functional distinctions between dorsal and ventral regions of the caudal PAG have been made, based upon work with both brain stimulation [Fardin et al., 1984a, b] and lesions [Liebman et al., 1970].

The opioid and nonopioid SPA regions of the midbrain affect the spinal cord through at least partially independent medullary pathways [Pietro et al., 1983]. The region of the NRM appears to serve an intermediary role for only the opioid form of midbrain SPA (see, however, [Basbaum and Fields, 1984]). Extensive lesions of the NRM disrupted SPA derived from the dorsal raphe region. Comparable lesions were without effect on SPA elicited from the more dorsal midline PAG. Bulbar regions distal to the NRM may serve as centrifugal relays for this latter system [Basbaum and Fields, 1984; Morton et al., 1984], or the recently discovered direct projections from the PAG to the spinal cord [Mantyh and Peschanski, 1982] could be involved.

There also is evidence of opioid and nonopioid forms of SPA when the medullary region that includes the NRM is directly stimulated. Akaike et al. [1978] found that SPA from the NRM itself was not affected by naloxone [Satoh et al., 1980, see, however, Oliveras et al., 1977; Zorman et al.; 1981], whereas naloxone did antagonize SPA from electrode placements just lateral to the NRM in the nucleus reticularis paragigantocellularis (NRPG). Lumbar intrathecal injections of naloxone can disrupt SPA from the NRPG [Zorman et al., 1982], suggesting a spinal location for the opiate receptors involved in this form of SPA. Interestingly, similar intrathecal administrations of naloxone are without effect on analgesia produced by *l*-glutamate injections into the NRPG [Satoh et al., 1983]. This differential antagonism, dependent on the method of brain stimulation, led to the speculation that electrical stimulation of fibers of passage are responsible for the opioid components of SPA from the NRPG [Satoh et al., 1983].

In summary, there are opioid and nonopioid forms of SPA. In the midbrain, electrical stimulation of the dorsal raphe nucleus and subjacent tegmentum produce analgesia that is at least partially opioid in nature. Stimulating more dorsally in the medial PAG, by contrast, elicits analgesia that does not rely on opioid release. In addition to a pharmacological dissociation, these midbrain regions have anatomically dissociable spinopedal medullary pathways. Electrical stimulation of adjoining regions in the medulla can also produce analgesic states with either opioid (NRPG) or nonopioid (NRM) characteristics.

Stress-Induced Analgesia

Exposure to an extensive array of stressful situations has been shown to diminish reactivity to noxious stimulation [Mayer et al., 1984; Terman et al., 1984]. This stress-induced analgesia (SIA) has been the focus of a great deal of research during recent years. Much of this interest is due to SIA's offering the first reliable, noninvasive means of activating the body's pain-inhibitory mechanisms. Particularly when compared with SPA, SIA research possesses other attractive features – it typically proceeds at a more rapid pace; it also has the appearance of being less technically demanding. Unfortunately, this final property can be an illusion.

Thus, for a variety of reasons, there is a wealth of data concerning the analgesic effects of numerous stress procedures and several models have been proposed to explain the underlying mechanisms of SIA. When considering even a single stressor, footshock, the field is not unanimous concerning the number of forms of SIA and the role opioids play in each. Without agreement on the preceding, the existing debate concerning which features of footshock stress are critical for selectively triggering the various forms of SIA is understandable [Bolles and Fanselow, 1980; Terman et al., 1984; Watkins and Mayer, 1982].

The first investigations of SIA using footshock provided contradictory indications of its biochemical nature. Akil et al. [1976] found that naloxone could antagonize SIA, whereas Hayes et al. [1978] found naloxone to be without effect. Our first work in this area appeared to resolve this discrepancy by demonstrating that the parameters of footshock exposure could define the biochemical nature of the resulting analgesia [Lewis et al., 1980]. Prolonged footshock (20–30 min of 1-second shock pulses every 5 s) elicited an opioid-form of SIA; briefer exposure (3 min of continuous shock) to the same shock intensity elicited nonopioid SIA [Lewis et al., 1980, 1981].

More recently, Terman et al. [1984] have further examined the role of shock parameters in SIA. They found that brief continuous footshock can elicit either opioid or nonopioid analgesia as a function of an interaction between the duration and intensity of footshock. Using a fixed shock duration, higher shock intensities resulted in nonopioid SIA, whereas lower intensities produced an opioid form. In a parallel fashion, with a fixed shock intensity, longer shock durations resulted in nonopioid SIA, whereas shorter durations produced opioid SIA. Based upon the preceding, Terman et al. [1984] proposed that increasing the severity (intensity or duration) of relatively brief footshock stress can change the neurochemical basis of the resulting SIA from opioid to nonopioid. Subsequent work has served to confirm the importance of severity in defining the opioid or nonopioid nature of the analgesia that follows brief exposures to footshock [Cannon et al., 1983; Terman et al., 1985], rotation [Cannon et al., 1986] and swimming [Terman et al., 1986].

The opioid analgesic states produced by brief and prolonged exposure to footshock share a number of properties. They demonstrate tolerance with repeated exposure to stress, cross-tolerance with repeated administrations of morphine, and cross-tolerance with each other (that is, repeated exposure to one reduces the analgesic effect of the other).

Nonopioid SIA exhibits none of the preceding characteristics [Lewis et al., 1981; Terman et al., 1983; Terman et al., 1984].

As mentioned above, repeated exposure to either of these opioid stress paradigms selectively reduced the effectiveness of the opioid form of SPA from dorsal raphe electrode placements. Repeated exposure to nonopioid stress parameters did not produce this cross-tolerance. Nonopioid SPA from more dorsal medial PAG electrode placements did not show cross-tolerance with any stress procedure [Terman et al., 1985].

The observation of cross-tolerance across all three of the opioid analgesic states examined (two elicited by stress and one by brain stimulation) suggests that a common population of opiate receptors is involved in these phenomena. The finding that all three analgesic states can be antagonized significantly by low (0.01–0.1 mg/kg) doses of naloxone or naltrexone is consistent with this hypothesis [Cannon et al., 1982; Lewis et al., 1980; Terman et al., 1984].

Despite the similarities that exist between the two opioid forms of SIA, a number of observations indicate that they should be regarded as independent phenomena. These differences include: (1) Only the opioid SIA produced by prolonged exposure to stress depends upon the integrity of the pituitary-adrenal axis [Lewis et al., 1980; Lewis et al., 1982; Terman, et al., 1984], and there are indications that enkephalin-like peptides from the adrenal medulla may mediate this form of SIA [Lewis et al., 1982]. (2) Only the analgesic response to prolonged stress is dependent upon forebrain mechanisms [Terman et al., 1984]. (3) There appears to be a learned component to the analgesic response to prolonged stress, and parallels have been drawn between this form of SIA and the phenomenon of learned helplessness [Maier et al., 1983]. (4) The analgesic response to prolonged, but not to brief, stress is abolished by pentobarbitol (Nembutal) anesthesia [Terman et al., 1984].

Nonopioid SIA does not rely upon the pituitary-adrenal axis [Lewis et al., 1981]. It also does not appear to involve a learned component. Nonopioid SIA still occurs in animals anesthetized with pentobarbitol and does not rely upon forebrain mechanisms [Terman et al., 1984]. Lesions in the region of the NRM have been found to disrupt nonopioid but not the hormonally dependent form of opioid SIA [Cannon et al., 1983]. The medullary relays for the two opioid forms of SIA have not been clearly determined. Research involving other paradigms has suggested a number of possibilities [Morton et al., 1984; Watkins et al., 1983].

In summary, variations in the temporal and intensity parameters of footshock can selectively activate three forms of SIA. (1) Prolonged exposure to footshock triggers an opioid analgesic state that relies upon hormonal mechanisms. This hormonal-opioid SIA depends upon an intact forebrain and cannot be observed in pentobarbitol anesthetized animals. (2) Briefer stress exposures can produce either opioid or nonopioid analgesic states dependent upon the severity of the stress experience. (3) Increasing the severity (duration or intensity) of stress produces a shift from opioid to nonopioid analgesia. Neither of the analgesic states that follow brief stress depend upon the forebrain or those hormonal systems examined to date. Both can be demonstrated under pentobarbitol anesthesia.

Finally, Watkins and Mayer [1982, 1984] have performed an extensive series of experiments defining the biochemical and anatomical substrates of the opioid and nonopioid analgesic states that they produced by selectively shocking the front or hind paws, respectively, of rats for 90 s. There are some intriguing parallels between these analgesic states and those that we have been studying following brief (1–4 min) footshock exposure [Mayer et al., 1971; Terman et al., 1984]. In fact, variations of shock intensity can produce both naloxone-sensitive and naloxone-insensitive SIA when either the front or hind paws of rats are shocked selectively [Cannon et al., 1984]. This pattern of results conforms to the findings of Terman et al., [1984], in which higher shock intensities produce analgesia that is nonopioid. While it is premature to speculate about the degree to which the data from these two lines of research can be integrated, the prospect of even a small reduction in the diversity of paradigms and models within this field is a welcome one.

Overview

This review reflects the emphasis that our field has placed on determining whether a given paradigm produces an analgesic state that depends upon opioid mechanisms. Fully addressed, this determination involves assessments of whether the analgesia is diminished by naloxone, shows tolerance, and exhibits cross-tolerance with other opioid or opiate manipulations. Overall, evaluating the roles of other neurotransmitters or neurohormones in endogenous analgesic mechanisms has been studied less often.

As we have seen, this concentration on the involvement of opioids in analgesia has led to several fruitful lines of research. It has provided a valuable benchmark from which to chart our rapidly developing field, and has allowed the identification of multiple forms of both SPA and SIA. These advances should make the task of defining the roles of nonopioid neurotransmitters in pain modulation considerably more manageable, and progress has been made toward this end. Such an 'opioid' frame of reference could become an impediment, however, were opioids to be viewed as functioning independently of other neurotransmitter systems or were 'nonopioid' to assume the characteristics of a term defining what something is, rather than what something is not.

Medicine has sought nontolerating, nonaddicting pain relief for centuries. To date, efforts to fulfill this quest by modifications of opiate compounds have proven futile. Contributing to this lack of success may be the prospect that tolerance and addiction are natural properties of the systems that opiates activate, rather than artifacts of exogenous activation. At least some nonopioid analgesic mechanisms do not appear to possess these clinically undesirable properties. Thus, a better understanding of the neuropharmacological and anatomical bases of nonopioid analgesic systems appears to be of particular significance.

References

Adams, J.E.: Naloxone reversal of analgesia produced by brain stimulation in the human. Pain *2:* 161–166 (1976).

Akaike, A.; Shibata, T.; Satoh, M.; Takagi, H.: Analgesia induced by microinjection of morphine into, and electrical stimulation of, the nucleus reticularis paragigantocellularis of rat medulla oblongata. Neuropharmacology *17:* 775–778 (1978).

Akil, H.; Liebeskind J.C.: Monoaminergic mechanisms of stimulation-produced analgesia. Brain Res. *94:* 279–296 (1975).

Akil, J.; Madden, J.; Patrick R.L.; Barchas, J.D.: Stress-induced increase in endogenous opiate peptides. Concurrent analgesia and its partial reversal by naloxone; in Kosterlitz, Opiates and endogenous opioid peptides, pp. 63–70 (Elsevier Amsterdam, 1976).

Akil, H.; Mayer, D.J.: Antagonism of stimulation-produced analgesia by p-CPA, a serotonin synthesis inhibitor. Brain Res. *44:* 692–697 (1972).

Akil, H.; Mayer, D.J.; Liebeskind, J.C.: Antagonism of stimulation-produced analgesia by naloxone, a narcotic antagonist. Science *191:* 961–962 (1976).

Akil, J.; Richardson D.E.; Barchas, J.D.; Li, C.H.: Appearance of β-endorphin-like immunoreactivity in human ventricular cerebrospinal fluid upon analgesic electrical stimulation. Proc. natn. Acad. Sci. USA *75:* 5170–5172 (1978).

Balagura, S.; Ralph, T.: The analgesic effect of electrical stimulation of the diencephalon and mesencephalon. Brain Res. *60:* 369–379 (1973).

Basbaum, A.I.; Fields, H.L.: Endogenous pain control mechanisms: Review and hypothesis. Ann. Neurol. *4:* 451–462 (1978).

Basbaum, A.I.; Fields, H.L.: Endogenous pain control systems. Brain stem spinal pathways and endorphin circuitry. A. Rev. Neurosci. *7:* 309–338 (1984).

Bolles, R.C.; Fanselow, M.S.: A perceptual-defensive-recuperative model of fear and pain. Behav. Brain Sci. *3:* 291–323 (1980).

Cannon, J.T.; Lewis, J.W.; Weinberg, V.W.; Liebeskind, J.C.: Evidence for the independence of brain stem mechanisms mediating analgesia induced by morphine and two forms of stress. Brain Res. *269:* 231–236 (1983).

Cannon, J.T.; Maro, N.; Telep, B.; Carty, G.: Differential effects of naloxone on analgesia and vocalizations induced by brief durations of pulsed footshock. Soc. Neurosci. Abst. *9:* 793 (1983).

Cannon, J.T.; Prieto, G.J.; Lee, A.; Liebeskind, J.C.: Evidence for opioid and nonopioid forms of stimulation-produced analgesia in the rat. Brain Res. *243:* 315–321 (1982).

Cannon, J.T.; Shavit Y.; Liebeskind J.C.: Rotation speed can determine whether the resulting stress-induced analgesia is blocked by naloxone. Ann. N.Y. Acad. Sci. *467:* 425–427 (1986).

Cannon, J.T.; Terman, G.W.; Lewis, J.W.; Liebeskind J.C.: Body region shocked need not critically define the neurochemical basis of stress analgesia. Brain Res. *323:* 316–319 (1984).

Carr, K.D.; Uysal, S.: Evidence of a supraspinal opioid analgesic mechanism engaged by lateral hypothalamic electrical stimulation. Brain Res. *335:* 55–62 (1985).

Dennis, S.G.; Choiniere, M.; Melzack, R.: Stimulation-produced analgesia in rats. Assessment by two pain tests and correlation with self-stimulation. Expl. Neurol. *68:* 295–309 (1980).

Fardin, V.; Olivaras, J.L.; Besson, J.M.: A reinvestigation of the analgesic effects induced by stimulation of the periaqueductal gray matter in the rat. I. The production of behavioral side effects together with analgesia. Brain Res. *306:* 105–123 (1984a).

Fardin, V.; Olivaras, J.L.; Besson, J.M.: A reinvestigation of the analgesic effects induced by stimulation of the periaqueductal gray matter in the rat. II. Differential characteristics of the analgesia induced by ventral and dorsal PAG stimulation. Brain Res. *306:* 125–139 (1984b).

Goodman, S.J.; Holcombe, V.: Selective and prolonged analgesia in monkey resulting from brain stimulation; in Bonica, Albe-Fessard, Advances in pain research and therapy, vol. 1, pp. 495–502 (Raven Press, New York, 1976).

Hardy, S.G.P.: Analgesia elicited by prefrontal stimulation. Brain Res. *339:* 281–284 (1985).

Hayes, R.L.; Bennett, G.J.; Newlon, P.G.; Mayer, D.J.: Behavioral and physiological studies of non-narcotic analgesia in the rat elicited by certain environmental stimuli. Brain Res. *155:* 69–90 (1978).

Hosobuchi, Y.; Adams, J.E.; Linchitz, R.: Pain relief by electrical stimulation of the central gray matter in humans and its reversal by naloxone. Science *197:* 183–186 (1977).

Hosobuchi, Y.; Rossier, J.; Bloom, F.E.; Guillemin, R.: Stimulation of human periaqueductal gray for pain relief increases immunoreactive β-endorphin in ventricular fluid. Science *203:* 279–281 (1979).

Lewis, J.W.; Baldrighi, G.; Watson, S.J.; Akil, H.: Electrical stimulation of nucleus tractus solitarius (NTS) causes opioid mediated analgesia in the rat. Soc. Neurosci. Abstr. *11:* 637 (1985).

Lewis, J.W.; Cannon, J.T.; Liebeskind, J.C.: Opioid and nonopioid mechanisms of stress analgesia. Science *208:* 623–625 (1980).

Lewis, J.W.; Chudler, E.H.; Cannon, J.T.; Liebeskind, J.C.: Hypophysectomy differentially affects morphine and stress analgesia. Proc. West. Pharmacol. Soc. *24:* 323–326 (1981).

Lewis, J.W.; Sherman, J.E.; Liebeskind, J.C.: Opioid and nonopioid stress analgesia: Assessment of tolerance and cross-tolerance with morphine. J. Neurosci. *1:* 358–363 (1981).

Lewis, J.W.; Tordoff, M.G.; Sherman, J.E.; Liebeskind, J.C.: Adrenal medullary enkephalin-like peptides may mediate opioid stress analgesia. Science *217:* 57–59 (1982).

Liebman, J.M.; Mayer, D.J.; Liebeskind, J.C.: Mesencephalic central gray lesions and fear-motivated behavior in rats. Brain Res. *23:* 353–370 (1979).

Maier, S.J.; Sherman, J.E.; Lewis, J.W.; Terman, G.W.; Liebeskind, J.C.: The opioid/nonopioid nature of stress-induced analgesia and learned helplessness. J. exp. Psychol. Anim. Behav. Proc. *9:* 80–90 (1983).

Mantyh, P.W.; Peschanski, M.: Spinal projections from the periaqueductal gray and dorsal raphe in the rat, cat and monkey. Neuroscience *7:* 2769–2776 (1982).

Mayer, D.J.; Hayes, R.L.: Stimulation-produced analgesia: Development of tolerance and cross-tolerance to morphine. Science *188:* 941–943 (1975).

Mayer, D.J.; Watkins, L.R.: Multiple endogenous opiate and nonopiate analgesia systems; in Kruger, Liebeskind, Advances in pain research and therapy, vol. 6, pp. 253–276 (Raven Press, New York 1984).

Mayer, D.J.; Wolfle, T.L.; Akil, H.; Carder, B.; Liebeskind, J.C.: Analgesia from electrical stimulation in the brainstem of the rat. Science *174:* 1351–1354 (1971).

Melzack, R.; Wall, P.D.: Pain mechanisms. A new theory. Science *150:* 971–979 (1965).

Morton, C.R.; Duggan, A.W.; Zhao, Z.Q.: The effects of lesion of medullary midline and lateral reticular areas on inhibition in the dorsal horn produced by periaqueductal gray stimulation in the cat. Brain Res. *301:* 121–130 (1984).

Oliveras, J.L.; Besson, J.M.; Guilbaud, G.; Liebeskind, J.C.: Behavioral and electrophysiological evidence of pain inhibition from midbrain stimulation in the cat. Expl Brain Res. *20:* 32–44 (1974).

Oliveras, J.L.; Guilbaud, G.; Besson, J.M.: A map of serotoninergic structures involved in stimulation producing analgesia in unrestrained freely moving cats. Brain Res. *164:* 317–322 (1979).

Oliveras, J.L.; Hosobuchi, Y.; Redjemi, F.; Guilbaud, G.; Besson, J.M.: Opiate antagonist, naloxone, strongly reduces analgesia induced by stimulation of a raphe nucleus (centralis inferior). Brain Res. *120:* 221–229 (1977).

Oliveras, J.L.; Redjemi, F.; Guilbaud, G.; Besson, J.M.: Analgesia induced by electrical stimulation of the inferior centralis nucleus of the raphe in the cat. Pain *1:* 139–145 (1975).

Pert, A.; Walter, M.: Comparison between naloxone reversal of morphine and electrical stimulation induced analgesia in the rat mesencephalon. Life Sci. *19:* 1023–1032 (1976).

Prieto, G.J.; Cannon, J.T.; Liebeskind, J.C.: N. raphe magnus lesions disrupt stimulation-produced analgesia from ventral but not dorsal midbrain areas in the rat. Brain Res. *261:* 53–57 (1983).

Reynolds, D.V.: Surgery in the rat during electrical analgesia induced by focal brain stimulation. Science *164:* 444–445 (1969).

Rhodes, D.L.: Periventricular system lesions in stimulation-produced analgesia. Pain *7:* 31–51 (1979).

Richardson, D.E.; Akil, H.: Pain reduction by electrical brain stimulation in man. I. Acute administration in periaqueductal and periventricular sites. J. Neurosurg. *47:* 178–183 (1977a).

Richardson, D.E.; Akil, H.: Pain reduction by electrical brain stimulation in man. II. Chronic self-administration in the periventricular gray matter. J. Neurosurg. *47:* 184–189 (1977b).

Satoh, M.; Akaike, A.; Nakazawa, T.; Takagi, H.: Evidence for involvement of separate mechanisms in the production of analgesia by electrical stimulation of the nucleus reticularis paragigantocellularis and nucleus raphe magnus in the rat. Brain. Res. *194:* 525–529 (1980).

Satoh, M.; Oku, R.; Akaike, A.: Analgesia produced by microinjection of *L*-glutamate into the rostral ventromedial bulbar nuclei of the rat and its inhibition by intrathecal α-adrenergic blocking agents. Brain Res. *261:* 361–364 (1983).

Terman, G.W.; Lewis, J.W.; Liebeskind J.C.: Opioid and nonopioid mechanisms of stress analgesia. Lack of cross-tolerance between stressors. Brain Res. *260:* 147–150 (1983).

Terman, G.W.; Morgan, M.J.; Liebeskind, J.C.: Opioid and nonopioid stress analgesia from cold water swim: Importance of stress severity. Brain Res. *372:* 167–171 (1986).

Terman, G.W.; Penner E.R.; Liebeskind, J.C.: Stimulation-produced and stress analgesia: Cross-tolerance between opioid forms. Brain Res. *360:* 374–378 (1985).

Terman, G.W.; Shavit, Y.; Lewis, J.W; Cannon, J.T.; Liebeskind, J.C.: Intrinsic mechanisms of pain inhibition: Activation by stress. Science *226:* 1270–1277 (1984).

Watkins, L.R.; Mayer, D.J.: The organization of endogenous opiate and nonopiate pain control systems. Science *216:* 1185–1192 (1982).

Watkins, L.R.; Young, E.G.; Kinscheck, I.B.; Mayer, D.J.: The neural basis of foot-shock analgesia. The role of specific ventral medullary nuclei. Brain Res. *276:* 305–315 (1983).

Yaksh, T.L.; Young, J.C.; Rudy, T.A.: An inability to antagonize with naloxone the elevated thresholds resulting from electrical stimulation of the mesencephalic central gray. Life Sci. *18:* 1193–1198 (1976).

Young, R.F.; Feldman, R.A.; Kroening, R.; Fulton, W.; Morris, J.: Electrical stimulation of the brain in the treatment of chronic pain in man; in, Kruger, Liebeskind, Advances in pain research and therapy, pp. 289–303 (Raven Press, New York 1984).

Zorman, G.; Belcher, G.; Adams, J.E.; Fields, H.L.: Lumbar intrathecal naloxone blocks analgesia produced by microstimulation of the ventromedial medulla in the rat. Brain Res. *236:* 77–84 (1982).

Zorman, G.; Hentall, I.D.; Adams, J.E.; Fields, H.L.: Naloxone-reversible analgesia produced by microstimulation in the rat medulla. Brain Res. *219:* 137–148 (1981).

J. Timothy Cannon, PhD, Department of Psychology, University of Scranton, Scranton, PA 18510 (USA)

Subject Index